T0341012

Number, Shape, and Symmetry

Number, Shape, and Symmetry

An Introduction to Number Theory, Geometry, and Group Theory

Diane L. Herrmann
Paul J. Sally, Jr.

CRC Press
Taylor & Francis Group
Boca Raton London New York

CRC Press is an imprint of the
Taylor & Francis Group, an **informa** business

AN A K PETERS BOOK

Cover photograph copyright © George W. Hart, http://georgehart.com

Eights by George W. Hart, 6″ x 6″ x 6″, paper. The underlying structure is based on five intersecting tetrahedra, with each color representing the six edges of one tetrahedron.

CRC Press
Taylor & Francis Group
6000 Broken Sound Parkway NW, Suite 300
Boca Raton, FL 33487-2742

© 2013 by Taylor & Francis Group, LLC
CRC Press is an imprint of Taylor & Francis Group, an Informa business

No claim to original U.S. Government works

Printed in the United States of America on acid-free paper
Version Date: 20120817

International Standard Book Number: 978-1-4665-5464-1 (Hardback)

This book contains information obtained from authentic and highly regarded sources. Reasonable efforts have been made to publish reliable data and information, but the author and publisher cannot assume responsibility for the validity of all materials or the consequences of their use. The authors and publishers have attempted to trace the copyright holders of all material reproduced in this publication and apologize to copyright holders if permission to publish in this form has not been obtained. If any copyright material has not been acknowledged please write and let us know so we may rectify in any future reprint.

Except as permitted under U.S. Copyright Law, no part of this book may be reprinted, reproduced, transmitted, or utilized in any form by any electronic, mechanical, or other means, now known or hereafter invented, including photocopying, microfilming, and recording, or in any information storage or retrieval system, without written permission from the publishers.

For permission to photocopy or use material electronically from this work, please access www.copyright.com (http://www.copyright.com/) or contact the Copyright Clearance Center, Inc. (CCC), 222 Rosewood Drive, Danvers, MA 01923, 978-750-8400. CCC is a not-for-profit organization that provides licenses and registration for a variety of users. For organizations that have been granted a photocopy license by the CCC, a separate system of payment has been arranged.

Trademark Notice: Product or corporate names may be trademarks or registered trademarks, and are used only for identification and explanation without intent to infringe.

Library of Congress Cataloging-in-Publication Data

Herrmann, Diane, 1953-
Number, shape, and symmetry : an introduction to number theory, geometry, and group theory / Diane L. Herrmann and Paul J. Sally, Jr.
p. cm.
Summary: "This textbook shows how number theory and geometry are the essential components in the teaching and learning of mathematics for students in primary grades. The book synthesizes basic ideas that lead to an appreciation of the deeper mathematical ideas that grow from these foundations. The authors reflect their extensive experience teaching undergraduate nonscience majors, students in the Young Scholars Program, and public school K-8 teachers in the Seminars for Endorsement of Science and Mathematics Educators (SESAME). "-- Provided by publisher.
Includes bibliographical references and index.
ISBN 978-1-4665-5464-1 (hardback)
1. Number theory--Textbooks. 2. Geometry--Textbooks. I. Sally, Paul. II. Title.

QA241.H44 2012
512.7--dc23

2012019333

Visit the Taylor & Francis Web site at
http://www.taylorandfrancis.com

and the CRC Press Web site at
http://www.crcpress.com

We dedicate this book to our fathers,
Kenneth Herrmann, the carpenter
and
Paul J. Sally, the bricklayer.

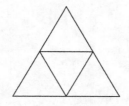

Contents

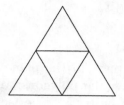

Preface

The topics of number theory and geometry are essential components in the teaching and learning of mathematics from the earliest experiences of students in the primary grades. These two subjects lay the foundation for the study of mathematics in the middle-grade, high-school, and university levels. However, as students proceed in linear fashion through the standard mathematics curriculum, they are less likely to encounter the rich ideas that grow directly from those foundations. Students are also less likely to see the common aspects of number theory and geometry. In the advanced study of university mathematics, these two topics return to their central roles. However, for the nonscientist, this contact is never reestablished. We do that in this book.

This book has emerged primarily from our work with three different groups of students. These groups are undergraduate nonscience majors at the University of Chicago, students in the grades 7–8 and 9–10 components of our Young Scholars Program (YSP) for mathematically talented students, and Chicago Public School K-8 teachers who have participated in Seminars for Endorsement in Science And Mathematics Education (SESAME), a staff development program. With all of these students, we attempt to develop serious mathematical ideas and an introduction to proof in mathematics. We do this through a careful treatment of number theory and geometry. In number theory, we begin with the Rules of Arithmetic (Axioms for the Integers) and develop all the basic ideas about divisibility, primes, and modular arithmetic. We then present applications of these topics. In addition, we introduce the abstract notion of a group and offer a number of examples, which is useful when we discuss symmetry groups in geometry. The final topics in number theory are rational numbers, real numbers, and ideas about infinity.

In geometry, we begin with polygons and polyhedra, including the construction of regular polygons and regular polyhedra. We study tessellation by looking at patterns in the plane, especially those made by regular polygons or sets of regular polygons. We determine the symmetry groups of these figures and patterns. The fact that groups arise significantly in both geometry and number theory is the unifying theme of this book.

We have successfully used this material with all three groups of students. More recently, this material has become an integral part of our preservice program for elementary teachers. For the YSP students, this material works because their contact with the ideas of number theory and geometry is still strong and leads easily to the topics and problems we discuss in this book. Number theory and geometry work for the K-8 mathematics teachers and our preservice teachers because our concrete study of the foundations helps them better understand the ideas they are trying to communicate to their own students. Finally, this material works for the nonscience college students because they appreciate, and even enjoy, the serious experience in mathematics that these subjects provide.

We begin the book in Chapter 0 with a simple problem that reflects the interplay between number and symmetry. This problem illustrates various aspects of the process of doing mathematics. We have used this problem in our teaching to show several things about mathematical inquiry. Among these are the idea that a problem in mathematics may have more than one solution, that one can reduce a problem by first eliminating certain cases, that geometry can aid in solving numerical problems, and finally that a given problem in mathematics can often be generalized in a direct and interesting way.

In Chapter 1, we begin our study of number theory with the Rules of Arithmetic, that is, the axioms for a commutative ring with 1. We continue with examples of mod 2 and mod 10 (one's digit) arithmetic. In Chapter 2, we derive some elementary consequences of these axioms and continue with the axioms for order. We provide careful mathematical proofs of our assertions, introducing the idea of a complete proof in a relatively simple context. These proofs are intended to be as much teaching tools as establishments of mathematical fact. That is, within the proof itself, we often include motivational discussion and clear directions of how to move from one assertion to the next in logical progression. As the reader gains some maturity in working with proofs, fewer and fewer details are necessary.

In Chapters 3, 4, and 5, we devote our attention to the fundamental aspects of divisibility in the integers. We give proofs of the central theorems about divisibility, namely, the Division Algorithm, the Euclidean Algorithm, and the Fundamental Theorem of Arithmetic. We follow this with some applications to divisibility, including Fibonacci numbers, sum and number of divisors, perfect numbers, rational arithmetic, and Egyptian fractions.

This discussion of divisibility leads naturally to congruences, one of the most important topics in number theory. Congruences and their applications are very concrete and make use of all the basic number theory facts developed earlier in the book. The study of congruences in Chapters 6 and 7 leads us to the abstract notion of a group and its properties. We present some interesting and surprising aspects of congruence arithmetic, from divisibility tests to UPC numbers on products. We end the number theory section of the book in Chapter 8 with a discussion of rational and irrational numbers and the concept of infinity.

Our geometry study begins in Chapter 9 and continues in Chapter 10 with the treatment of polygons, including regular polygons and their construction. By looking at the symmetries of regular polygons in Chapter 11, we return to the idea of groups. We present groups in which the binary operation is not commutative. In Chapter 12, we introduce permutations as a way of describing symmetries, which leads naturally to a discussion of permutation groups. In Chapter 13, we turn our attention to polyhedra, including regular polyhedra and their constructions and symmetry groups. Chapter 14 presents a concise introduction to graph theory, an important idea in modern mathematics. Paralleling the last chapter of the number theory section, in Chapter 15, we turn to the idea of the infinite in geometry. This includes tessellations, frieze patterns, and wallpaper groups. We end the book with a discussion of two topics that connect the themes of number theory and geometry. Chapter 16 includes the relationship between the Fibonacci numbers and the golden ratio and also between constructible numbers and constructible polygons. In geometry, we assume knowledge of elementary plane geometry, which we include in the Appendix. Here the reader will find some of the information usually contained in a first text for high-school Euclidean geometry.

We know that mathematics is best learned by doing. Therefore, as in any mathematics book, the problems and exercises that we include constitute an essential feature for learning the subject. We provide three avenues for students to develop their understanding. First, the text itself includes Practice Problems that test a student's grasp of the material as it is presented. These problems are generally rather simple; we recommend that students keep a pencil and paper handy as they read the text so that the Practice Problems can be solved as soon as they are encountered. The second category of problems presented in the text is that of Challenge Problems. These are intended to be more difficult than the Practice Problems, and their solutions may require considerable ingenuity. Finally, at the end of each chapter, we present a large set of exercises. Some of these exercises test the students' understanding of the material in the text. Others are designed to extend the textual material and project students into more sophisticated aspects of number theory and geometry.

This book first emerged from a two-quarter course for nonscience majors at the University of Chicago. That course covered number theory in the first quarter and geometry in the second. We expect that, by expanding on the textual material through projects contained in the exercises, this text could constitute a two-semester course. At Chicago, we have also met the need to present this material in a single quarter. The selected topics for a coherent and interesting one-quarter (or one-semester) course include

▷ Chapters 0–4,

▷ Chapter 6,

▷ Chapters 9–14.

Chapter 5, Variations on a Theme, and Chapter 15, Tessellations, contain material that can be assigned as projects for student investigation and presentation in class. A one-quarter class would also omit Chapters 8 and 16, which deal with ideas of the infinite.

We would like to thank those who provided help for getting this text in print. First, we had support from the College of the University of Chicago. This support allowed us to hire course assistants, without whose help the completion of the text would have taken considerably more time. Second, we owe a tremendous debt of gratitude to Ashley Reiter Ahlin, Emily Peters, and Alex Zorn, who not only prepared the manuscript, but also provided valuable suggestions and feedback about the organization and content of the book. We also thank Barbara Csima, John Zekos, David Schmitz, Ryan Reich, and Kevin Tucker for their help in writing and collecting exercises, and for typing part of the text. Those who used the preliminary notes in the Math 110 course at Chicago, including Chris Degni, Chris French, Kaj Gartz, Chris Hallstrom, Moon Duchin, and

Arunas Liulevicius, as well as the tutors and students in their courses, have also provided corrections and suggestions. Finally, we would like to thank many years of students in the 7–8 and 9–10 components of YSP as well as Chicago Public Schools middle-grade teachers in SESAME and preservice teachers who attended our classes and affirmed for us each year the value of presenting this exciting material to enthusiastic learners.

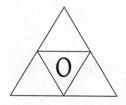

Warm-up: The Triangle Game

Let's start by investigating a problem we call the *Triangle Game*. The Triangle Game shows that a mathematics problem may not have just a single answer and that some solutions to a problem can lead to other solutions as well as intriguing new questions in a direct and interesting way. The Triangle Game illustrates a fundamental mathematical idea: it shows that a problem can be presented to elementary school students in the third or fourth grade and can be solved mainly by trial and error. As the problem is approached in a more sophisticated way, we can use fundamental tools from algebra to reduce the trial-and-error approach to the actual solution.

We begin by drawing an equilateral triangle. Now we draw a dot on the midpoint of each side and at each vertex, or corner (see Figure A).

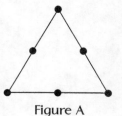

Figure A

Figure B

We place each of the numbers 1 through 6 on one of the dots, in any order. For example, let's start by putting 1 at the top and going around the triangle clockwise, as in Figure B.

If we sum the numbers along each side, we get $1 + 2 + 3 = 6$ on the right, $3 + 4 + 5 = 12$ across the bottom, and $5 + 6 + 1 = 12$ on the left. The object of the Triangle Game is to find all the ways to place the six numbers on the dots in such a way that the sum along each of the three sides, or "side sum," is the same. (Thus, Figure B is not a solution.)

Practice Problem 0.1. *Before reading further, try to experiment and find some solutions to the Triangle Game.*

When we explained how to start the game, we began by putting a 1 at the top, but the side sums were different. Let's start with 1 at the top. We want to experiment with computation to develop an algorithm for finding solutions to the game.

If we start with the numbers 1, 2, and 3 clockwise on the vertices, and carefully place 4, 5, and 6 on the midpoints, we get a solution

1

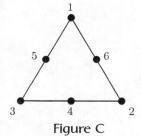

Figure C

to the problem with a triangle whose side sum is 9 (see Figure C).

Many students might stop playing the game now that they have found a solution to the problem. But as mathematicians, we are interested in the other questions that this solution raises, such as:

1. Is this the only way to get 9 as the side sum?

2. Can any other numbers be side sums?

3. Can 8 be a side sum? Or 13?

4. Can there be two different ways to place the numbers to get the same side sum in the game?

By experimenting, we hope to find a complete solution to the Triangle Game.

Practice Problem 0.2. *Show that if you put a 1, 5, and 3 from top to bottom along the left side of the triangle, there is only one way to solve the Triangle Game, and that is the solution we already know.*

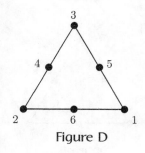

Figure D

We could try to develop an algorithm to solve the Triangle Game in several ways. Let's see what happens if we put 1 at a different vertex, say, the lower left. We could now proceed in either of two ways. One would be to start with a 1 at the lower left corner and redo all the computations to fill out the triangle. The other way would be to notice that if we rotate our original solution triangle (Figure C) 120° counterclockwise, we get the 1 at the lower left vertex and have a ready-made solution (see Figure E).

Notice that if we rotate Figure C 120° clockwise, the side sum is unchanged, but the numbers are in different places on the picture of the triangle (see Figure D).

We could also rotate Figure C another 120° clockwise, or 240° in all, and get Figure E, which we found before by rotating Figure C 120° counterclockwise.

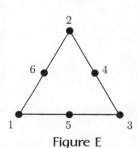

Figure E

There's another way to get a side sum of 9 using the original solution (Figure C) by flipping or reflecting it through a line that goes through any vertex and the opposite midpoint. For example, if we draw such a line through the top of our original triangle (redrawn as Figure F, left), and flip or reflect around this line, we get another different-looking triangle (Figure F, right), still with a side sum of 9.

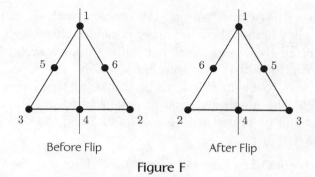

Figure F

There are two other flips of the original solution, one through the line from the bottom right vertex to the opposite side (Figure G), and one through the line from the bottom left vertex to the opposite side (Figure H).

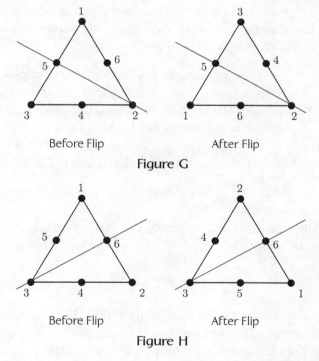

Figure G

Figure H

So we now have six different-looking triangles, all with a side sum of 9. Even though the solutions look different, the same set of numbers is on the corners and on the midpoints. In all these triangles, the numbers 1, 2, and 3 are on vertices and 4, 5, and 6 are on the midpoints.

Although we have now "solved" the problem with six different triangles, we can ask whether there are other possible solutions. This is where we need to bring some organizational unity to our search. We look for ideas, or connections, that lead to other solutions.

Here's a way to begin this kind of thinking: to place the numbers on the triangle, the number 6 must go somewhere, and the two smallest numbers that can be paired with 6 are 1 and 2. So the smallest possible side sum is 9.

Practice Problem 0.3. *Using the same ideas, show that the largest possible side sum is 12.*

Through this deductive process, we can conclude that the only possible side sums are 9, 10, 11, and 12.

Practice Problem 0.4. *Using methods of experimentation and deduction, show that there is a solution for each of the side sums 9, 10, 11, and 12. Also show that all other solutions with each side sum are related to your solution by rotation or reflection of the triangle.*

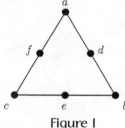

Figure I

Actually, if we use some elementary algebraic techniques, we can refine this process to something less experimental. Start with the letters a, b, c, d, e, f to represent the integers from 1 to 6 in some order. We place the letters a, b, c at the vertices and fill in the midpoints with the letters d, e, f (Figure I).

Assuming that we have a solution of the Triangle Game, we can write

$$(a + d + b) + (b + e + c) + (c + f + a) = 3S,$$

where S is the side sum. The sum of the integers from 1 to 6 is 21, so this leads to the equation

$$a + b + c + 21 = 3S.$$

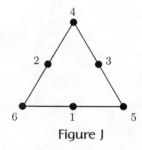

Figure J

From this, we can conclude immediately that $a + b + c$ is divisible by 3. Starting with the smallest values, $a = 1$, $b = 2$, $c = 3$, we get $6 + 21 = 3S$. This tells us that to have a side sum 9, we must have 1, 2, and 3 as the vertices. Taking the next step, the largest values of a, b, and c are $4, 5, 6$. This leads to the equation $15 + 21 = 3S$, or $S = 12$. This is the largest possible value of S. A solution with $4, 5$, and 6 on the vertices and 12 as a side sum is pictured in Figure J.

Practice Problem 0.5. *For the value $S = 10$, we have $a + b + c = 9$. Find three distinct values of the integers from 1 to 6 that sum to 9. Is there more than one case? Which of these cases can be used to create a side sum of 10?*

Is there a way to get one of these solutions from another one? Think about how we might generate a solution with a different side sum from the triangle with side sum 9. One possibility is to use an idea called duality, which exchanges the number x at a vertex with the number $7 - x$. For example, in the triangle with side sum 9 we would interchange the numbers 1 and 6, 2 and 5, and 3 and 4. This results in a triangle with side sum 12 (see Figure K).

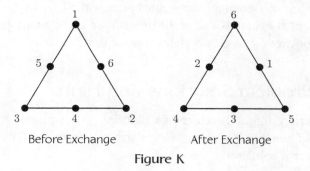

Before Exchange   After Exchange

Figure K

Practice Problem 0.6. *Try to do this exchange with the triangles with side sum 10 and side sum 11 to see what happens.*

We may finally conclude that in order to completely solve the Triangle Game, we need to produce only one solution with side sum 9 and one with side sum 10, and then use reflections, rotations, and duality to get all the others. These mathematical ideas thus streamline the way to arrive at a complete solution.

Why do people enjoy problems such as the Triangle Game? There are many reasons that different people find it appealing. One is that it has one easy solution. Another is that it doesn't have just one solution. There are even some arrangements that we can argue are impossible. There is some interplay in this problem between the numbers on the figure and the way the figure moves around with rotations and reflections. This gives a nice relation between the numerical and geometrical ideas that characterize mathematics. In addition, in this problem we can get a small number of basic solutions and use ideas such as symmetry and duality to generate all the solutions to the problem. Finally, this problem leads to generalizations and therefore new problems. For example, what would happen if we played the Square Game, where we used the numbers 1 through 8 on the vertices and midpoints of the edges of a square? Why stop there? What about a pentagon or hexagon game? Can we play a similar game in three dimensions? All of these questions are interesting to ponder.

The Triangle Game illustrates what we are trying to do in this book. Even though this text does not emphasize computational algebra or prove an overwhelming number of theorems, it does generate a significant number of mathematical ideas. We will work carefully with these ideas to give a rigorous treatment of the underlying mathematical concepts. We hope the reader will engage these mathematical ideas in a positive way and learn about mathematical proof in number theory and geometry and the relation between these two rich areas of mathematics.

Practice Problem Solutions and Hints

0.1. One strategy might be to try to place three numbers on a side that add up to 9. Then see if you can figure out where to put the other three numbers to get a solution.

0.2. This starting position forces the number 6 to be on the midpoint of the side where 1 is a vertex. Then the places for 4 and 2 are determined.

0.3. The number 1 must go somewhere. The largest numbers that can be paired with 1 are 5 and 6. So the largest possible side sum is 12.

0.4. Solutions with sums of 10, 11, and 12 are shown below.

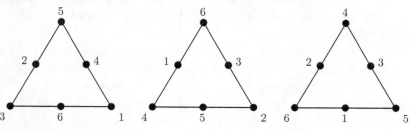

0.5. The possible solutions are $1+2+6 = 9$, $1+3+5 = 9$, and $2+3+4 = 9$. Only the second of these yields a solution with side sum 10. Try it!

0.6. Look at the triangles in Practice Problem 0.4.

Exercises

0.1. The Cross Game is played by placing each of the numbers $\{1, 2, 3, 4, 5\}$ in one of the boxes on the cross so that the sum on each of the two diagonals is the same.

(a) Find a solution with a sum of 9. Explain why any other solution with a sum of 9 must have the same number in the center square.

(b) Find all solutions that have a sum of 9. For each one, describe a rotation or a flip of the cross that changes your solution in (a) into each of the other possible solutions.

(c) Why must every solution (having any sum) have an odd number in the middle?

(d) Find all possible "row sums" that occur in solutions to the Cross Game.

0.2. Suppose that the rules of the Cross Game are changed to use products: Using the numbers $\{2, 3, 4, 5, 6\}$, fill in the grid so that you obtain the same products in each of the two rows.

(a) Find a solution to this Cross Game.

(b) Why must 5 be the center number in every solution?

(c) If instead you use the numbers $\{1, 2, 3, 4, 5\}$, are there any solutions to the game?

(d) Do any other sequences of five consecutive positive numbers yield a solution?

0.3. Find all the possible ways of placing the numbers $\{5, 6, 7, 8, 9, 10\}$ on the edges and corners of a triangle to get the same sum on each side. How are these related to the solutions of the original Triangle Game?

0.4. The following exercises develop the idea of duality from an algebraic standpoint:

(a) Suppose you have a solution to the Triangle Game using the integers 1 to 6 with side sum S. Show that if you replace each of the integers 1 to 6 with its negative, you get a triangle with side sum $-S$ along each edge.

(b) Suppose you have a solution to the Triangle Game using the integers 1 to 6 with side sum S. Show that if you pick a constant C and add it to each of the integers 1 to 6 at their positions on the triangle, you get a triangle with side sum $S + 3C$ along each edge.

(c) Show that parts (a) and (b) justify the concept of duality.

0.5. The Square Game is played by placing the numbers $\{1, 2, 3, 4, 5, 6, 7, 8\}$ on the midpoint of each edge and vertex of a square so that the sum along each side is the same. Each of the following squares is a possible start to the Square Game. Either complete it or show that no solution is possible.

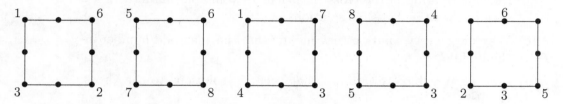

0.6. Use an algebraic approach similar to that used in the Triangle Game to analyze the Square Game.

(a) Show that if a, b, c, d are placed on the vertices of a square and e, f, g, h are placed on the midpoints, you get $a + b + c + d + 36 = 4S$.

(b) Determine that the possible side sums for the Square Game are 12, 13, 14, and 15.

0.7. None of the arrangements below can lead to a solution of the Square Game. Is there a way to show this is true, other than trying all possible combinations of the remaining numbers?

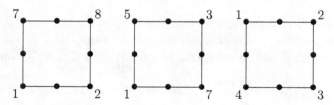

0.8. Find all solutions to the Square Game.

0.9. A number is a *lower bound* for a set of solutions if all solutions are greater than or equal to that number. Similarly, a number is an *upper bound* for a set of solutions if all solutions are less than or equal to that number. Find a lower bound and an upper bound for the side sums of a Pentagon Game if we use the numbers $\{1, 2, \ldots, 10\}$.

0.10. Is there a solution to the Pentagon Game for every number between the bounds you found in Exercise 0.9? **Hint:** Look specifically at 15.

0.11. Suppose n is any odd number greater than or equal to 3. Can you find a solution to the n-gon game using the numbers $\{1, 2, \ldots, 2n\}$?

0.12. Suppose n is any even number greater than or equal to 4. Can you find a solution to the n-gon game, using the numbers $\{1, 2, \ldots, 2n\}$?

0.13. A *magic square* is an $n \times n$ grid in which each of the numbers $\{1, 2, 3, \ldots, n^2\}$ is used once and the sum of each row, column, and diagonal is the same. For example,

16	3	2	13
5	10	11	8
9	6	7	12
4	15	14	1

is a 4×4 magic square whose rows, columns, and diagonals add up to 34.

(a) Find all possible 3×3 magic squares.

(b) Prove that you have found all of the possible 3×3 magic squares.

(c) Find some more 4×4 magic squares. What is their row sum? Why is there only one possible row sum?

(d) Find a formula for the row sum of an $n \times n$ magic square.

(e) Now suppose you are working with a rectangular grid, and there are no restrictions on which numbers you can use (all real numbers are fair game and numbers can be repeated). Can you fill in a 4×5 grid so that each row and each column adds up to 0?

(f) Can you fill in a 4×5 grid with nonzero numbers so that each row and column adds up to 10?

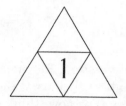

The Beginnings of
Number Theory

1.1 Setting the Table: Numbers, Sets, and Functions

Games such as the Triangle Game are intriguing to people who enjoy mathematics. These games build on elementary arithmetic skills and provide problems that have multiple solutions and even give rise to more problems than the initial simple question. They presume only a knowledge of basic arithmetic and a facility with elementary calculation.

Numbers and Number Systems

Before we start with the formal aspects of arithmetic and number theory, let's discuss informally some of the terms and ideas encountered in elementary school. First, we need to identify different number systems so that we know precisely which numbers we are using in our problem solving attempts. The first set of numbers we meet are the *natural numbers* $\mathbb{N}$. These are the ones we use to count objects: $1, 2, 3, 4, 5, \ldots$. Next come the *integers*, which include 0 and the negatives of all the natural numbers. The letter $\mathbb{Z}$ stands for the integers, and comes from the German word for number, *zahlen*. The natural numbers allow us to do addition and multiplication, and the integers allow us to do both of these as well as subtraction. But when we get to division, we need the *rational numbers*. The rational numbers are a rich source of number theory questions. These numbers, sometimes called fractions, are those that can be expressed as the quotient of two integers a and b, written $\frac{a}{b}$, where b is not allowed to be 0 and the fraction is reduced to lowest terms. We label the rational numbers $\mathbb{Q}$, for quotient.

Finally, we need the *real numbers*, $\mathbb{R}$, which include the rational numbers and also irrational numbers such as π and $\sqrt{2}$. These numbers will surface when we do geometry. For now, think of the real numbers as the set of all decimals.

Sets

Mathematicians use the word *set* to mean the same thing as the word does in ordinary English, namely, a collection of objects. We assume that the objects, or things, that make up a set come from some "big" set that is large enough to include all the objects we might need. This big set is sometimes called the "universe" or the "universal set." For now, we denote this big set by the letter X.

Just as we use numerals to represent numbers, we can use symbols to represent sets and their members. If S is a set, we write $s \in S$ to indicate membership in S, and we say that *s is an element of S*. Because we think of a set as being determined by its members, we can specify a set S simply by listing its elements. If the set is finite—for example, the first four months of the year—we write $S = \{$January, February, March, April$\}$. If the set is large, maybe even infinite, one way to describe the set is to list the first few elements and describe the set to the reader. For example, we can write the set of the natural numbers $\mathbb{N}$ as $\mathbb{N} = \{1, 2, \ldots\}$. Here, those three dots mean that the pattern continues as indicated.

Practice Problem 1.1. *Use set notation to write the following sets:*

(a) S, whose members are the first five letters of the alphabet.

(b) $\mathbb{Z}$, whose members are the integers.

There is a second way to specify the members of a set that satisfy some condition. For example, we may want to use set notation to write the set S consisting of all academy award winners. Here, we want to specify just those people from the universe X who have won an academy award. In this example, the universe X could be the set of human beings or X could be the set of all people who have lived since 1900. There is a way to write the set of all objects $x \in X$ that satisfy some condition $P(x)$. The proposition $P(x)$ just represents a statement that is either true or false, depending on the value of x. In our academy award example, $P(x)$ is the statement "x has won an academy award." $P(x)$ is true if person x has won an academy award, and false otherwise. In abbreviated notation, we write this set S as $S = \{x \in X \mid x$ has won an academy award$\}$, and we read this as "the set of x in X such that x has won an academy award."

Practice Problem 1.2. *Use the abbreviated set notation and identify the universe X and the statement $P(x)$ for the following sets:*

(a) *S is the set of all even numbers between, but not including, 3 and 50.*

(b) *S is the set of all presidents of the United States who have been senators.*

(c) *S is the set of all positive integers less than or equal to 8.*

There is one special set that has no members at all. We denote this set, called the *empty set*, by the symbol $\emptyset$.

Example. The set $S = \{$All women who were president of the United States prior to the year 2008$\}$ is the empty set.

Other ideas about sets include how to combine them and how to make new sets from ones you already have. In this text, we focus on a few of the ideas that make our geometry and number theory discussions clearer. One of these ideas is that of a subset of a set S. We say that A is a *subset* of S if every element of A is also an element of S (that is, if $x \in A$, then also $x \in S$). We write this as $A \subset S$. Additionally, we say that A is a *proper subset* of S if $A \subset S$ and $A \neq S$.

Example. If $S = \{$all even numbers$\}$ and $A = \{$all multiples of 4$\}$, then A is a subset of S.

Example. A set S is always a subset of itself. The empty set $\emptyset$ is always a subset of any set S, and is a proper subset of S if S is not the empty set.

To state our ideas precisely, from now on, we assume that all sets are subsets of some universal set X that contains all the elements we will ever deal with.

Practice Problem 1.3. *Write out the subsets of the set $S = \{1, 2, 3\}$.* **Hint:** *There should be eight subsets.*

We can combine two sets by creating their union. The *union* of A and B, written $A \cup B$, consists of all elements of A and all elements of B. That is, $A \cup B = \{x \mid x \in A$ or $x \in B\}$.

Example. If $A = \{1, 2, 3\}$ and $B = \{2, 4, 6\}$, then $A \cup B = \{1, 2, 3, 4, 6\}$. Note that the 2 is not mentioned twice in $A \cup B$.

Another way to combine two sets is to create their intersection. The *intersection* of A and B, written $A \cap B$, is the set of all elements that are elements of both A and B. That is, $A \cap B = \{x | x \in A \text{ and } x \in B\}$.

Example. If $A = \{1, 2, 3\}$ and $B = \{2, 4, 6\}$, then $A \cap B = \{2\}$.

Practice Problem 1.4. *Suppose* $A = \{1, 2, 3, 4, 5, 6\}$, $B = \{2, 4, 6, 8\}$, *and* $C = \{1, 3, 6\}$.

(a) *Write out the elements in the following sets:* $A \cup B$, $A \cap B$, $B \cup C$, $B \cap C$, $A \cup C$, *and* $A \cap C$.

(b) *Does the set* $A \cup B \cup C$ *make sense?*

(c) *Show that* $A \cup \emptyset = A$ *and* $A \cap \emptyset = \emptyset$.

There is a special set associated with a particular set S. This new set, called the *power set* of S and denoted $\mathcal{P}(S)$, consists of all subsets of the set S. The answer to Practice Problem 1.3 is the members of $\mathcal{P}(S)$ for the set $S = \{1, 2, 3\}$.

Another idea that is familiar from work with the coordinate plane and graphing is the Cartesian product of two sets. The coordinate plane is the set of all pairs (x, y) of real numbers, where x denotes the "first coordinate" and y the "second coordinate." We call the symbol (x, y) an ordered pair because order does make a difference. The point $(1, 2)$ on the coordinate plane is quite different from the point $(2, 1)$. We can make a general definition for the Cartesian product of two sets A and B.

Definition. Suppose A and B are sets. The *Cartesian product* of A and B, written $A \times B$, is defined as the set $A \times B = \{(a, b) | a \in A \text{ and } b \in B\}$.

A way to describe the Cartesian product of A and B is the set of all ordered pairs where the first element comes from the set A and the second element comes from the set B.

Example. If $A = \{1, 2, 3\}$ and $B = \{a, b\}$, then $A \times B = \{(1, a), (2, a), (3, a), (1, b), (2, b), (3, b)\}$.

Practice Problem 1.5. *For* $A = \{1, 2, 3\}$ *and* $B = \{a, b\}$, *write out the elements of the Cartesian product* $B \times A$. *Are the sets* $A \times B$ *and* $B \times A$ *the same or different?*

More information about operations on sets can be found in the Exercises.

Functions

The concept of function is another idea that is important in mathematics. Informally, a function from a set A to a set B is a correspondence between elements of A and elements of B so that each element of A is assigned to exactly one element of B. This is the same idea of a function we encounter in other mathematics courses, where the sets A and B are the real numbers and there is ordinarily a rule or an explicit formula that describes the correspondence.

Example. The function $f(x) = x^2$ assigns to each real number x its square, x^2.

Notice in this example that the numbers $x = 2$ and $x = -2$ have the same value assigned by this function, namely, 4. This isn't against the rules!

We need a more formal definition of a function.

Definition. Suppose A and B are nonempty sets. A *function* from A to B is a subset of $A \times B$ such that each element of A occurs exactly once as a first coordinate.

Practice Problem 1.6. *Which of the following sets of ordered pairs are functions from A to B, where $A = \{a, b, c\}$ and $B = \{1, 2, 3, 4\}$?*

$$S = \{(a, 2), (a, 3), (b, 4), (c, 1)\};$$
$$T = \{(a, 2), (b, 2), (c, 2)\};$$
$$R = \{(a, 1)\};$$
$$P = \{(a, 1), (b, 2), (c, 3)\}.$$

Most people do not think of ordered pairs when they think of functions. The informal idea of a function as a correspondence satisfying certain properties often proves more useful. Typically, we use the letter f, as we did when we said $f(x) = x^2$. To indicate that f is a function from A to B, we use the notation $f : A \to B$.

Next, we need a language to describe functions and their related sets.

Definition. Suppose f is a function from a set A to a set B. The set A is called the *domain* of f and the set B is called the *range* of f. The set of elements $f(A) = \{f(a)|a \in A\}$ is the *image* of f. If $f(A) = B$, then we say that f is *onto* B. Along with the property of being a function, f may also have the property that each element in the image of f corresponds to exactly one element in the domain. A function with this property is called *one-to-one*. This means that f is one-to-one if whenever $f(a) = f(a')$, then $a = a'$.

Practice Problem 1.7. *Determine which of the functions in Practice Problem 1.6 are one-to-one.*

Practice Problem 1.8. *What is the image of each of the functions in Practice Problem 1.6?*

There are many ways to get "new functions from old." If domains agree and it makes sense to add, subtract, or multiply elements of the range, we can combine two functions to get a new one. Dividing functions may require some additional restrictions, however.

Example. Suppose f is the function $f(x) = x^2$ and g is the function $g(x) = x + 1$, where the domain of f and g is $\mathbb{R}$. Then we can create a new function $(f + g)(x)$ by adding f and g. So $(f + g)(x) = x^2 + x + 1$. If we wanted to create the quotient $\frac{f(x)}{g(x)}$, we would have to be careful to exclude any zero denominators so that the quotient $\frac{x^2}{(x+1)}$ makes sense. Therefore, we must exclude $x = -1$ from the domain of $\frac{f}{g}$.

Practice Problem 1.9. *What would the function $f - g$ be? How about the function $f \cdot g$? What are their domains?*

There is another way to combine functions that is useful.

Definition. Suppose A, B, and C are sets and f and g are functions, where $f : A \to B$ and $g : B \to C$. The *composition* of f and g is a function $g \circ f : A \to C$ defined by $(g \circ f)(a) = g(f(a))$.

Example. Suppose f is the function $f(x) = x^2$ and g is the function $g(x) = x + 1$, both with domain $\mathbb{R}$. Then $(g \circ f)(x) = g(x^2) = x^2 + 1$. The composition $(f \circ g)(x) = f(x + 1) = (x + 1)^2$. Both $g \circ f$ and $f \circ g$ also have domain $\mathbb{R}$. Note that $f \circ g \neq g \circ f$.

Practice Problem 1.10. *Let $f(x) = x - 7$, $g(x) = x^2$, and $h(x) = \frac{1}{x}$. What is $f \circ g$? What is $g \circ f$? What is $f \circ h$? What is $h \circ h$? What are the domain restrictions in each case that make the composition make sense?*

Math Words

Let's go back to the number systems we started with and think about the associated mathematical vocabulary words familiar from school. The first words that pop into people's minds have to do with operations on numbers: addition, multiplication, subtraction, and division. When we say "numbers," we are talking about real numbers unless we specify otherwise. Let's focus on the topic of addition first. What are some properties of addition? There is a fancy name for the fact that the order in which you add two numbers is irrelevant. This is the *commutative property of addition:*

$$a + b = b + a.$$

You can also add a column of numbers without worrying how to group them in a particular order. This is the *associative property of addition:*

$$(a + b) + c = a + (b + c).$$

These same two properties, commutativity and associativity, also hold for multiplication:

$$a \cdot b = b \cdot a,$$

$$(a \cdot b) \cdot c = a \cdot (b \cdot c).$$

The operations of addition and multiplication are related by a third property, the *distributive property*, which we know from early work in algebra:

$$a \cdot (b + c) = (a \cdot b) + (a \cdot c).$$

The distributive law is the property that really makes arithmetic work.

Practice Problem 1.11. *Assuming the commutative property, explain why you can write the distributive property in the following way:*

$$(b + c) \cdot a = (c \cdot a) + (b \cdot a).$$

Other words are associated with addition and multiplication. Two of these words are "identity" and "inverse." Each of our operations has an identity element. An *identity* for an operation is an element that, when combined with any element, gives that second element back again. In the operation of addition, 0 is the identity since for any number a,

$$a + 0 = a,$$

and because of the commutative law,

$$0 + a = a.$$

In the operation of multiplication, 1 plays the role of the identity since

$$a \cdot 1 = a,$$

and because of the commutative law,

$$1 \cdot a = a.$$

If an operation has an identity, then we can ask whether elements have inverses in the system related to that operation. In each operation, if we take an element a, then the *inverse* of a is the element we combine with a to get the identity for that operation. Let's focus first on addition. For the *additive* inverse of a number a, we need a number that, when added to a, gives the additive identity, or 0. We label this inverse "$-a$" or "the opposite of a." Sometimes we call the additive inverse of a "minus a." We have

$$a + (-a) = 0,$$

and because of the commutative law,

$$(-a) + a = 0.$$

Example. What are the additive inverses of some sample numbers?

1. The number 3 has additive inverse -3, since $3 + (-3) = 0$.

2. The number -7 has additive inverse $-(-7)$, since $(-7) + (-(-7)) = 0$.

Example 2 may look a bit awkward, and you may recall from elementary arithmetic that $-(-7) = 7$. This magic transformation of two negatives into a positive takes a little more discussion to understand completely; so we return to it later.

To get the *multiplicative* inverse of a number a, we need a number that, when multiplied by a, gives the multiplicative identity, or 1. We label this inverse "a^{-1}," and we have

$$a \cdot a^{-1} = 1,$$

and because of the commutative law,

$$a^{-1} \cdot a = 1.$$

In elementary arithmetic, the symbol a^{-1} is often written $\frac{1}{a}$. Sometimes we call the multiplicative inverse of a "a inverse."

Example. What are the multiplicative inverses of some sample numbers?

1. The number -3 has multiplicative inverse $-\frac{1}{3}$, since $(-3) \cdot (-\frac{1}{3}) = 1$.
2. The number $\frac{2}{7}$ has multiplicative inverse $\frac{7}{2}$, since $\frac{2}{7} \cdot \frac{7}{2} = 1$.
3. The number 1 is its own multiplicative inverse.
4. The number -1 has multiplicative inverse -1, since $(-1) \cdot (-1) = 1$.

Here are two more observations about multiplicative inverses. First, without a multiplicative identity, a multiplicative inverse cannot exist since the definition of an inverse depends on there being an identity. Second, what is the multiplicative inverse of 0? Students learn in elementary school that they cannot divide by 0, which is to say that 0 does not have a multiplicative inverse. This isn't just an arbitrary rule; there are valid mathematical reasons why it is so. Don't just accept this as a rule to be followed; reading further in this book will help to understand why this is a necessary consequence of the properties of addition and multiplication.

Practice Problem 1.12. *Can you figure out a mathematical reason why you can't divide by 0?*

Let's go back to the other operations for a minute. What about the properties of commutativity and associativity for, say, subtraction?

Practice Problem 1.13. *Give examples of numbers that make the following equations true:*

(a) $a - b \neq b - a$.

(b) $(a - b) - c \neq a - (b - c)$.

From the result of Practice Problem 1.13, we can conclude that subtraction has neither the commutative nor the associative property.

This problem illustrates a point: when we state a property of numbers for an operation, it has to hold for all numbers. To show that a certain property doesn't hold, we have only to find one set of numbers for which it fails. We can then call this set of numbers a counterexample. In mathematics, exceptions to rules, or counterexamples, are often a rich source of good ideas and questions.

1.2 Rules of Arithmetic

When we pursue mathematics in a formal way, it is customary to establish a set of rules, or "axioms," that govern the behavior of the set of objects we are studying. This, in turn, allows us to deduce properties of the system. As we pointed out in Chapter 0, these rules are established on the basis of experience. Our objective is to impose as few rules as possible and deduce all the properties of the system directly from these rules. The properties of addition and multiplication, discussed earlier in this chapter, will become part of the axioms or rules of arithmetic. These rules will not only govern the arithmetic in the usual number systems, such as $\mathbb{Z}$ and $\mathbb{Q}$, but also in other number systems that arise in subsequent chapters.

We want these rules to hold for specific operations on sets of numbers. Keep in mind that to work with an operation, we must always specify on what set of elements it operates. An operation is called *binary* if it combines two elements of a set to get an element of the universal set X. The "bi" part of the word means that we will be taking just two things at a time, not three or more. Most of our operations in elementary arithmetic are binary. As we saw earlier, addition on $\mathbb{N}$ is really a binary operation, and multiplication on $\mathbb{Q}$ is another binary operation.

Example. A nonmathematical operation is combining colors in a painter's palette. Combining two colors at a time is a binary operation. For example, red and yellow yield orange, and blue and red give purple. However, combining the three colors red, yellow, and blue all at once to get brown is not binary.

If we start with a binary operation on a set, the set is called *closed* under this binary operation if, whenever we combine two elements from that set, the result is an element of the same set. The set $\mathbb{N}$ is closed under addition since whenever we add two natural numbers, we get a third natural number. The set $\mathbb{N}$ is not closed under subtraction because, although we sometimes get a natural number, sometimes we don't. For example, if we subtract 7 from 2, we get -5, which is in $\mathbb{Z}$ but is not in $\mathbb{N}$. We have to go outside $\mathbb{N}$ to get the answer for $2 - 7$, so $\mathbb{N}$ is not closed under the operation of subtraction.

Example. Think about painting again. If the set is just the primary colors of blue, red, and yellow, this set is not closed under combining colors, since the secondary colors are outside the set. If we use the same operation of combining colors on the set of all shades of color, the set will be closed under the operation of combining colors.

It is important to specify both the set and the operation to check the binary and closure properties. Closure depends on the set and operation. For example, we have already seen that $\mathbb{N}$ is not closed under subtraction, but if we use $\mathbb{Z}$ instead, the set is closed under subtraction. Similarly, the integers are not closed under the operation of division. To get a set that is closed under division, we must use the nonzero rational numbers so that all the fractions are included.

Next, we list the Rules of Arithmetic. In these rules, we always assume the operations are binary and that the set of numbers is closed under the binary operations. We use $+$ to denote addition and $\cdot$ to denote multiplication. We call these rules by the more formal term axiom.

Axiom A1 (Commutativity of Addition). For any two elements a and b from the set, $a + b = b + a$.

Axiom A2 (Associativity of Addition). For any three elements a, b, and c from the set, $a + (b + c) = (a + b) + c$.

Axiom A3 (Additive Identity). There is a unique element 0 that is the identity for addition. That is, if a is any element of the set, $0 + a = a$ and $a + 0 = a$.

When we say the element 0 is *unique*, we mean that no other element in the system can have the same properties as 0.

Axiom A4 (Additive Inverse). If a is any element of the set, then there is a unique corresponding element $-a$ that is the additive inverse of a. That is, $a + (-a) = 0$ and $(-a) + a = 0$.

Axiom M1 (Commutativity of Multiplication). For any two elements a and b from the set, $a \cdot b = b \cdot a$.

Axiom M2 (Associativity of Multiplication). For any three elements a, b, and c from the set, $a \cdot (b \cdot c) = (a \cdot b) \cdot c$.

Axiom M3 (Multiplicative Identity). There is a unique element $1 \neq 0$ that is the identity for multiplication. That is, if a is any element from the set, $1 \cdot a = a$ and $a \cdot 1 = a$.

Axiom M4 (Multiplicative Inverse). If a is any nonzero element of the set, then there is a corresponding unique element a^{-1} that is the multiplicative inverse for a. That is, $a \cdot a^{-1} = 1$ and $a^{-1} \cdot a = 1$.

Axiom D (Distributivity). If a, b, and c are any three elements from the set, $a \cdot (b + c) = (a \cdot b) + (a \cdot c)$.

We have already looked at these rules for our familiar number systems. Axioms A1 and A2 hold whether we are doing arithmetic in $\mathbb{N}$, $\mathbb{Z}$, $\mathbb{Q}$, or $\mathbb{R}$. The number 0 is an additive identity in three of these four number systems. But one of these systems, $\mathbb{N}$, fails to include 0, so it has no additive identity. The term *whole numbers* is sometimes used to refer to the set of natural numbers along with 0. The set of whole numbers does have an additive identity. Also, since the numbers in $\mathbb{Z}$, $\mathbb{Q}$, and $\mathbb{R}$ have additive inverses, Axiom A4 holds in each of these systems. But since there is no additive identity in $\mathbb{N}$, it makes no sense to look for additive inverses for any of the elements. Thus, Axioms A3 and A4 fail in $\mathbb{N}$.

Similarly, Axioms M1, M2, and M3 work just fine in all these number systems. Here, the number 1 plays the multiplicative identity role in all of them. However, both $\mathbb{N}$ and $\mathbb{Z}$ have trouble with Axiom M4. To get multiplicative inverses, we need to include fractions. Remember that we are not looking for a multiplicative inverse for 0 in any of these systems, since Axiom M4 only requires multiplicative inverses only for nonzero elements. So Axiom M4 holds in $\mathbb{Q}$ and $\mathbb{R}$.

1.3   A New System

Let's make a new operation $\oplus$ called "parity sum," which tells whether the sum of two integers is odd or even. We know from elementary arithmetic that if we add two even numbers, we get an even number. If we add two odd numbers, we also get an even number. If we mix them and add an odd number to an even number, we get an odd number. If we let our set consist of the two words "Even" and "Odd," this set is closed under the binary operation parity sum. If we make an addition table for our parity sum, we get Table 1.1.

Let's check the addition rules for this operation. Commutativity and associativity are easy to see by just thinking about how addition of numbers works, so Axioms A1 and A2 hold in this system. What about an identity for this operation? A close check will show that Even is behaving

⊕	Even	Odd
Even	Even	Odd
Odd	Odd	Even

Table 1.1. Parity sum.

⊙	Even	Odd
Even	Even	Even
Odd	Even	Odd

Table 1.2. Parity multiplication.

like an identity here, since if we add Even to Even, we get Even, and if we add Even to Odd, we get Odd. So in this operation, Even plays the role of 0, or the additive identity.

What about inverses? Again, a check shows that Even is its own inverse. The same is true of Odd. So we have found a new system that is very simple and in which all four of the addition rules, Axioms A1, A2, A3, and A4, hold.

Let's try introducing "parity multiplication," or ⊙ on this set of Even and Odd, with Table 1.2 as the multiplication table.

Here, as for the rules for addition, Axioms M1 and M2 clearly hold. Do we have something that acts like a multiplicative identity? There aren't many things to try. Even ⊙ Even = Even makes it look like Even might work, but Even ⊙ Odd = Even shows that Even doesn't leave Odd alone. What about trying Odd as the identity? This time, Odd ⊙ Odd = Odd, and Odd ⊙ Even = Even, so Odd must be the multiplicative identity.

Thus, in this system of parity multiplication, Axioms M1, M2, and M3 hold. To check Axiom M4, we need to look for multiplicative inverses only for the nonzero elements in the set. Since Even plays the role of the additive identity 0, we need to check only that Odd has a multiplicative inverse. Since Odd ⊙ Odd = Odd, we can see that Odd is its own inverse! So we have discovered a new system in which Axioms A1–A4 and M1–M4 hold.

Practice Problem 1.14. *Check that Axiom D, Distributivity, holds in this new system.*

1.4   One's Digit Arithmetic

Let's try another number system that builds on the familiar operations we already know. We begin by thinking about "one's digit arithmetic." To introduce this system, we first describe the operation of one's digit addition. In one's digit addition, we will be adding integers in the usual way, but instead of getting the normal answer, we will use just the one's digit as the answer. That is, to get an answer for $a + b$ in one's digit arithmetic, we add $a + b$ in the regular way, and then our answer is the one's digit of the result.

Example. Let's add some numbers in this system. If we add $2 + 3$, we still get 5. If we add $9 + 6$ in normal arithmetic, we get 15. In one's digit arithmetic, however, we only want to have the digit in the one's place as the answer. Therefore, in this system, $9 + 6 = 5$.

Practice Problem 1.15. *What is $9 + 8$? What is $6 + 7$?*

What answers do we get when we add numbers in this system? Since the only possible numbers that can appear as the one's digit are 0, 1, 2, 3, 4, 5, 6, 7, 8, and 9, these are the only numbers we can use in the system. So, for one's digit arithmetic, we take $\{0, 1, 2, 3, 4, 5, 6, 7, 8, 9\}$ as our set of numbers. Table 1.3 is a table of addition in this system.

Now let's make some observations about Table 1.3. Take a look at the table now and see how many things you notice about it.

Practice Problem 1.16. *Take a look at the diagonals from the lower left to the upper right, for example. Do you see a pattern? Compare columns and rows. How many of each number are present in each row and column? Are there any repeats in a row or column? Do you see a pattern in how to get from one row to the next? Is there any symmetry in the table?*

+	0	1	2	3	4	5	6	7	8	9
0	0	1	2	3	4	5	6	7	8	9
1	1	2	3	4	5	6	7	8	9	0
2	2	3	4	5	6	7	8	9	0	1
3	3	4	5	6	7	8	9	0	1	2
4	4	5	6	7	8	9	0	1	2	3
5	5	6	7	8	9	0	1	2	3	4
6	6	7	8	9	0	1	2	3	4	5
7	7	8	9	0	1	2	3	4	5	6
8	8	9	0	1	2	3	4	5	6	7
9	9	0	1	2	3	4	5	6	7	8

Table 1.3. One's digit addition.

+			a			b			
a			$a+a$			$a+b$			
b			$b+a$			$b+b$			

Table 1.4. Axiom 1 in one's digit arithmetic.

Let's consider this new system together with the Rules of Arithmetic that we stated for ordinary addition. First, is this system closed under the binary operation +? Since we are using only the set of numbers 0 through 9 in our new system, we see that combining any two of them gives a third element of this same set. There are no "strangers" in the table, so the set is closed under the binary operation of addition.

We next consider whether Axioms A1–A4 hold in this new system. Let's begin with Axiom A1, the Commutative Property of Addition. In ordinary arithmetic, we know that Axiom A1 holds, since $a + b = b + a$. So when we check $(a + b)$'s one's digit, it will be the same as $(b + a)$'s one's digit. So Axiom A1 does hold in our new system.

Think about how Table 1.4 illustrates this fact. If we draw a diagonal line from the upper left to the lower right, Table 1.4 is symmetric across that diagonal line. In other words, $(a + b)$ appears as a reflection of $(b + a)$ in the table. So the symmetry we noticed in the table actually represents the mathematical fact that the operation of addition in one's digit arithmetic is commutative.

Now let's think about Axiom A2, Associativity of Addition. One way to check if this is true for our new system would be to check every one of the 1,000 possibilities for the associative law. Let's try one now.

Practice Problem 1.17. *Show that* $(9 + 8) + 7 = 9 + (8 + 7)$ *in one's digit arithmetic.*

Instead of doing lots of computation, we want to streamline the process of establishing the associative property for our new system. To do this, we will use the associative property of addition of integers. In ordinary arithmetic, $(a + b) + c = a + (b + c)$ for any choices of a, b, and c, so the one's digit of $(a + b) + c$ is the same as the one's digit of $a + (b + c)$. This observation shows that Axiom A2 holds in this new system.

Do we have an identity for addition? The number 0 is a good candidate, and a look at the very first row of Table 1.3 shows that if you add 0 to any number, you get that number back. The same principle is illustrated in the first column of this table. So we do have an additive identity, and it is the familiar number 0.

Finally, we need to check for additive inverses. Here we begin to have something new to think about. There are no negative numbers in Table 1.3. So where will we get the inverses? Axiom A4 actually says that for any number a, we have $-a$ as an additive inverse if $-a$ satisfies $a + (-a) = 0$.

Let's focus our attention on the number 2. Is there a number in Table 1.3 that when added to 2 gives 0 back? A look across the row that begins with 2 shows that 8 has this property. Is 8 the additive inverse for 2? In fact, a look down the column that begins with 2 shows that $8 + 2 = 0$. Thus, $2 + 8 = 8 + 2 = 0$.

This is a new concept. How can this make sense? Think about the additive inverse of 2 in ordinary arithmetic. It is -2. In our new system, the number 8 is playing the role of 2's additive inverse, so here we identify the number 8 with -2. Is 8 really negative 2? We need to be careful here. The minus sign does not necessarily signify a negative number. The minus sign indicates that a number, $-a$, when added to the number a gives 0. In the Rules of Arithmetic, the minus sign simply denotes the additive inverse and nothing else. For example, in ordinary arithmetic, $-(-7)$ is the additive inverse of the number -7, and we know $-(-7)$ is a positive number.

Practice Problem 1.18. *Find* -3, *the additive inverse of 3, in one's digit arithmetic.*

How did we tell by looking at Table 1.3 where 2's additive inverse is? If we look across the row for addition by 2, there is a 0 in the row under the column for 8. This indicates that $2 + 8 = 0$, so that 2 and 8 are additive inverses for each other. So one way to find the additive inverse of a number in the table is to look across its row for a 0 and then read off the number at the top of that column. As we identified the additive inverse of 2 with 8 in the one's digit addition table, the same thing is true for all the numbers in the table. In general, to check for Axiom A4 in a table, check that the identity occurs in every row and every column. In fact, since the axiom says inverses are unique, we must check that the identity occurs exactly once in each row and column. Axiom A4 holds, since every element has exactly one additive inverse.

Practice Problem 1.19. *Find the additive inverses for each number in Table 1.3. Which, if any, are their own additive inverses?*

Is it unsettling that in this system $-3 = 7$ and $-5 = 5$? These are just the familiar Rules of Arithmetic working in a different setting. They do take some getting used to!

We have established that all the rules for addition hold in our system of one's digit arithmetic. We have created a binary operation on the set of numbers 0 through 9 that is commutative and associative, and this set is closed under the binary operation. There is an additive identity and an additive inverse for each element of the set.

Let's now move on to multiplication in our system. To multiply in this system, we use one's digit multiplication; that is, just doing regular multiplication and then using the one's digit of the result as the answer. In other words, the product of two numbers a and b will be the one's digit of the usual product ab.

Example. $7 \cdot 2 = 4$ and $3 \cdot 8 = 4$.

Practice Problem 1.20. *What is $6 \cdot 6$? What is $3 \cdot 7$? What is $4 \cdot 5$?*

Again, the only entries we have in a multiplication table for this system are the numbers $0, 1, \ldots, 9$. It takes a little more effort to fill out the multiplication table, but it does give good practice with all those memorized multiplication facts! Table 1.5 is a multiplication table in this system.

Just as we did for the addition table, we now make observations about the multiplication table.

Practice Problem 1.21. *Before reading further, make a list of your own observations and see whether you have any that are different from the ones listed next.*

·	0	1	2	3	4	5	6	7	8	9
0	0	0	0	0	0	0	0	0	0	0
1	0	1	2	3	4	5	6	7	8	9
2	0	2	4	6	8	0	2	4	6	8
3	0	3	6	9	2	5	8	1	4	7
4	0	4	8	2	6	0	4	8	2	6
5	0	5	0	5	0	5	0	5	0	5
6	0	6	2	8	4	0	6	2	8	4
7	0	7	4	1	8	5	2	9	6	3
8	0	8	6	4	2	0	8	6	4	2
9	0	9	8	7	6	5	4	3	2	1

Table 1.5. One's digit multiplication.

Some observations about the one's digit multiplication table follow.

1. There are a whole lot of 0 entries in the table. The entire first row and column, plus a sprinkling of other places are filled with zeroes.

2. As in the addition table, corresponding rows and columns read the same. For example, the row that begins with 3 and the column that begins with 3 each have the same entries in exactly the same order.

3. There is symmetry across the diagonal that goes from the upper left to lower right. (What mathematical property does this illustrate?)

4. Some rows read in the reverse order of other rows. Compare, for example, just the nonzero entries of the 3 row and 7 row.

5. Several of the rows have repeating patterns. Some repeat in blocks of five numbers (the 4 row, for example) and some repeat in blocks of two (the 5 row, for example).

6. Unlike the addition table, numbers are repeated in rows and columns.

7. Some columns and rows don't have every number.

Other ideas about this table also may seem significant.

Let's check whether Axioms M1–M4 hold for this system. First, is the multiplication commutative? Since $a \cdot b = b \cdot a$ in ordinary multiplication, the one's digit will be the same for each answer. So Axiom M1, Commutativity of Multiplication, holds in one's digit arithmetic. The diagonal symmetry in the table neatly illustrates this fact.

Number	Inverse
0	None
1	1
2	None
3	7
4	None
5	None
6	None
7	3
8	None
9	9

Table 1.6. Multiplicative inverses in one's digit arithmetic.

Just as associativity holds for addition, Axiom M2 holds in this system because it is a consequence of associativity of ordinary multiplication in the integers.

Is there a multiplicative identity? The number 1 is an obvious choice, and since multiplication by 1 does not change the one's digit of any number, it is clear that 1 is the identity for multiplication, so Axiom M3 holds.

Since we have an identity, it makes sense to look for multiplicative inverses for the nonzero entries in the table. Now we run into some difficulty. Checking 1 is easy to start, since $1 \cdot 1 = 1$, so 1 is its own inverse. The next number, 2, is a problem. If we look for a 1 in the 2 row of the table, we are out of luck. There is no number in this system that will multiply by 2 and give an answer of 1. Thus, in this system, Axiom M4 cannot hold because we have found a counterexample: at least one number does not have a multiplicative inverse. However, it is still possible that some numbers do have multiplicative inverses.

Which are the numbers that do have multiplicative inverses? This is the same as asking: in which rows does a 1 appear? Look at each row and figure out which numbers have multiplicative inverses. Table 1.6 lists these numbers and their inverse.

Why do the numbers 1, 3, 7, and 9 have inverses, but all the other numbers do not? A first guess might be that these are all odd numbers. But a closer look shows that even though 5 is odd, it has no multiplicative inverse.

Practice Problem 1.22. *Can you come up with a reason why the numbers 1, 3, 7, and 9 are the only ones that have multiplicative inverses in this system?*

When we looked at the additive inverses, we noticed that 8 played the role of -2, since that's the notation we are using to denote additive inverses. Let's think about multiplicative inverses in arithmetic. For example, what is the number 3's ordinary multiplicative inverse? The fraction $\frac{1}{3}$ is the number that when multiplied by 3 gives the ordinary multiplicative identity of 1. In one's digit arithmetic, 7 is the number that when multiplied by 3 gives the identity, or 1. So in this system, we might call 7 by the name $\frac{1}{3}$. How do we interpret a fraction such as $\frac{4}{7}$? A fraction like this is interpreted as the product of 4 and $\frac{1}{7}$, provided the fraction $\frac{1}{7}$ makes sense. It is interesting to think about what fractions exist and don't exist in the one's digit system.

Practice Problem 1.23. *Make a list of fractions that do exist in one's digit arithmetic. Make another list of fractions that do not exist in this new system.*

If we think about both the addition table and the multiplication table for the one's digit arithmetic system, we can determine whether Axiom D, the distributive law, holds.

Practice Problem 1.24. *Try some entries from the tables to see whether you think Axiom D holds in this system. Try showing $a \cdot (b+c) = (a \cdot b) + (a \cdot c)$ for the following values of $a, b,$ and c:*

a	b	c
5	6	7
8	4	9
7	3	8

Because $a \cdot (b + c) = (a \cdot b) + (a \cdot c)$ in ordinary arithmetic, the one's digit of both sides must be the same, so the distributive law holds. To sum up, our one's digit arithmetic system satisfies some, but not all, of the Rules of Arithmetic. In particular, Axioms A1–A4, M1–M3, and D hold, but Axiom M4 does not.

Let's isolate the numbers that do have multiplicative inverses and write a table just for them (Table 1.7).

·	1	3	7	9
1	1	3	7	9
3	3	9	1	7
7	7	1	9	3
9	9	7	3	1

Table 1.7. One's digit multiplication of numbers with multiplicative inverses.

What can we notice about Table 1.7? One striking thing that this table displays is that this set is closed under multiplication since there are no "strangers" in the table. We already know that multiplication is commutative and associative. There is a multiplicative identity, and every element has a multiplicative inverse. So here is a system in which Axioms M1–M4 hold. Observe, however, that $\{1, 3, 7, 9\}$ is not closed under one's digit addition.

Mathematicians have a name for a system in which there is only one operation for which certain rules hold. The name for such a system is an *abelian group*.

Definition. An *abelian group* is a system closed under a binary operation that is associative, commutative, has an identity, and has an inverse for every element.

If the operation is addition, then the system is an abelian group if the four Axioms A1–A4 hold. If the operation is multiplication, then the system is an abelian group if the three Axioms M1–M3 hold and Axiom M4 is replaced by an axiom stating that every element has a multiplicative inverse. For example, $\mathbb{Z}$ is an abelian group under addition, and the set $\{1, 3, 5, 7\}$ in the one's digit multiplication is an abelian group. We talk more about groups in Chapters 6 and 11.

Example. The numbers $\{0, 1, 2, 3, 4, 5, 6, 7, 8, 9\}$ with one's digit addition form an abelian group.

Practice Problem 1.25. *What other subsets of* $\{0, 1, 2, 3, 4, 5, 6, 7, 8, 9\}$ *with one's digit addition as the operation form an abelian group? What other subsets of* $\{0, 1, 2, 3, 4, 5, 6, 7, 8, 9\}$ *with one's digit multiplication as the operation form an abelian group?*

Many systems that we work with will have not one operation, but two. For example, $\mathbb{N}$, $\mathbb{Q}$, and $\mathbb{Z}$ all have addition and multiplication. Mathematicians have a name for these systems too.

Definition. A system with two binary operations, $+$ and $\cdot$, in which Axioms A1–A4, M1–M3, and D hold is called a *commutative ring with 1*.

Practice Problem Solutions and Hints

1.1.
  (a) $S = \{a, b, c, d, e\}$.
  (b) $\mathbb{Z} = \{0, 1, -1, 2, -2, 3, -3, \ldots\}$.

1.2.
  (a) $S = \{x \in \mathbb{Z} \mid x$ is even, $x > 3$ and $x < 50\}$.
  (b) $S = \{x \in X \mid x$ was a president of the United States and x was a senator $\}$.
  (c) $S = \{x \in \mathbb{Z} \mid 1 \leq x \leq 8\}$.

1.3. The subsets of $\{1, 2, 3\}$ are $\emptyset$, $\{1\}$, $\{2\}$, $\{3\}$, $\{1, 2\}$, $\{1, 3\}$, $\{2, 3\}$, and $\{1, 2, 3\}$.

1.4.
  (a) $A \cup B = \{1, 2, 3, 4, 5, 6, 8\}$, $A \cap B = \{2, 4, 6\}$, $B \cup C = \{1, 2, 3, 4, 6, 8\}$, $B \cap C = \{6\}$, $A \cup C = \{1, 2, 3, 4, 5, 6\}$, and $A \cap C = \{1, 3, 6\}$.
  (b) $A \cup B \cup C$ is the set $\{1, 2, 3, 4, 5, 6, 8\}$.
  (c) $A \cup \emptyset = \{x \mid x \in A$ or $x \in \emptyset\} = \{x \mid x \in A\} = A$. $A \cap \emptyset = \{x \mid x \in A$ and $x \in \emptyset\} = \{x \mid x \in \emptyset\} = \emptyset$.

1.5. $B \times A = \{(a, 1), (a, 2), (a, 3), (b, 1), (b, 2), (b, 3)\}$. Since these are ordered pairs, $A \times B$ and $B \times A$ are different sets.

1.6. T and P are functions.

1.7. P is one-to-one.

1.8. The image of T is $\{2\}$. The image of P is $\{1, 2, 3\}$.

1.9. The domain of $f-g = x^2-x-1$ is $\mathbb{R}$, and the domain of $f \cdot g = x^3+x^2$ is also $\mathbb{R}$.

1.10. The domain of $f \circ g = x^2 - 7$ and $g \circ f = (x-7)^2$ is $\mathbb{R}$. The domain of $f \circ h = \frac{1}{x} - 7$ and $h \circ h = x$ is $\{x \mid x \in \mathbb{R}, x \neq 0\}$.

1.11. By commutativity of multiplication, $a \cdot b = b \cdot a$ and $a \cdot c = c \cdot a$. By commutativity of addition, we can add these two terms in either order. So, the right-hand side of this equation is the same as the right-hand side of the original equation.

1.12. If 0 had a multiplicative inverse, say, x, then $x \cdot 0$ would have to be 1. But $x \cdot 0 = 0$ for any number x.

1.13.

(a) If $a = 7$ and $b = 3$, $a - b = 4$, but $b - a = -4$.

(b) If $a = 7$, $b = 3$, and $c = 1$, $(a-b) - c = (4) - 1 = 3$, but $a - (b-c) = 7 - (2) = 5$.

1.14. Check that $E \cdot (O+O) = E \cdot O + E \cdot O$, $E \cdot (E+E) = E \cdot E + E \cdot E$, and then check the other six combinations.

1.15. The sum $9 + 8$ is 17 in real numbers, so in one's digit arithmetic, $9 + 8$ is 7, and likewise, $6 + 7$ is 3.

1.16. Among the things you might notice:

(a) Rows and columns shift one number from one row or column to the next.

(b) Every row and column has each number exactly once; there are no repeats.

(c) There's symmetry across the diagonal from top left to bottom right, but not across the other diagonal.

1.17. The sum $9 + 8$ is 7, then $7 + 7$ is 4. The sum $8 + 7$ is 5, then $9 + 5$ is also 4.

1.18. The number -3 is 7 in one's digit arithmetic since $7 + 3 = 0$ and $3 + 7 = 0$.

1.19. The additive inverses are $-0 = 0$, $-1 = 9$, $-2 = 8$, $-3 = 7$, $-4 = 6$, $-5 = 5$, $-6 = 4$, $-7 = 3$, $-8 = 2$, $-9 = 1$. The numbers 0 and 5 are their own additive inverses.

1.20. We get $6 \cdot 6$ is 6, $3 \cdot 7$ is 1, and $4 \cdot 5$ is 0.

1.21. There are many observations to make. Some are:
 (a) This table has rows and columns with repeated elements.
 (b) There is symmetry on the diagonal from the top left to the bottom right.

1.22. These numbers share no common factors with 10.

1.23. The fractions $\frac{1}{3}$, $\frac{2}{3}$, $\frac{2}{7}$, and $\frac{4}{9}$ are examples of those that do exist; $\frac{1}{2}$, $\frac{1}{4}$, $\frac{1}{5}$, and $\frac{1}{8}$ are examples of those that do not.

1.24. The product $5 \cdot (6+7)$ is $5 \cdot 3$ is 5; $(5 \cdot 6) + (5 \cdot 7)$ is $0+5$ is 5; $8 \cdot (4+9)$ is $8 \cdot 3$ is 4; $(8 \cdot 4) + (8 \cdot 9)$ is $2+2$ is 4; $7 \cdot (3+8)$ is $7 \cdot 1$ is 7; $(7 \cdot 3) + (7 \cdot 8)$ is $1 + 6$ is 7.

1.25. In addition, the numbers $\{0, 5\}$ form an abelian group, as do do the numbers $\{0, 2, 4, 6, 8\}$, and the number $\{0\}$. In multiplication, the numbers $\{1, 9\}$ form an abelian group, as does the number $\{1\}$.

Exercises

1.1. Write the following sets in set notation.
 (a) The days of the week.
 (b) The members of your immediate family.
 (c) The integers between and including 3 and 9.
 (d) The primary colors.
 (e) All two-digit numbers written with a 7.
 (f) Integers whose square is a number between 50 and 60.

1.2. Write the following sets in $\{x \in X \mid P(x)\}$ notation.
 (a) Countries that are part of the United Nations.
 (b) People born in 1950.

(c) All vegetarians.

(d) Numbers whose square is written with a 5.

1.3. Let $A = \{1, 3, 5, 7, 8, 11\}$ and $B = \{4, 5, 6, 7, 8\}$. What is $A \cup B$? What is $A \cap B$?

1.4. Let $A = \{x \in \mathbb{Z} \mid x > 8\}$ and $B = \{y \in \mathbb{Z} \mid y \text{ is even.}\}$. What is $A \cup B$? What is $A \cap B$?

1.5. Let $A = \{1, 2, 3\}$ and $B = \{1, 2, 3, 4, 5\}$. What is $A \times B$? What is $B \times A$?

1.6. Let A and B be as above. Which of the following represent a function from A into B?

$$F = \{(1, 2), (2, 3), (3, 4), (4, 5)\};$$
$$G = \{(1, 4), (2, 4), (3, 3)\};$$
$$H = \{(2, 1), (1, 3), (2, 3), (3, 5)\};$$
$$I = \{(1, 5), (3, 3)\};$$
$$J = \{(1, 6), (2, 7), (3, 8)\};$$
$$K = \{(3, 3), (2, 4), (1, 5)\}.$$

1.7. If S is a set with n elements, show that the power set of S has 2^n elements.

1.8. The set $\{(1, 11), (2, 15), (3, 10), (4, 13), (5, 12), (6, 14), (7, 13)\}$ is a function.

(a) What is the domain of this function?

(b) What is its range?

(c) Is it one-to-one? Onto?

1.9. For integers x, $f(x) = 2$.

(a) Is f a function?

(b) What is the domain of f?

(c) Is f one-to-one?

(d) Can you choose the range of f so that f is onto?

1.10. Let sets A, B, and C have functions $f : A \to B$, $g : B \to C$, $h : C \to A$, and $j : C \to B$.

(a) Which pairs of these functions can be composed? What is the domain and range of each composite function?

(b) Can you compose some of these functions to get a function from A to A?

1.11. Let $A = \{2, 4, 6\}$ and $B = \{1, 2, 3\}$. Define $f : A \to B$ by saying f sends 2 to 1, 4 to 2, and 6 to 3. Then define $g : B \to A$ by saying g sends 1 to 6, 2 to 6, and 3 to 4.

(a) What is the range of $f \circ g$?

(b) What are $(f \circ g)(1)$, $(f \circ g)(2)$, and $(f \circ g)(3)$?

(c) What is the range of $g \circ f$?

(d) What are $(g \circ f)(2)$, $(g \circ f)(4)$, and $(g \circ f)(6)$?

1.12. Let f and g be functions from $\mathbb{R}$ to $\mathbb{R}$, with $f(x) = \frac{x}{1+x^2}$ and $g(x) = x^2 - 5$. What are $f + g$, $f - g$, and $f \cdot g$? What is $\frac{f}{g}$ and what is its domain?

1.13. Find the additive and multiplicative inverses in $\mathbb{R}$ of the following real numbers:

(a) 2.

(b) 7.

(c) -1.

(d) -3.

(e) 0.

(f) $2 + \frac{1}{2}$.

(g) 0.01.

(h) π.

(i) $\frac{1}{2} - \frac{1}{3}$.

1.14. In the following proof that $(a + b)(c + d) = (ac + bc) + (ad + bd)$, identify which of the axioms of arithmetic is being used at each step:

$$(a + b)(c + d) = a(c + d) + b(c + d)$$
$$= (ac + ad) + (bc + bd)$$
$$= ac + (ad + (bc + bd))$$
$$= ac + ((ad + bc) + bd)$$
$$= ac + ((bc + ad) + bd)$$
$$= ac + (bc + (ad + bd))$$
$$= (ac + bc) + (ad + bd).$$

1.15. Use the axioms of arithmetic to show that

(a) $a(b + (c + d)) = (ab + ac) + ad$.

(b) $a((b + c)d) = (ab)d + a(cd)$.

1.16. For each of the following sets, tell whether the set is closed under addition and whether the set is closed under multiplication. Justify your answers either by giving an explanation of why the set is closed or an example that shows it is not closed.

(a) The real numbers.

(b) The natural numbers.

(c) $\{0, \frac{1}{2}, 1, \frac{3}{2}, \ldots\}$.

(d) The odd integers $\{\ldots, -3, -1, 1, 3, \ldots\}$.

(e) The irrational numbers.

(f) The negative real numbers.

(g) The set of real numbers between 0 and 1.

(h) The set $\{-1, 0, 1\}$.

1.17. Are the following operations associative? Are they commutative? Do they have identities?

(a) Division on the nonzero elements in $\mathbb{Q}$.

(b) The *exponentiation* of two nonzero integers, defined by $a \circ b = a^b$.

(c) The *concatenation* of two numbers $a_1 a_2 \ldots a_n$ and $b_1 b_2 \ldots b_m$, where a_k and b_k are the digits of the numbers, is $a_1 a_2 \ldots a_n b_1 b_2 \ldots b_m$. We use the symbol $\diamond$ to show concatenation. So, for example, $12 \diamond 7 = 127$. Which of the above properties does concatenation have?

1.18. Find an example that shows that in $\mathbb{Q}$, division does not distribute over addition. That is, find rational numbers a, b, and c for which $a \div (b + c) \neq (a \div b) + (a \div c)$. For which a, b, and c does $a \div (b+c) = (a \div b) + (a \div c)$?

1.19. Which of the following sets have the property that the multiplicative inverse of every number in the set is also in the set? For each set that doesn't have this property, give an element whose inverse is not in the set.

(a) The natural numbers.

(b) The irrational numbers.

(c) The rational numbers other than 0.

(d) The positive real numbers.

1.20. Suppose that $a \square b$ is defined to be the arithmetic mean (the average) of a and b. Is the operation $\square$ commutative? Is it associative? Is the set of integers closed under $\square$? Is there an identity for the operation $\square$? If so, are there inverses?

1.21. We define a new operation on the integers by $a \star b = ab + a + b$.

(a) Is $\star$ commutative? Justify your answer.

(b) Is $\star$ associative? Justify your answer.

(c) Does multiplication distribute over $\star$?

(d) Does $\star$ distribute over addition?

(e) Is there an identity for the $\star$ operation? If so, are there inverses?

1.22. If S is a set that contains the number 1, is closed under addition, and has additive inverses, what other numbers must be in the set S?

1.23. Suppose S is a set of integers.

(a) If S is closed under multiplication and contains the number 2, what other numbers must be in the set S?

(b) If S also has multiplicative inverses, what other numbers must be in the set S?

1.24. Let $\mathbb{Z} \times \mathbb{Z}$ be the set of all ordered pairs of integers. In other words, $\mathbb{Z} \times \mathbb{Z} = \{(a, b) \mid a, b \text{ are integers}\}$. Define the addition and multiplication of two elements of $\mathbb{Z} \times \mathbb{Z}$ as follows:

$$(a, b) + (c, d) = (a + c, b + d),$$

$$(a, b) \cdot (c, d) = (ac, bd).$$

(a) Compute $(-3, 7) + (5, 2)$ and $(-3, 7) \cdot (5, 2)$.

(b) Show that addition and multiplication on $\mathbb{Z} \times \mathbb{Z}$ are associative and commutative.

(c) Which element of $\mathbb{Z} \times \mathbb{Z}$ is the identity for addition? Which element of $\mathbb{Z} \times \mathbb{Z}$ is the identity for multiplication?

(d) What is the additive inverse of (a, b)?

(e) Which elements (a, b) have multiplicative inverses?

(f) Show that the distributive axiom is satisfied in $\mathbb{Z} \times \mathbb{Z}$.

1.25. The set of *Gaussian integers*, written as $\mathbb{Z}[i]$,[1] is the set $\mathbb{Z} \times \mathbb{Z}$ (see Exercise 1.24) with the following two binary operations:

$$(a, b) \oplus (c, d) = (a + c, b + d),$$
$$(a, b) \odot (c, d) = (ac - bd, ad + bc).$$

(a) Compute $(-3, 7) \oplus (5, 2)$ and $(-3, 7) \odot (5, 2)$.

(b) Show that $\oplus$ is an associative and commutative binary operation on $\mathbb{Z}[i]$. Find the identity element in $\mathbb{Z}[i]$ with respect to $\oplus$ and find the inverse element of (a, b) with respect to $\oplus$.

(c) Show that $\odot$ is an associative and commutative binary operation on $\mathbb{Z}[i]$.

(d) Find the identity element in $\mathbb{Z}[i]$ with respect to $\odot$.

(e) Verify that $\odot$ distributes over $\oplus$ in $\mathbb{Z}[i]$.

1.26. Let a, b and c be integers, not all of which are 0, such that $a^2 + b^2 = c^2$.

(a) Why must c be nonzero?

(b) Show that $X = \frac{a}{c}$ and $Y = \frac{b}{c}$ satisfy the equation $X^2 + Y^2 = 1$.

1.27. Let $S^1(\mathbb{Q})$ represent the set of all pairs of *rational* numbers (x, y) for which $x^2 + y^2 = 1$. Define multiplication of elements of $S^1(\mathbb{Q})$ as follows:

$$(x_1, y_1) \odot (x_2, y_2) = (x_1 x_2 - y_1 y_2, x_1 y_2 + x_2 y_1).$$

---

[1]The symbol $\mathbb{Z}[i]$ refers to the set of complex numbers of the form $a + bi$, where a and b are in $\mathbb{Z}$. We study these in Chapter 8.

(a) Show that $(\frac{3}{5}, \frac{-4}{5})$ and $(\frac{-12}{13}, \frac{-5}{13})$ are elements of $S^1(\mathbb{Q})$.

(b) Compute $(\frac{3}{5}, \frac{-4}{5}) \odot (\frac{-12}{13}, \frac{-5}{13})$.

(c) Show that $S^1(\mathbb{Q})$ is closed under the binary operation $\odot$. **Hint:** This involves showing that if (x_1, y_1) and (x_2, y_2) are pairs of rational numbers that each satisfy $x^2 + y^2 = 1$, then $(x_1, y_1) \odot (x_2, y_2)$ is another pair of rational numbers that satisfies $x^2 + y^2 = 1$.

(d) Show that $\odot$ is commutative.

(e) Show that $\odot$ is associative.

(f) Show that $(1,0)$ is the identity element of $S^1(\mathbb{Q})$ with respect to $\odot$.

(g) Show that the inverse of (x, y) with respect to $\odot$ is $(x, -y)$.

1.28. Even-Odd arithmetic is often used to model electrical circuits. Consider the two following situations.

(a) In one room, there are two light switches. The light is off if both switches are down or both switches are up. Fill in the following table to show under what conditions the light is on or off.

	Switch #1 Down	Switch #1 Up
Switch #2 Down		
Switch #2 Up		

Which table in Even-Odd arithmetic illustrates this situation?

(b) In another room, a lamp comes on when a wall switch is up and a knob on a lamp is turned. Create a similar table for this situation and tell which Even-Odd table illustrates this.

1.29. Calculate the following in one's digit arithmetic.

(a) $1 + 2 + 3 + 4 + 5 + 6 + 7$.

(b) $\frac{4}{7}$.

(c) $2^{54} \cdot 5^{123}$.

(d) $1 \cdot 2 \cdot 3 \cdot 4 \cdot 5 \cdot 6 \cdot 7 \cdot 8 \cdot 9 = 9!$

1.30. The following equalities are true in ordinary arithmetic. Are they true in one's digit arithmetic?

(a) $-(-a) = a$.

(b) $\frac{1}{\frac{1}{a}} = a$, provided $\frac{1}{a}$ exists.

(c) $\frac{-1}{a} = \frac{1}{-a} = -\frac{1}{a}$, provided $\frac{1}{a}$ exists.

(d) $(-a)(-b) = ab$.

1.31. Which numbers are perfect squares in one's digit arithmetic? Which numbers are perfect cubes?

1.32. Verify that for all the numbers in one's digit arithmetic, $x^5 = x$.

1.33. Explain why the following reformulation of one's digit arithmetic is valid: to add two numbers in one's digit arithmetic, add them by using ordinary addition of integers, and then find the remainder of the sum after division by 10. Does the same idea work with multiplication in place of addition?

Exercises 1.34 to 1.39 are about the game *Clock Arithmetic:* Consider the numbers $\{0, 1, 2, 3, 4, 5, 6, 7, 8, 9, 10, 11\}$ arranged in their normal places on a clock, with 0 occupying the position normally occupied by 12. Call this set C_{12}.

1.34. To add two numbers a and b from C_{12}, start at a o'clock and move forward b hours.

(a) Fill in the addition table in C_{12}:

+	0	1	2	3	4	5	6	7	8	9	10	11
0												
1												
2												
3												
4												
5												
6												
7												
8												
9												
10												
11												

(b) Is addition in C_{12} commutative? What feature of the addition table tells you this?

(c) Choose three different sets of three elements of C_{12} and verify that associativity of addition is satisfied for these elements.

(d) What is the identity of C_{12} with respect to addition?

(e) Fill in the following table with the additive inverse of the indicated elements:

+	0	1	2	3	4	5	6	7	8	9	10	11
inverse												

1.35. In $\mathbb{N}$, the product $a \cdot b$ is defined to be the result of b added to itself a times. So, $2 \cdot 3$ is $3 + 3$, and $3 \cdot 2$ is $2 + 2 + 2$. Define the multiplication $a \cdot b$ in C_{12} in an analogous way: starting at 0 o'clock, proceed clockwise a times through successive b-hour increments. For example, in C_{12}, $2 \cdot 8 = 4$ because 4 o'clock is two 8-hour increments after 0 o'clock.

(a) Fill in the following multiplication table in C_{12}.

·	0	1	2	3	4	5	6	7	8	9	10	11
0												
1												
2												
3												
4												
5												
6												
7												
8												
9												
10												
11												

(b) Is multiplication in C_{12} commutative? What feature of the multiplication table tells you this?

(c) Choose three different sets of three elements of C_{12} and verify that associativity of multiplication is satisfied for these elements.

(d) What is the identity of C_{12} with respect to multiplication?

(e) Fill in the following table with the multiplicative inverse of the indicated elements, or with "N" if an inverse does not exist:

·	0	1	2	3	4	5	6	7	8	9	10	11
inverse												

1.36. Create a multiplication table just for the elements of C_{12} that have a multiplicative inverse. When restricted to this subset of C_{12}, is multiplication still closed? Discuss some differences between this table and the similar table constructed for the elements in one's digit arithmetic that have multiplicative inverses (Table 1.6).

1.37. Choose three elements, a, b, and c, in C_{12} and verify that $a \cdot (b+c) = a \cdot b + a \cdot c$. Do this for two other sets of elements of C_{12}.

1.38. Explain why the following reformulation of addition and multiplication in C_{12} is valid: to add or multiply two numbers in C_{12}, add or multiply them using ordinary addition or multiplication of integers, and then find the remainder of the sum or product after division by 12.

1.39. If n is in $\mathbb{N}$, then an nth root of a number x is a number r such that $r^n = x$. A square root is a 2nd root, and a cube root is a 3rd root.

(a) Find all the square roots of 0 in C_{12}.

(b) Find all the square roots of 1 in C_{12}.

(c) Find all the square roots of 10 in C_{12}.

(d) Find the three numbers in C_{12} that do not have cube roots in C_{12}.

(e) Find all 4th roots of 1 in C_{12}.

1.40. In the set $\mathbb{Z} \times \mathbb{Z}$ (see Exercise 1.24), find all the square roots of (1,1). Write a multiplication table for these elements. From the table, determine which of Axioms M1–M4 this set satisfies.

1.41. Let S be a set and let $*$ be a closed binary operation on S that satisfies the following axioms.

Axiom I. $*$ is associative.

Axiom II. There is a unique element e in S that is the $*$-identity.

Axiom III. For any x in S, $x * x = e$.

Prove that $*$ is necessarily commutative. **Hint:** For any pair of elements y and z, show that $y * z$ and $z * y$ are each a $*$-inverse of $z * y$.

1.42. Given that the following table satisfies Axioms A1–A4, complete as much of the table as possible.

+	a	b	c	d	e
a		c			b
b		a			d
c		e			
d	a	b	c		
e					

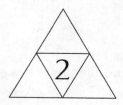

Axioms in Number Theory

In this chapter, we are trying to accomplish at least three things. First, we introduce the axiomatic method. This involves proving theorems by using axioms so that the theorems apply to any system satisfying those axioms. Second, we want to put familiar facts on a solid basis. Finally, we are aiming at an axiomatic characterization of the integers $\mathbb{Z}$.

To do problems in number theory, we want to be able to solve equations. To solve a simple equation such as $2x + 1 = 7$, we need to make use of all the rules we have at our disposal. After a course in elementary algebra, we can solve this equation mentally by subtracting 1 and dividing by 2 on both sides of the equation to get $x = 3$. Let's slow down and take this solution apart to look carefully at each of the steps involved in getting the answer $x = 3$. Try to think about solving this equation by using only our established Rules of Arithmetic for the two operations of addition and multiplication.

First, there's the matter of notation. At this stage, we employ a familiar device. In any equation, we assume that multiplication precedes addition. For example, the expression $ab + ac$ means that you should multiply a by b and a by c, and then add the two resulting numbers. With this convention, the distributive law will read $a(b + c) = ab + ac$. In an algebraic expression such as $2x + 1$, we multiply 2 times x first and then add 1. This is quite different from the expression $2(x + 1)$.

To solve our equation, we need to move the 1 from the left-hand side of the equation to the right. We use Axiom A4 to add (-1) to both sides of the equation to get

$$(2x + 1) + (-1) = 7 + (-1).$$

By using associativity and the properties of additive inverses and the additive identity, we get

$$2x + [1 + (-1)] = 6, \qquad \text{(Axiom A2)}$$
$$2x + 0 = 6, \qquad \text{(Axiom A4)}$$
$$2x = 6. \qquad \text{(Axiom A3)}$$

Next, in order to isolate the x, we need to "get rid of" the coefficient 2. Algebraically, we would just need to divide both sides of this equation by 2. However, remember that we have only the two operations, addition and multiplication, at our disposal for now. So we need to use the multiplicative inverse of 2, or 2^{-1}, to "cancel" the 2 on the left-hand side of the equation. We get

$$(2^{-1}) \cdot (2x) = (2^{-1}) \cdot (6).$$

Here's how it works using the axioms:

$$[(2^{-1}) \cdot (2)] \cdot x = [(2^{-1}) \cdot 2] \cdot (3), \qquad \text{(Axiom M2)}$$
$$1 \cdot x = 1 \cdot (3), \qquad \text{(Axiom M4)}$$
$$x = 3. \qquad \text{(Axiom M3)}$$

It would be useful to be able to do some of these algebraic manipulations without starting from the axioms for every argument. Mathematicians use theorems to state facts that are consequences of the axioms. Likewise, we can apply a theorem to each situation that satisfies an axiom without going through the justification of the axiom again. Theorems are useful tools for working with equations. In this chapter, we derive some theorems that are consequences of the Rules of Arithmetic. Some of these are straightforward and easy to prove; others require more subtle reasoning and clever use of proof techniques. We begin with one that is very natural and is also easy to prove.

2.1   Consequences of the Rules of Arithmetic

Cancelation for Addition

Suppose we start with two numbers, a and b. Take a third number, c, and add it to each of them. Now we have two new numbers, $a + c$ and

$b + c$. If we know that $a + c = b + c$, what conclusion can we draw? A first thought is probably that $a = b$, and that is correct. However, this kind of work on an equation is not one of the Rules of Arithmetic that we have established. Therefore, we need to justify this by proving it as a theorem that is a consequence of the Rules of Arithmetic. Fortunately, this is not very difficult.

Theorem 2.1 (Cancelation Law for Addition). *Suppose a, b, and c are elements of an abelian group under addition, and $a + c = b + c$. Then $a = b$.*

Proof: Start with the equation we know is true, namely,

$$a + c = b + c.$$

Since c is a number, by Axiom A4 it has an additive inverse, $-c$. We add $-c$ to both sides of the equation to get

$$(a + c) + (-c) = (b + c) + (-c).$$

Now, using Axiom A2, associativity of addition, we regroup and get

$$a + [c + (-c)] = b + [c + (-c)].$$

By Axiom A4, $c + (-c) = 0$, so we have

$$a + 0 = b + 0.$$

Finally, by Axiom A3, this becomes

$$a = b. \qquad \qquad \square$$

Example. In the equation $2x + 1 = 7$, we can use cancelation for addition to move the 1 and isolate the $2x$. Formally, we can write $2x + 1 = 6 + 1$, and now use Theorem 2.1 to cancel the 1s.

Properties of –1 and 0

Some facts about numbers that we know from elementary school arithmetic seem very reasonable. When we try to think about why they are true, the reasons may be a little difficult to explain. We now want to focus on two of these facts, namely,

$$0 \cdot a = 0,$$
$$(-1) \cdot a = -a.$$

These are two facts that we use all the time. Let's take a closer look at the first one that says 0 times any number is 0.

Why should this statement be true? Elementary texts sometimes provide only a nonmathematical justification to help make sense for students. Some texts present this fact as just another rule without any proof or connection to the Rules of Arithmetic. Why do we need a proof at all? Think back to the first time we talked about the number 0, the additive identity in addition. All its properties so far have had to do with the operation of addition. However, this new equation brings the additive identity into contact with the operation of multiplication, and we have no rules to tell us how 0 behaves with respect to multiplication. So there *really* is something to prove here.

We have to show that the additive identity has a special property that relates to multiplication, namely, that if we multiply a number by 0, the result is 0. Interestingly, if we go back to the axioms, the proof is straightforward.

Theorem 2.2. *If a is an element of a commutative ring, then $a \cdot 0 = 0$.*

Proof: We need to rely on the only fact we know about 0—that it is the additive identity. So we start with an identity,

$$a \cdot 0 = a \cdot 0.$$

Since we are trying to show something about the quantity $a \cdot 0$, we can use Axiom A3 to write

$$a \cdot 0 = a \cdot (0 + 0).$$

Now, using Axiom D on the right-hand side and Axiom A3 on the left, this becomes

$$(a \cdot 0) + 0 = (a \cdot 0) + (a \cdot 0).$$

Now we can use Theorem 2.1 to cancel and get $0 = a \cdot 0$.   □

This proof is important because it establishes a pattern for proving facts that we take for granted in elementary arithmetic. In this proof, we used only the axioms, or the Rules of Arithmetic, to establish a new role for the additive identity in the operation of multiplication. It is not surprising to realize that we needed to use the property of 0 as the additive identity to establish the theorem. Since we are also working with the relationship between addition and multiplication, it is necessary to use Axiom D, Distributivity. That is the only axiom that relates the two operations on numbers.

Let's now look at the second number fact, that $(-1) \cdot a = -a$. This, too, is a statement about the additive inverse of a. It gives us a new expression for $-a$ that involves multiplying the number a by the additive inverse of 1. Some of the same axioms we used in the previous proof are used in the argument for this theorem.

Although this equation may seem obvious, since we use it so often in elementary algebra, it helps to think about what this equation says in words: "When you multiply any number by the additive inverse of the multiplicative identity, you get the additive inverse of the original number."

Theorem 2.3. *If a is an element of a commutative ring, then $(-1) \cdot a = -a$.*

Proof: What do we know about $(-a)$? The only thing we know is that it is the additive inverse of a. Since the theorem says that this new number, $(-1) \cdot a$, is equal to $-a$, what we need to show is that our new number also behaves like a's additive inverse. In other words, if we add this new number to a, we should get 0. Let's see whether that happens when we add these two numbers. We start with $a + (-1) \cdot a$ and use the axioms to see whether this expression is equal to 0. With an eye toward using distributivity, we rewrite the simple term a as $1 \cdot a$ by using Axiom M3. We thus get

$$a + [(-1) \cdot a] = [1 \cdot a] + [(-1) \cdot a],$$

and using Axiom D on the right-hand side, we get

$$a + [(-1) \cdot a] = [1 + (-1)] \cdot a.$$

Since 1 and -1 are additive inverses, by Axiom A4 we have

$$a + [(-1) \cdot a] = 0 \cdot a.$$

But now we have at our disposal Theorem 2.2, which says that

$$0 \cdot a = 0.$$

So we have proved that

$$a + [(-1) \cdot a] = 0,$$

and since additive inverses are unique, by Axiom A4, that means that $(-1) \cdot a$ must be the additive inverse of a, as desired.   □

Notice that never in the proof did we use the term "negative a." Keep in mind that when we see a minus sign, it only indicates an additive inverse, not a negative number.

The proofs of these two statements illustrate how mathematics is done. That is, in proving a "new statement," rather than using only the axioms, we rely on both the axioms and "old statements," or theorems that we have already proved.

Practice Problem 2.1. *In a commutative ring, it is a direct consequence of the axioms that $-0 = 0$. Use Theorems 2.2 and 2.3 to provide an alternate proof.*

Let's continue working with the relationships between additive inverses and multiplication by presenting some theorems about them.

Theorem 2.4. *If a and b are any two numbers, then $(-a) \cdot b = -(ab)$.*

Proof: Similar to Theorem 2.3, this is a claim about the form of the additive inverse of the product $a \cdot b$. We need to show that the number $(-a) \cdot b$ is just another way of writing $-(ab)$. So again we begin by adding $(-a) \cdot b$ and ab to see whether their sum is 0. We start with

$$(-a) \cdot b + ab$$

and use distributivity to write

$$(-a) \cdot b + ab = [(-a) + a] \cdot b.$$

By Axiom A4, we get

$$(-a) \cdot b + ab = 0 \cdot b.$$

Again, by Theorem 2.2, since $0 \cdot b = 0$, we have

$$(-a) \cdot b + ab = 0,$$

and since additive inverses are, unique by Axiom A4, then $(-a) \cdot b$ must be the additive inverse of ab, as desired. □

It follows from Theorem 2.4 and commutativity of multiplication that $a \cdot (-b) = -(ab)$ as well.

Practice Problem 2.2. *Prove that if a and b are any two numbers, then $(-a) \cdot (-b) = ab$.*

Keep in mind that we are working with additive inverses, not "negative" numbers. However, when we finally do define what a negative number is, this will be the fact we need in order to prove that the product of two negative numbers is positive.

Cancelation for Multiplication

Suppose we have three numbers, a, b, and c, and that we know the equation $ab = ac$ is true. Is it then true that $b = c$? Recall that this is not true in one's digit arithmetic, where $2 \cdot 8 = 2 \cdot 3$, but $8 \neq 3$. Even in regular arithmetic, if $a = 0$, then b and c may not be equal. For systems in which every nonzero element has a multiplicative inverse, we can use the same idea that we used for cancelation for addition to establish that if $ab = ac$ and $a \neq 0$, then $b = c$.

Practice Problem 2.3. *Suppose a, b, and c are elements of an abelian group under multiplication. If $ab = ac$, show that $b = c$.*

Now, by Practice Problem 2.3 we know that if Axiom M4 holds in our number system, we have the property of cancelation for multiplication. What happens in a system where we do not have multiplicative inverses? If we are working with the integers and have the equation $2a = 2b$, then it is still true that $a = b$. To show this, our instinct is to divide the equation by 2 and get $a = b$. Unfortunately, this assumes 2^{-1} exists. But 2 does not have a multiplicative inverse in $\mathbb{Z}$. Therefore, using the Rules of Arithmetic, we cannot prove that cancelation for multiplication holds in $\mathbb{Z}$. So if we want the integers as a number system to have cancelation for multiplication, we must assign it as an axiom.

Axiom C (Cancelation Law for $\mathbb{Z}$). If a, b, and c are in $\mathbb{Z}$, $a \neq 0$, and $ab = ac$, then $b = c$.

Example. In our earlier work with the equation $2x + 1 = 7$, we had $2x = 6$, or $2x = 2 \cdot 3$. If we want a solution in the integers, we can now apply the cancelation law, and immediately we have $x = 3$.

Cancelation for multiplication is more complicated in one's digit arithmetic, in which some numbers have multiplicative inverses but others do not. If we go back to our simple algebraic equation, $2x + 1 = 7$, we can still get to the equation $2x = 6$ by using additive inverses, just as we did above. But in one's digit arithmetic, there is no multiplicative inverse for 2. Can we "cancel" as we did for $\mathbb{Z}$? That still seems to work, since in one's digit arithmetic, if $2x = 6$, $x = 3$ is still a solution. But there is another solution, namely, if $2x = 6$, $x = 8$ is also a solution.

Will including the cancelation axiom in one's digit arithmetic get us out of this difficulty? Unfortunately, the answer is no. If we use what we have just deduced, we know that $2 \cdot 3 = 2 \cdot 8$, but certainly 3 is not 8. This would contradict the cancelation axiom. The problem here is precisely with the nonzero numbers that do not have multiplicative inverses in one's digit arithmetic.

The set of numbers $\{2, 4, 5, 6, 8\}$ are the nonzero numbers in one's digit arithmetic that do not have multiplicative inverses. Each of them can be paired with at least one number in one's digit arithmetic so that the product of these two numbers is 0.

Example. For 2, we have $2 \cdot 5 = 0$ in one's digit arithmetic.

Practice Problem 2.4. *Find all pairs of numbers from the set $\{2, 4, 5, 6, 8\}$ whose product is 0 in one's digit arithmetic.*

Such numbers are called *zero divisors*.

Definition. A nonzero element a in a commutative ring R is called a *zero divisor* if there is a nonzero element b in R such that $ab = 0$.

This is *not* the rule we recall from elementary arithmetic, in which if the product of two numbers is 0, one of the numbers must be 0. Notice that the number 0 is not a zero divisor. It is not a coincidence that the proof of this arithmetic fact is based upon cancelation for multiplication.

Theorem 2.5. *Let a and b be any two numbers in a system for which the cancelation property for multiplication, Axiom C, holds. If $ab = 0$, then either $a = 0$ or $b = 0$.*

Proof: If $a = 0$, we're all set. If $a \neq 0$, then from Theorem 2.2, we know that $a \cdot 0 = 0$. So by substituting, $ab = 0$ implies $ab = a \cdot 0$. Now we use cancelation to get $b = 0$.   □

The moral of the story is this: if we want to use the cancelation property for multiplication in a number system in which there aren't multiplicative inverses for every nonzero element, then we need to include an axiom about cancelation for multiplication in the rules of arithmetic for that system.

We give these number systems a name, integral domain.

Definition. An *integral domain* is a number system with two binary operations, $+$ and $\cdot$, in which Axioms A1–A4, M1–M3, D, and cancelation hold.

We can conclude that a number system is an integral domain in two different ways. One way is for Axiom M4 to hold and then, by Practice Problem 2.3, cancelation comes free. This means $\mathbb{Q}$ and $\mathbb{R}$ are integral domains. The other way is to include cancelation as an axiom. Thus, because of Axiom C, the integers $\mathbb{Z}$ are also an integral domain.

We have seen at least one system in which Axioms A1–A4, M1–M3, and D hold and cancelation does not hold—one's digit arithmetic. Hence, one's digit arithmetic is not an integral domain; however, it is a commutative ring.

Subtraction and Division

Until now in our development of the Rules of Arithmetic, we have used only two operations, addition and multiplication. We need to define the other two common arithmetic operations, subtraction and division, that are suggested by the notation of the axioms.

Definition (Subtraction). If a and b are any two elements of a commutative ring, $a - b = a + (-b)$.

Definition (Division). If a and b are any two elements of a commutative ring, and if b has a multiplicative inverse, $\frac{a}{b} = a(b^{-1})$.

These are two natural definitions. We have included some exercises at the end of this chapter for practice. The definition of division leads directly to another of those arithmetic facts that seem as natural as breathing but are serious consequences of the axioms.

Theorem 2.6. *Zero has no multiplicative inverse, so division by zero is not defined.*

The method of proof we use here is called *proof by contradiction*. In this method, we assume that the conclusion of the theorem is false. Then we argue, using clear, logical steps, and hope that we encounter a contradiction, or a statement that doesn't make sense. If we do get a contradiction, then we know that our original assumption, that the theorem is false, must have been an incorrect assumption. This will establish that the theorem is true. This may seem like a roundabout way to prove a theorem, but it is a useful method that is mathematically sound.

Proof: Suppose that 0 did have a multiplicative inverse, call it z. By Axiom M4, we must have $0 \cdot z = 1$. However, by Theorem 2.2, $0 \cdot z = 0$. Since $0 \neq 1$ by Axiom M3, this is a contradiction, and so the element z cannot exist. □

2.2 Inequalities and Order

So far we have just been working with equalities; now we begin to work with inequalities. As we saw in our first encounter with the Rules of Arithmetic, there are several "facts" that we assume about inequalities, but whose proofs may elude us. Why, for example, when we start with an inequality and multiply both sides by a negative number, does the inequality symbol "flip"? What does it mean for a number to be "negative"? In this section, we add four "order axioms" to the Rules of Arithmetic. These are additional properties we want to have in some of the number systems with which we are working. In particular, we restrict our use of these axioms to $\mathbb{Z}$, $\mathbb{Q}$, and $\mathbb{R}$ for reasons that will appear later. So in this section, when we refer to numbers, we mean elements of $\mathbb{Z}$, $\mathbb{Q}$, or $\mathbb{R}$. These four order axioms introduce new mathematical words, which we include here.

Before we state the order axioms, however, we need to have something similar to the binary operations that characterized Axioms A1–A4 (addition) or Axioms M1–M4 (multiplication). For order, we introduce a *binary relation* "<," which reads as "is less than." The four axioms satisfied by < are the following.

Axiom O1 (Trichotomy). For any two numbers, a and b, one and only one of the following statements is true: $a < b$, $a = b$, or $b < a$.

Axiom O2 (Transitivity). For any three numbers, a, b, and c, if $a < b$ and $b < c$, then $a < c$.

Axiom O3 (Addition for Inequalities). For any three numbers, a, b, and c, if $a < b$ then $a + c < b + c$.

Axiom O4 (Multiplication for Inequalities). For any three numbers, a, b, and c, if $a < b$ and $0 < c$, then $ac < bc$.

We are in familiar territory here. The first order axiom says if we compare two numbers, either they are equal or one is less than the other. The second axiom says that the inequality is preserved when we compare numbers, and the third says the inequality is preserved if we add the same number to both sides. To make the fourth axiom sound more familiar, let's formally introduce more of the familiar terms from elementary arithmetic.

We say that a is *greater than* b if $a > b$. We introduce the symbol ">" to indicate this relation. So the inequalities $b < a$ and $a > b$ stand for the same thing. We say that a is *less than or equal to* b if $a < b$ or $a = b$, and we write this as $a \leq b$. We define $a \geq b$ similarly. We also use the terms "positive" and "negative" in their usual senses. A number a will be defined as *positive* if $a > 0$. A number b will be defined as *negative* if $b < 0$.

Thus, the fourth order axiom says that multiplying an inequality by a positive number preserves the inequality.

Keep in mind that a number with a minus sign in front of it might not be negative.

Example. The number $-(-3)$ is positive.

Let's state and prove some of the familiar facts about inequalities.

Theorem 2.7. *If $a < b$ and $c < d$, then $a + c < b + d$.*

Proof: We start with the inequality $a < b$, and add c to both sides. Then, by Axiom O3, we have

$$a + c < b + c.$$

Now we take the second inequality, $c < d$, and add b to both sides. Again, by Axiom O3, we have

$$c + b < d + b.$$

By using Axiom A1, we can rewrite this as

$$b + c < b + d.$$

Comparing the two inequalities that resulted from Axiom O3, we have $a+c < b+c$ and also $b+c < b+d$. So by Axiom O2, we have $a+c < b+d$.□

Practice Problem 2.5. *Use the result of Theorem 2.7 to show that the sum of two positive numbers is positive and the sum of two negative numbers is negative.*

Theorem 2.8. *If $a < b$, then $-a > -b$.*

Caution. Remember that this is a theorem about additive inverses, and not about negative numbers.

Proof: We begin by adding both $(-a)$ and $(-b)$ to the inequality $a < b$. We get

$$a + [(-a) + (-b)] < b + [(-a) + (-b)].$$

By using commutativity and associativity, we can rewrite this as

$$[a + (-a)] + (-b) < [b + (-b)] + (-a).$$

By using Axiom A4 and the subtraction notation introduced in the last section, we get

$$0 - b < 0 - a,$$

or by Axiom A3,

$$-b < -a;$$

that is,

$$-a > -b. \qquad\qquad \square$$

Practice Problem 2.6. *Prove that if $-a < -b$, then $a > b$.*

Theorem 2.9 states that when we multiply an inequality by a negative number, the sign reverses.

Theorem 2.9. *If $a < b$ and $c < 0$, then $ac > bc$.*

Proof: Since $c < 0$, by Theorem 2.8 we have $-c > -0$, or $-c > 0$, since $0 = -0$. Now we can use Axiom O4 to get

$$a(-c) < b(-c).$$

From Theorem 2.4, this is the same as

$$-(ac) < -(bc).$$

From Practice Problem 2.6, this means

$$ac > bc. \qquad \square$$

Practice Problem 2.7. *Explain how the proof of Theorem 2.9 shows that the product of two negative numbers is positive.*

We now prove another of those "facts" that we all believe is true, namely, squares of numbers other than zero are positive.

Theorem 2.10. *If $a \neq 0$, then $a^2 > 0$.*

Proof: By Axiom O1, either $a > 0$ or $a < 0$. If $a > 0$, then by Axiom O4, $a \cdot a > 0 \cdot a$, or $a^2 > 0$. If $a < 0$, then by Practice Problem 2.7, $a^2 > 0$. $\square$

We follow this with an even more obvious notion.

Corollary 2.11. $1 > 0$.

Proof: We know $1 \neq 0$, so by Theorem 2.10, $1 = 1^2 > 0$. $\qquad \square$

We have labeled this result a "corollary." In mathematics, a corollary is a result that follows easily from a theorem. The exercises at the end of the chapter present more problems for practice with the order axioms.

Order and Other Number Systems

Before we get too complacent and think that every familiar number system obeys these new order axioms, let's think about order in one's digit arithmetic. Do the order axioms make sense in this system? If we assume they hold, let's see what happens. There are only ten numbers in this system. From Corollary 2.11, we start with $0 < 1$. By Axiom O3, if we add 1 to both sides of this inequality, we get $1 < 2$. By Axiom O2, this means $0 < 2$. Continuing in this way, we will get $0 < 3$, until finally $0 < 9$. Axiom O3 then shows that $0+1 < 9+1$, or $1 < 0$, which is a contradiction to Axiom O1. This shows that we cannot impose an ordering on the one's digit number system and have a consistent set of rules for arithmetic.

Practice Problem 2.8. *What familiar or new number systems obey the set of order axioms? Consider* $\mathbb{Q}$, $\mathbb{R}$, *and the Even-Odd arithmetic system.*

We are closing in on a careful description of the integers, $\mathbb{Z}$. This is a very special number system for us. The integers $\mathbb{Z}$ satisfy all the Rules of Arithmetic except for the existence of multiplicative inverses. They do obey cancelation for multiplication, which makes up to some extent for the absence of multiplicative inverses. So we add the order axioms to the characterization of the integers. At the moment, then, we call the integers $\mathbb{Z}$, or any other integral domain with the order axioms, an *ordered integral domain*.

Well-Ordering

There is one more property that makes the integers $\mathbb{Z}$ special. This property is called *well-ordering*, and it guarantees that if we take a nonempty set of positive integers, there is a smallest number in the set. This is not a particularly surprising fact. For any set of positive integers, we should be able to pick out the smallest one.

Example. Think of all the positive even integers. The smallest element of this set is the number 2.

The formal statement of the well-ordering axiom for $\mathbb{Z}$ is the following.

Axiom (Well-Ordering for $\mathbb{Z}$). Any nonempty set of positive integers has a smallest element.

Why should this be such a special property? Let's think about $\mathbb{Q}$, the rational numbers. If we take a set of positive rational numbers, the well-ordering axiom may not be true. Of course, some sets of positive rational numbers do have smallest elements.

Example. The set $\{\frac{1}{2}, \frac{2}{3}, \frac{3}{4}, \ldots, \frac{n}{n+1}, \ldots\}$ has a smallest element. It is $\frac{1}{2}$.

However, many sets of rational numbers do not have a smallest element.

Example. The set $\{\frac{1}{2}, \frac{1}{3}, \frac{1}{4}, \ldots, \frac{1}{n}, \ldots\}$ is a set of smaller and smaller positive rational numbers, each one closer and closer to 0. In this set, there is no smallest rational number.

Practice Problem 2.9. *Is there a smallest positive rational number? Think about how to prove your answer.*

In practice, we can apply the well-ordering axiom not only to the set of positive integers, but also to the set of all integers greater than any fixed integer m.

Example. If A is a nonempty subset of $S = \{n \in \mathbb{Z} | n > m\}$, then the well-ordering axiom implies that A has a smallest element. This works because by subtracting m, we can replace S with the set of all positive integers.

Finally, the well-ordering axiom leads us to Theorem 2.12, another of those "obvious" facts that doesn't seem to need a proof.

Theorem 2.12. *There are no integers between 0 and 1.*

Proof: This is a proof by contradiction. Assume there is at least one integer between 0 and 1. In fact, make a set of all the integers that are between 0 and 1. By our assumption, this set is a nonempty set of positive integers. By the well-ordering axiom, this set has a smallest element. Call it a. Then $0 < a$ and $a < 1$. We also know that a^2 is a positive integer. Since $a < 1$, by Axiom O4, $a^2 < a$. But then a^2 is an integer between 0 and 1 that is smaller than a, our smallest element. This is a contradiction, so our assumption was false, and the theorem is proved. $\square$

Practice Problem Solutions and Hints

2.1. By Theorem 2.3, $-0 = (-1) \cdot 0$. By Theorem 2.2, $(-1) \cdot 0 = 0$. So $-0 = 0$.

2.2. By Theorem 2.4, $(-a)(-b) = -[a(-b)]$. This is the additive inverse of $[a(-b)]$. But again, by Theorem 2.4, $[a(-b)] = -(ab)$. So $-[a(-b)] = ab$ by the uniqueness of additive inverses.

2.3. Suppose $ab = ac$. Multiply by a^{-1}:

$$(a^{-1})(ab) = (a^{-1})(ac).$$

So $(a^{-1}a)b = (a^{-1}a)c$ by Axiom M2; $1 \cdot b = 1 \cdot c$ by Axiom M4; and $b = c$ by Axiom M3.

2.4. In one's digit arithmetic, $2 \cdot 5$ is 0. So are $4 \cdot 5$, $6 \cdot 5$, and $8 \cdot 5$.

2.5. If $0 < a$ and $0 < b$, then $0 + 0 < a + b$, so $0 < a + b$ and $a + b$ is positive. If $a < 0$ and $b < 0$, then $a + b < 0 + 0$, so $a + b < 0$ and $a + b$ is negative.

2.6. Suppose $-a < -b$. Then $-(-a) > -(-b)$ by Theorem 2.8. But $-(-a) = a$ and $-(-b) = b$ by Theorem 2.4, so $a > b$.

2.7. If $a < 0$ and $b < 0$, Theorem 2.9 says $ab > 0$.

2.8. $\mathbb{Q}$ and $\mathbb{R}$ obey the order axioms; Even-Odd does not.

2.9. Think about a smallest rational number z. How big would $\frac{1}{2}z$ be, compared to z? **Warning:** Zero is not a positive rational number.

Exercises

For these exercises, assume we are working with elements of a commutative ring unless otherwise stated.

2.1. Using the axioms or the results of this chapter, prove that $-(a+b) = (-a) + (-b)$ for all numbers a and b.

2.2. Using the axioms or the results of this chapter, prove that $(a \cdot b)^{-1} = b^{-1} \cdot a^{-1}$ for all numbers a and b that have a multiplicative inverse. Conclude that if a and b have multiplicative inverses, then so does $a \cdot b$.

2.3. The following tables describe the operations $a \bigcirc b$ and $a \triangle b$. Determine whether each of them satisfies the cancelation property. If either does not, give specific letter values for x, y, and z such that $xy = xz$ but $y \neq z$.

$\bigcirc$	a	b	c	d	e	f
a	e	f	a	b	c	d
b	c	c	b	f	d	b
c	a	e	c	b	e	f
d	d	b	d	a	b	e
e	a	d	e	c	e	a
f	e	a	f	d	a	c

$\triangle$	a	b	c	d	e	f
a	e	a	d	f	b	c
b	a	b	c	d	e	f
c	d	c	a	e	f	b
d	f	d	e	b	c	a
e	b	e	f	c	a	b
f	c	f	b	a	d	e

2.4. Prove the following by using the definitions of subtraction and division.

(a) $(a - b) + (c - d) = (a + c) - (b + d)$.

(b) $(a - b) - (c - d) = (a + d) - (b + c)$.

(c) $(a - b)(c - d) = (ac + bd) - (ad + bc)$.

(d) If $a - b = c - d$, then $a + d = b + c$.

(e) $\frac{a}{b} \div \frac{c}{d} = \frac{ad}{bc}$.

(f) $\frac{a}{b} + \frac{c}{d} = \frac{ad + bc}{bd}$.

(g) If $\frac{a}{b} = \frac{c}{d}$ then $ad = cb$.

2.5. Division of real numbers is not associative in general. For which real numbers a, b, and c does $a \div (b \div c) = (a \div b) \div c$?

2.6. Suppose S is a set with addition and multiplication that satisfy the Rules of Arithmetic except possibly Axiom M4. Also suppose there are no zero divisors in S. Prove that cancelation for multiplication holds in S.

2.7. Define addition and multiplication on $\mathbb{Z} \times \mathbb{Z}$ as:

$$(a, b) + (c, d) = (a + c, b + d),$$
$$(a, b) \cdot (c, d) = (ac, bd).$$

(a) Find an example that shows that $\mathbb{Z} \times \mathbb{Z}$, under these operations, does not satisfy cancelation for multiplication.

(b) Describe all the zero divisors in $\mathbb{Z} \times \mathbb{Z}$. (Recall that in $\mathbb{Z} \times \mathbb{Z}$, a zero divisor is any element (a, b) in the set for which there exists an element $(c, d) \neq (0, 0)$ having the property that $(a, b) \cdot (c, d) = (0, 0)$.)

2.8. Suppose that S is a set that has addition and multiplication satisfying the Rules of Arithmetic except possibly Axiom M4. Let Z be the set of all zero divisors of S, and let T be the set of all elements of S that are not zero divisors.

(a) Prove that if x is an element of S that has a multiplicative inverse, then x is a member of T.

(b) Show that T is closed under multiplication.

(c) Find the subsets Z and T when $S = C_{12}$ (see the definition of C_{12} before Exercise 1.34).

2.9. Let S be a set and let $*$ be an associative binary operation under which the set S is closed. Let e be the $*$-identity. Let a and b be two elements of S that have the same $*$-inverse. Prove that $a = b$.

2.10. Let $<$ be the ordinary order on the set of real numbers.

(a) Prove that if $0 < a < b$ and $0 < c < d$, then $ac < bd$.

(b) Let n be a positive real number, and suppose that $n = ab$, where a and b are positive real numbers. Show that either $a \leq \sqrt{n}$ or $b \leq \sqrt{n}$.

2.11. Suppose an order is defined on a set that satisfies Axioms O1–O4. Prove the following for elements $a, b, x,$ and y of the set.

(a) If $a - x < a - y$, then $x > y$.

(b) If $x > y$, then $a - x < a - y$.

(c) If $0 < a < b$, then $a^2 < b^2$.

(d) If $0 > a > b$, then $a^2 < b^2$.

(e) $a^2 + b^2 \geq 2ab$.

2.12. Prove the following for elements of $\mathbb{Q}$, with the usual order:

(a) If $0 < \frac{a}{b} < \frac{c}{d}$ and $b > 0$ and $d > 0$, then $ad < bc$.

(b) If $0 < \frac{a}{b} < \frac{c}{d}$ and $b > 0$ and $d > 0$, then $\frac{a}{b} < \frac{a+c}{b+d} < \frac{c}{d}$.

2.13. For a real number a, define the *absolute value* of a, written as $|a|$, as the largest number of the set $\{a, -a\}$.

(a) Show that an equivalent definition for absolute value is

$$|a| = \begin{cases} a & \text{for } a \geq 0, \\ -a & \text{for } a < 0. \end{cases}$$

(b) Show that $|a| \geq 0$ and $|a| = 0$ only when $a = 0$.

(c) Show that $|ab| = |a||b|$.

(d) Show that $|a + b| \leq |a| + |b|$.

2.14. Which of these sets have a least element?

(a) The set of odd natural numbers.

(b) The set of positive rational numbers.

(c) The set of negative rational numbers.

(d) The set of rational numbers greater than π.

(e) Any finite subset of the real numbers.

2.15. Define the following order on the set $\mathbb{Z} \times \mathbb{Z}$: $(a, b) < (c, d)$ if either $a < c$ or $a = c$ and $b < d$. This is referred to as the *dictionary order* on $\mathbb{Z} \times \mathbb{Z}$.

(a) Show that there are infinitely many elements (x, y) in $\mathbb{Z} \times \mathbb{Z}$ satisfying the inequalities $(0, 0) < (x, y) < (1, 1)$.

(b) Show that Axioms O1–O3 are satisfied for this ordering.

(c) Give an example that shows that Axiom O4 is not satisfied for this ordering.

(d) Is the well-ordering axiom satisfied for $\mathbb{Z} \times \mathbb{Z}$ with the dictionary order?

2.16. Let S be a set on which there is an addition defined that satisfies Axioms A1–A4. Assume also that there is an ordering $<$ defined on S that satisfies Axioms O1, O2, and O3. Two elements, a and b, in S are said to be *consecutive* if there is no element x in S satisfying $a < x < b$.

(a) Show that if a and b are consecutive, then so are $a + c$ and $b + c$, where c is any element of S.

(b) Conclude that n and $n + 1$ are consecutive for any n in $\mathbb{Z}$.

(c) Consider the set of rational numbers $\mathbb{Q}$ with the usual order. Show that no two elements are consecutive.

(d) Consider the set $\mathbb{Z} \times \mathbb{Z}$ with the dictionary order. When are (a, b) and (c, d) consecutive?

2.17. Let S be the set {rock, scissors, paper}. Define an order $<$ on the set S such that scissors $<$ rock, rock $<$ paper, and paper $<$ scissors. Show that $<$ is not transitive.

2.18. Suppose that we define a different order on the rational numbers as follows: to compare two numbers, first write each one as a fraction in lowest terms with a positive denominator; then $\frac{a}{b} < \frac{c}{d}$ if $a < c$.

(a) Fill in each blank with $<$, $>$, or an N if neither inequality would make a true statement:

   (i) $\frac{1}{2}$ ___ $\frac{2}{3}$.
   (ii) $\frac{-2}{6}$ ___ $\frac{-2}{5}$.
   (iii) $\frac{3}{9}$ ___ $\frac{2}{10}$.
   (iv) 1 ___ $\frac{1}{5}$.
   (v) $\frac{-6}{-14}$ ___ 0.

(b) Which of the order axioms does this ordering satisfy? Give a counterexample for each axiom that is not satisfied.

2.19. The following ordering of $\mathbb{N}$ (known as Sarkovskii's ordering of $\mathbb{N}$), denoted $>>$, is important in the study of dynamical systems and chaos theory.

$$3 >> 5 >> 7 >> 9 >> \ldots >> 6 >> 10 >> 14 >> 18 >> \ldots$$
$$>> 12 >> 20 >> 28 >> 36 >> \ldots >> 8 >> 4 >> 2 >> 1.$$

In words, this means that all odds are more than twice all odds, which are more than four times all odds, which are more than eight times all odds, and so on. The powers of 2 are less than all the other numbers. Every natural number can be found exactly once on this list. In this ordering, 3 is the largest number and 1 is the smallest.

(a) Fill in each blank with either $<<$ or $>>$ to make a true statement:

   (i) 7 __ 12.
   (ii) 9 __ 20.
   (iii) 24 __ 20.
   (iv) 48 __ 56.
   (v) 32 __ 31.

(b) Give an example to show that $<<$ does not satisfy Axiom O3.

(c) Show that the following variation of Axiom O4 is satisfied for $<<$:
   Axiom O4′: If $a << b$ with neither a nor b being a power of 2, and
   if c is any natural number, then $ac << bc$.

2.20. The integers satisfy a property known as *mathematical induction*.
This is a familiar topic in high school textbooks.

(a) The *First Principle of Mathematical Induction* is stated as follows.
   Suppose S is a subset of $\mathbb{N}$ with the following properties:

   (i) The number 1 is in S.
   (ii) If n is in S, then $n+1$ is in S.

   Using well-ordering, prove $S = \mathbb{N}$.

(b) The *Second Principle of Mathematical Induction* is stated as follows.
   Suppose S is a subset of $\mathbb{N}$ with the following properties:

   (i) The number 1 is in S.
   (ii) If n is in S and if every natural number k, where $k \leq n$, is in
        S, then $n+1$ is in S.

   Using well-ordering, prove $S = \mathbb{N}$.

2.21. Show, using mathematical induction, that $2^{k-1} \leq k!$ for all natural
numbers k, by doing parts (a) and (b).

(a) Let S be the set of natural numbers for which the inequality is true.
   Show that 1 is in S.

(b) Now conclude, using the induction hypothesis and the theorems
   about inequalities, that

$$2^n = 2 \cdot 2^{n-1} \leq 2 \cdot n! \leq (n+1) \cdot n! = (n+1)!$$

   That is, if n is in S, then $n+1$ is in S.

2.22. Suppose that a commutative ring has a binary relation $<$ that satisfies Axioms O1–O4. Prove that this commutative ring satisfies cancelation for multiplication and is therefore an integral domain.

2.23. Let R be a commutative ring. Suppose that P is a subset of R with the following properties:

P1: For any element $a \in R$, one and only one of the following holds: $a = 0, a \in P$, or $-a \in P$.

P2: P is closed under addition and multiplication.

Define an order relation $<$ on R by saying that for $a, b \in R$, $a < b$ if $b - a \in P$. Show that $<$ satisfies Axioms O1–O4.

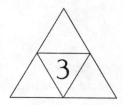

Divisibility and Primes

One of the earliest concepts that comes up when people discuss numbers is the idea of a prime number. Usually, if students are asked to describe what a prime number is, the definition comes out as: "A prime is a number that can only be divided by itself and 1." Since this will be an important definition, let's take some time to be sure the formal definition is completely precise. In fact, before we can define a prime number, we have to decide what we mean by "divided by."

3.1 Divisibility

When we first talk about division in the integers, we usually mean "divides evenly." That means we can divide a number into several equal parts with nothing left over. For example, we can agree that 7 divides 21, but does not divide 23.

Definition. Start with two integers a and b. We say that a *divides* b if there is an integer k so that $ak = b$. We will write this as $a \mid b$.

Example. We claimed 7 divides 21. How does this fit the definition? Here, $a = 7$ and $b = 21$. What is the integer k for the definition? Simple arithmetic shows that if $k = 3$, then $7 \cdot 3 = 21$. Also, since we cannot find an integer k so that $7k = 23$, we can say that 7 does not divide 23.

Example. What about 0? In order for $0 \mid n$ to be true, we would need an integer k so that $0 \cdot k = n$. Since $0 \cdot k$ is 0 for every possible k, the only integer that 0 divides is 0 itself.

Practice Problem 3.1. *Which integers divide 0?*

If a divides b, we call a a *factor*, or a *divisor*, of b. What are the other factors of 21? From elementary arithmetic, we would come up with a short list of the numbers 1, 3, 7, and 21. These are all the *positive* divisors of 21. Are there any negative divisors? Yes, in fact, -1, -3, -7, and -21 are also divisors of 21. Let's check one of them.

Example. Let's show that -3 divides 21. Here, $a = -3$ and $b = 21$. What integer k makes the equation $-3k = 21$ true? If $k = -7$, we know that $(-3)(-7) = 21$.

Practice Problem 3.2. *List all the divisors, positive and negative, of the number* 24.

There are some elementary properties of divisibility that we will find useful as we pursue the study of number theory. In fact, we might say that divisibility and its consequences are what number theory is all about.

Let's look at some divisibility facts (theorems) and their proofs. In these simple proofs, we usually need to write down what the definition tells us and then follow a set of logical steps through the argument.

Theorem 3.1. *For any three integers, a, b, and c, if $a \mid b$ and $b \mid c$, then $a \mid c$.*

Proof: By the definition of divisibility, to show that $a \mid c$, we need to find an integer n such that $an = c$. But how do we find n? We start by taking a closer look at what we know to be true. Let's take the two "if" parts of the statement and translate them into what the definition says they mean. For a to divide b, there must be some integer k that makes the equation $ak = b$ true. Likewise, since we know that b divides c, we know there must be an integer, say, m, that makes the equation $bm = c$ true.

The definition gave us two equations. The first relates a and b, the second relates b and c. If we put these two together somehow, we could get an equation that relates a and c. The idea is to use the first equation, $ak = b$, to substitute for b in the second equation. Then $bm = c$ becomes $(ak)m = c$.

Now, if we use Axiom M2, associativity of multiplication, we can rewrite this as

$$a(km) = c.$$

This is what we want. Since both k and m are integers, and integers are closed under multiplication, km must also be an integer. We can call it any name we like. So let's rename km as n. Then our equation reads

$$an = c,$$

which is precisely what we need to show in order to prove that a divides c. □

We have gone through the proof of Theorem 3.1 very carefully to illustrate the kind of thinking we need to prove facts in number theory. As we give more and more proofs, we will be able to shorten them.

Theorem 3.2. *For any three integers, a, b, and c, if $a \mid b$ and $a \mid c$, then $a \mid (b + c)$.*

Proof: Again, we first use the definition to write down what the "if" statements tell us. Since a divides b, there is an integer k that makes the equation $ak = b$ true. Since a divides c, there is an integer m that makes the equation $am = c$ true. We want to combine these two equations to show that a divides $(b + c)$. In other words, we must find an integer n that makes the equation $an = b + c$ true.

Using the first two equations, we write an expression for $(b + c)$,

$$b + c = ak + am.$$

Now, if we use Axiom D, the Distributive Property, we can rewrite the equation as

$$b + c = a(k + m).$$

Since integers are closed under addition, $(k + m)$ is also an integer. We can call it n. Then our equation becomes

$$b + c = an,$$

which is what the definition requires for a to divide $(b + c)$. □

Theorem 3.3. *For any three integers a, b, and c, if $a \mid b$ and $a \mid c$, then $a \mid (b - c)$.*

Practice Problem 3.3. *Prove Theorem 3.3.*

Theorem 3.4. *For any two integers, a and b, if $a \mid b$, then for any integer c, $a \mid bc$.*

Proof: For a to divide b, there must be an integer k that makes the equation $ak = b$ true. Now, if we just multiply both sides of this equation by c, we get

$$(ak)c = bc.$$

By Axiom M2, Associativity of Multiplication, this is the same as

$$a(kc) = bc.$$

Since integers are closed under multiplication, kc is still an integer, which we can rename n. Then our equation reads

$$an = bc,$$

which is what the definition requires for a to divide bc. □

Challenge Problem. Think about the converse of Theorem 3.4. If we take integers $a, b,$ and c, and if we know $a \mid bc$, what can we conclude? Must $a \mid b$ or $a \mid c$? Or is there an example for which $a \mid bc$, but a does not divide b and does not divide c?

Theorem 3.5. *For any three integers, $a, b,$ and c, if $a \mid b$ and $a \mid c$, and if m and n are any two integers, then $a \mid (bn + cm)$.*

Proof: Since $a \mid b$, by Theorem 3.4, $a \mid bn$. Similarly, we get $a \mid cm$. So then $a \mid (bn + cm)$ by Theorem 3.2. □

Practice Problem 3.4. *Show that Theorem 3.5 implies Theorem 3.3.*

These five theorems about divisibility have all used equations in their proofs. There is also one important fact about divisibility that involves an inequality.

Theorem 3.6. *If b is a positive integer and $a \mid b$, then $a \leq b$.*

Proof: We suppose $a > b$ and look for a contradiction. Since $a \mid b$, there is an integer k such that $b = ak$. We know that $b > 0$, and since $a > b$, we also know that $a > 0$. It follows that k is an integer greater than 0, and hence $k \geq 1$. But then, by Axiom O4, $ak > bk$ and $bk \geq b$, so $ak > b$ by Axiom O2. This is a contradiction. Therefore, $a \leq b$, since Axiom O1 tells us that either $a < b$, $a = b$, or $a > b$. □

3.2   Greatest Common Divisor

We are interested in finding out whether two numbers have any of the same positive divisors. All positive integers are divisible by 1, so 1 is always in the list of positive common divisors of two integers. In this section, when we discuss common divisors, we are talking about *positive* common divisors.

Definition. Two integers, a and b, are said to have a *common divisor d* if $d \mid a$ and $d \mid b$. The largest of all the common divisors of a and b is called the *greatest common divisor* of a and b. We write $(a, b) = d$ if d is the greatest common divisor of a and b. We sometimes write GCD for greatest common divisor. For now, we restrict our attention to the positive numbers a and b.

Examples. What are the greatest common divisors of some pairs of integers?

1. The list of common divisors of 24 and 36 is $1, 2, 3, 4, 6,$ and 12. Therefore, $(24, 36) = 12$.

2. The list of common divisors of 15 and 45 is 1, 3, 5, and 15. So $(15, 45) = 15$.

3. The only common divisor of 25 and 33 is 1, so $(25, 33) = 1$.

Practice Problem 3.5. *If a and b are two positive integers and $a \mid b$, show that $a = (a, b)$.* **Hint:** *First show that a divides both a and b. Then think about how to show that a is the* largest *integer dividing both a and b.*

An important special case of GCDs occurs when a and b have no common factors other than 1.

Definition. If the greatest common divisor of two integers a and b is 1, then a and b are said to be *relatively prime.*

Practice Problem 3.6. *Find all the integers between 25 and 30 that are relatively prime to 10.*

As was the case with divisibility, some facts about GCDs will be useful in some of our later discussions. We prove two here and leave a third as a Challenge Problem.

Theorem 3.7. *For any positive integers a and b:*

1. *Suppose $(a, b) = d$. If $a = md$ and $b = nd$, $(m, n) = 1$.*

2. *For any integer c, $(a + cb, b) = (a, b)$.*

Proof:

1. Assume that e is a positive integer that divides both m and n. Then, by the definition of divisibility, there must be an integer p so that $ep = m$ and an integer q so that $eq = n$. Multiplying by d, we get $dep = dm = a$ and $deq = dn = b$. So the integer de divides a, and it also divides b. This means that de is a common divisor of a and b. Since d was the greatest of all the common divisors, de must be no larger than d. This forces e to be 1. Thus, $(m, n) = 1$.

2. For this part, we show that all the common divisors of a and b are the same as all the common divisors of $(a + cb)$ and b. That way, the largest divisor will be the same for both pairs of numbers.

   Start with a number f that is a common divisor of $(a + cb)$ and b. By Theorem 3.5, f will divide $(a + cb) - cb$, or a. So f is a common divisor of a and b. Now, start with a number e that divides both a and b. Then, by Theorem 3.5, e also divides $(a + cb)$. So e will be in the set of common divisors of $(a + cb)$ and b. $\qquad\square$

Theorem 3.8. *Suppose c and d are positive integers and $c = dq + r$, where q and r are also integers. Then $(c, d) = (d, r)$.*

Challenge Problem. Prove Theorem 3.8.

Exercises for practice with divisibility and GCDs are included at the end of this chapter. Now we can get back to what started this discussion, namely, what is a prime number?

3.3   Primes

Definition. A *prime number p* is a positive integer greater than 1 whose only positive divisors are 1 and p.

For reasons that will become clear later, we want to exclude 1 from the list of prime numbers. We can begin a list of primes by just doing some mental arithmetic. Here are the first few:

$$2, 3, 5, 7, 11, 13, 17, 19, 23, 29, 31, 37, 41, 43.$$

Right away, we begin to notice some things about the list. First, there is only one even number, 2, on the list. This is because all the other even integers are divisible by 2. The even integers greater than 2 are examples of composite numbers.

Definition. The positive integers greater than 1 that are not primes are called *composite numbers*.

Theorem 3.9. *Every composite number is divisible by some prime.*

Proof: Suppose n is a composite number so that $n > 1$. Since n is not prime, n has a positive divisor besides 1 and n. Call this divisor a. Since $a \neq 1$, $1 < a$. Since $a \neq n$, by Theorem 3.6, $a < n$. So $1 < a < n$. Since $a \mid n$, there is an integer k such that $ak = n$. Again, $k > 1$ since $a < n$, and $k < n$ since $a > 1$. So $1 < k < n$.

If a or k is prime, then n is divisible by a prime. If, instead, neither is prime, then both a and k are composite. Let's factor one of these divisors of n, say, a. By Theorem 3.1, any divisor of a will also be a divisor of n. Since a is not prime, a has a positive divisor besides 1 and a. Call this divisor b. Since $b \neq 1$, $1 < b$. Since $b \neq a$, by Theorem 3.6, $b < a$. So $1 < b < a$. Since $b \mid a$, there is an integer h such that $bh = a$. Again, $1 < h < a$. If b or h is prime, then a, and therefore n, are divisible by a prime. If neither b nor h is prime, we repeat this process.

This process yields a set of factors of n that are greater than 1. Call this set S. Our goal is to show that there is a prime in S. By the well-ordering axiom, since S is a nonempty set of positive integers, S has a smallest element. Call this element p. Since p is in S, $p > 1$. This number p is either prime or composite. What if p is composite? Then p has a positive divisor q besides 1 and p. Then $q < p$, and q is an element of S. This is a contradiction, since p is the smallest element of S. So p cannot be composite and must be prime. Thus, n has a positive prime divisor.□

Looking at the beginning of the list of prime numbers, we see that there are three primes in a row among the first odd numbers. Then there is a gap, and then two come together, then another gap, then another two. One of the first questions we can ask about primes has to do with this pattern of having two or three odd primes in a row. We call the set of primes 3, 5, and 7 *triple primes*. Odd primes that come next to one another in counting, such as 17 and 19, or 29 and 31, are called *twin primes.* How many triple primes are there? How many twin primes? How many primes in all are there?

One of the most intriguing things about working with primes is that these kinds of simple questions can lead to dramatically different answers. In the case of triple primes, we can answer that {3, 5, 7} is the only triple prime set. We can also prove that there are an infinite number of primes. However, the question of how many twin primes there are is unanswered.

Let's begin by making a convincing mathematical argument to show that there are an infinite number of primes. This argument is ancient. In fact, Euclid[1] used this argument to prove this fact thousands of years ago. In this discussion, we use some familiar terms that we will discuss more formally in later sections.

Theorem 3.10. *There are infinitely many primes.*

Before we start the proof of this theorem, let's think about what it says. It states that the number of primes is not finite. That means that there is not a largest prime number. If there were, we would be able to count all the primes up to and including that largest prime, and the total would be a finite number. We use this idea to establish the proof of the theorem.

Proof: We use the method of proof by contradiction, so we assume the theorem is false. For the theorem to be false, there would have to be a finite number of primes. We thus number the primes, starting with the first one, and call it p_1. The next would be p_2, the next p_3, and so on. The largest prime will have a number also; call the last prime p_n.

To reach a contradiction, we need to find another prime number that is not included in this list. So, we want a number that is not divisible by any of the prime numbers p. We multiply all the prime numbers together and then add 1, and call this new number N. Then an equation for N would be

$$N = p_1 p_2 \cdots p_n + 1.$$

We need to figure out whether N is a prime or composite number. By Theorem 3.9, if N is composite, it will be divisible by some prime. Let's

---

[1] Euclid (ca. 365–300 B.C.) wrote the first well-organized treatment of the mathematics the Greeks used. Euclid taught at the Library of Alexandria about 2,300 years ago. His work, *Elements*, comprises 13 books that include theorems and proofs in number theory and geometry. He was younger than Plato but older than Archimedes, and he is one of the most influential mathematicians of all time.

see what divides N evenly. If we divide N by p_1, we get a remainder of 1. So p_1 is not a divisor of N. If we try to divide N by p_2, we also get a remainder of 1. So p_2 is not a divisor of N. If we try this with all the primes in order, each will leave a remainder of 1 when it divides N. This is true even for our biggest prime, p_n. So we have shown that N is not divisible by any of the primes. Therefore, since no prime divides N, N itself must be a prime number. But how can N be a prime number? It is bigger than the largest prime p_n. This is a contradiction, which means that our original assumption, that there were only a finite number of primes, must have been wrong. Thus, there must be an infinite number of primes, as the theorem says.   □

How do we know which numbers are primes? The first thing most students try is to take a number and check whether anything less than that number divides it evenly. This is a good first try. Let's think about a number that may or may not be prime, and figure out whether it is a prime or composite. Take the number 199. To start, we would begin dividing 199 by the smaller prime numbers and see what happens. For example, 199 is not divisible by 2 since it is odd.

Practice Problem 3.7. *Figure out if* 199 *is prime by dividing it by primes smaller than* 199.

If you tried this Practice Problem, where did you stop? Did you check all the primes less than 199? Or was your first thought that if you check division by all the primes halfway to 199, that would be enough? You may have become weary of dividing 199 by so many numbers. Is there a shortcut to all this dividing?

Yes, there is. If we think about division with integers, positive factors come in pairs (sometimes with two of a kind). For example, the factors of 10 are 1 and 10, and 2 and 5, and the factors of 12 are 1 and 12, 2 and 6, and 3 and 4. So once we get halfway to the number, we will have accounted for all the possible factors. In fact, we can stop checking once we get to a prime whose square is greater than the number.

Example. Let's factor 100. We get the pairs 1 and 100, 2 and 50, 4 and 25, 5 and 20, and 10 and 10. Notice that once we get to the square root of 100, there are no larger numbers that will be factors that aren't already in our list.

In our example of trying to figure out if 199 is prime or composite, we need to check the primes only up to 13. Why can we stop at 13? The next prime is 17, and $17^2 = 289$. This means that if 17 were to divide 199, some smaller prime would divide 199. Since the largest prime less than 17 is 13, we know we can stop there.

Theorem 3.11. *If an integer N is composite, it will be divisible by some integer less than or equal to $\sqrt{N}$.*

Practice Problem 3.8. *Prove Theorem 3.11.*

It is useful to have a list of primes up to 1,000. Make a list by using the *Sieve of Eratosthenes*.[2] Here is how the sieve works: Make a list of numbers from 2 to N. Cross out 4, and then every even number after 4. In this way, we are eliminating all composite numbers that have a factor of 2. The integer 3 is the next prime. Starting at 9, cross out every third number. Now we've deleted all the composite numbers with a factor of 3. Our next prime number is 5, so start at 25 and cross out every fifth number to eliminate multiples of 5. In general, for a prime p, start at p^2 and cross out every pth number.

Example. Table 3.1 is the result of using the Sieve of Eratosthenes on the integers between 1 and 100.

1	2	3	4	5	6	7	8	9	10
11	12	13	14	15	16	17	18	19	20
21	22	23	24	25	26	27	28	29	30
31	32	33	34	35	36	37	38	39	40
41	42	43	44	45	46	47	48	49	50
51	52	53	54	55	56	57	58	59	60
61	62	63	64	65	66	67	68	69	70
71	72	73	74	75	76	77	78	79	80
81	82	83	84	85	86	87	88	89	90
91	92	93	94	95	96	97	98	99	100

Table 3.1. The Sieve of Eratosthenes applied to the numbers 1 through 100.

---

[2]Eratosthenes (ca. 276–194 B.C.) was born in Cyrene, a Greek colony near Egypt. He was the tutor of Ptolemy II's son in Athens. He was the chief librarian in Alexandria, where he was a scholar in many fields. He is famous for measuring the earth.

Practice Problem 3.9. *Use the Sieve of Eratosthenes to make a list of primes up to 1,000.*

This kind of analysis will tell us whether a given number is prime or composite. It may take a long time, particularly if the number is large.

Formulas for Primes

Since there is an infinite number of primes, it is an interesting problem to see whether there is a way to find a formula that will result in a prime number. We do not know whether such a formula exists, but it has led to very interesting mathematics.

Example. Here's a procedure that seems to result in a prime number:

1. Take a positive integer.
2. Add one to that number.
3. Multiply the first two numbers together.
4. Add 17 to the result. This should be a prime number.

Does this work? Let's try an example with 2 as the first integer. Then the second integer is 3. The product of these two numbers is 6. If we add 17, the result is 23, which is a prime number.

We could write a formula for this procedure. If n stands for our original number, then $n + 1$ is the second number. Our possible prime p will then be the number

$$p = (n)(n + 1) + 17.$$

Practice Problem 3.10. *Show that this formula gives a prime for $n = 1, 2, \ldots, 15$.*

Unfortunately, this procedure breaks down when $n = 16$, since the resulting "prime" is divisible by 17. This formula does work for a while, though. A similar formula would be

$$p = (n)(n + 1) + 41.$$

This too yields primes for a while but breaks down when $n = 40$.

A lawyer named Pierre de Fermat[3] thought he had a good formula for primes. His proposed formula uses powers of 2.

Definition. A *Fermat number* is any number of the form $2^{2^n} + 1$, where n is a nonnegative integer. If a Fermat number is prime, it is called a *Fermat prime.*

As we did with the other formulas, let's check to see which of these Fermat numbers are primes. For $n = 0$, we get 3, a prime. For $n = 1$, we get 5, which is a prime, and for $n = 2$, we get 17, which is also prime. For $n = 3$ we get 257.

Practice Problem 3.11. *Verify that* 257 *is prime.*

For $n = 4$, we get 65,537, which is also a prime. However, for $n = 5$ the resulting Fermat number, 4,294,967,297, is composite because it factors as $641 \cdot 6{,}700{,}417$. In fact, all other Fermat numbers that have been computed are composite. The question of whether any other Fermat numbers are prime is unanswered. The Fermat primes play a surprising role in geometry that is presented in Chapter 16.

A contemporary of Fermat also had a formula for finding primes. Mersenne[4] was a teacher of Descartes and a Franciscan friar. His formula, like Fermat's, involves powers of 2.

Definition. A *Mersenne number* is any number of the form

$$2^m - 1,$$

where m is a natural number. If a Mersenne number is prime, it is called a *Mersenne prime.*

---

[3]Pierre de Fermat (1601–1665) was born into a wealthy family and studied the classics and languages before becoming a lawyer. He did not like to publish his mathematical discoveries, but wrote about them in letters to his contemporaries. Fermat was one of the cofounders of analytic geometry and also worked in the field of probability. He is perhaps most famous for writing a comment in the margin of a book of Diophantus, stating that he had a proof of a statement in the text, but that the proof would not fit in the margin. This theorem, called Fermat's Last Theorem, states that the equation $a^n + b^n = c^n$ has no positive integer solutions a, b, c for any $n > 2$. It took mathematicians nearly 400 years to come up with a proof of this theorem. The proof, due to Andrew Wiles in 1995, took seven years to finish.

[4]Marin Mersenne (1588–1648) was a priest and teacher who corresponded extensively with scholars in Europe, including Fermat. He studied number theory, and was particularly interested in the search for prime numbers. In 1644, he claimed to have a list of the primes of the form $2^p + 1$.

We currently know of 47 Mersenne primes. The largest one has more than twelve million digits.

Practice Problem 3.12. *Compute the first ten Mersenne numbers and determine which are prime.*

Twin Primes and Triple Primes

We asked earlier how many sets of twin and triple primes there are. Let's consider triple primes first. We know that the set $\{3, 5, 7\}$ is a set of three consecutive odd numbers, all of which are prime. Is there another such set? One place to start might be to look at twin primes and look for a third to make the twins a triplet. If we try this, we notice something interesting. Look at 11 and 13, for example. The odd number before 11 and the odd number after 13 are both divisible by 3.

Practice Problem 3.13. *Look at the following sets of twin primes and write down your observations about trying to find a third prime to make a triplet.*

(a) 29 *and* 31.

(b) 41 *and* 43.

The set of small primes $\{3, 5, 7\}$ turns out to be the only set of triple primes. Our argument to prove this fact includes using an observation about divisibility by 3. Let's write the fact as a theorem and prove the result.

Theorem 3.12. *The set of primes $\{3, 5, 7\}$ is the only set of triple primes.*

Proof: What we need to show is that if we take three other odd numbers in a row, one of them will be a composite number divisible by 3. So we first write three odd numbers in a row. If we name the first number in our list n, then the next two odd numbers would be $n+2$ and $n+4$. Since we do not want $\{n, n+2, n+4\} = \{3, 5, 7\}$ we must have $n > 3$.

We first note that if $3 \mid n$, then since n is bigger than 3, n is a composite number, and our list will not be a list of triple primes.

If 3 does not divide our number n evenly, then the remainder when we divide n by 3 will be either 1 or 2. How can we write this symbolically? Either $n = 3k+1$ or $n = 3k+2$ for some integer k. We'll think about each of these possibilities separately. Call the possibilities case 1 and case 2.

Case 1 is where $n = 3k + 1$. Remember that n was the first number in the list of possible triple primes. The other two numbers were $n + 2$ and $n + 4$. In case 1, what would be an equation for the second number, $n + 2$?

$$n + 2 = (3k + 1) + 2$$
$$= 3k + 3$$
$$= 3(k + 1)$$

According to our definition of divisibility, this says that $3|(n+2)$. So $n+2$ is composite and our list is therefore not a list of triple primes.

Case 2 is where $n = 3k + 2$. This time, let's look at the third number in the list of possible triple primes. That would be $n + 4$. Let's write an equation for $n + 4$.

$$n + 4 = (3k + 2) + 4$$
$$= 3k + (2 + 4)$$
$$= 3k + 6$$
$$= 3(k + 2)$$

According to our definition of divisibility, this says that $3|(n+4)$. So $n+4$ is composite and our list is therefore not a list of triple primes.

To sum up, if we try to list three odd numbers in a row, one of them will be divisible by 3. The only way for such a set of numbers to be all primes is for one of the numbers to be 3. Therefore, the set $\{3, 5, 7\}$ is the only set of triple primes. $\square$

What about twin primes? It has been conjectured that the list of twin primes, like the list of primes themselves, is infinite. In the search for large prime numbers, twin primes keep appearing on the list. However, unlike the theorem about triple primes, the twin prime conjecture is considerably more complicated and has not been resolved.

Other Conjectures about Primes

It is interesting to think about primes and what other facts may be true about them. Christian Goldbach[5] made a conjecture. He looked at sums

---

[5]Christian Goldbach (1690–1764) was born in Konigsburg, Prussia, and became a professor of mathematics at the Imperial Academy of St. Petersburg. He also corresponded with the mathematicians of his day.

of prime numbers and noticed an interesting fact. Certainly, if you add two odd numbers, you get an even number. (We talked about that in Section 1.3 on Even-Odd arithmetic.) So, if you add two odd primes, you get an even number. Goldbach turned this question around and asked whether every even integer bigger than 2 could be written as the sum of two primes.

Let's try this for some small even numbers. Try the number 10. If you think about it for a minute, you can get 10 as the sum of two primes in two different ways, since $3 + 7 = 10$ and $5 + 5 = 10$.

Practice Problem 3.14. *Try to write n as the sum of two primes. How many different ways can you find to do this?*

(a) $n = 20$.

(b) $n = 30$.

(c) $n = 100$.

Goldbach first stated his conjecture in a letter to Leonard Euler.[6] People have tried to verify this conjecture ever since. No one has yet to find an even number that cannot be written as the sum of two primes. However, this does not prove that Goldbach's conjecture is true. This is another open question about prime numbers.

Practice Problem Solutions and Hints

3.1. All of them!

3.2. $\pm 1, \pm 2, \pm 3, \pm 4, \pm 6, \pm 8, \pm 12, \pm 24$.

3.3. Since $a \mid b$, there exists an integer k such that $ak = b$. Since $a \mid c$, there exists an integer m such that $am = c$. So $b - c = ak - am = a(k - m)$. Since $k - m$ is an integer, $a \mid b - c$.

3.4. To get Theorem 3.3, let $m = 1$ and $n = -1$ in Theorem 3.5.

---

[6]Leonard Euler (1707–1783) was a famous Swiss mathematician. He is one of the most prolific mathematicians who ever lived. He abandoned plans to pursue theology as a vocation and instead studied extensively in the sciences, including physics and astronomy. He wrote more than 800 books and papers that already fill more than 80 volumes. Although he was blind for the last 17 years of his life, he continued to pursue mathematics. He had 13 children.

3.5. $a \mid a$ since $a = 1 \cdot a$, and $a \mid b$ by assumption. So a is a common divisor of a and b. If $d \mid b$ and $d > a$, then by Theorem 3.6, d cannot be a divisor of a. So $a = (a, b)$.

3.6. The integers 27 and 29 are relatively prime to 10. Integers 25 and 10 have a common factor of 5; 30 and 10 have a common factor of 10; and 26 and 28 have a common factor of 2 with 10. So $(25, 10) = 5$; $(26, 10) = 2$; $(27, 10) = 1$; $(28, 10) = 2$; $(29, 10) = 1$; and $(30, 10) = 10$.

3.7. Check by dividing. Start with 2, 3, 5, 7, and so on. Continue until you've checked all the primes less than $\sqrt{199}$, as discussed next.

3.8. See Exercise 2.10(b) for a hint about how to start this problem.

3.9. This table lists all the primes less than 1,000, and has twin primes in bold.

2	3	5	7	11	13	17	19	23	29
31	37	**41**	**43**	47	53	**59**	**61**	67	**71**
73	79	83	89	97	**101**	**103**	**107**	**109**	113
127	131	**137**	**139**	**149**	**151**	157	163	167	173
179	181	**191**	**193**	**197**	**199**	211	223	**227**	**229**
233	**239**	**241**	251	257	263	**269**	**271**	277	**281**
283	293	307	**311**	**313**	317	331	337	**347**	**349**
353	359	367	373	379	383	389	397	401	409
419	**421**	**431**	**433**	439	443	449	457	**461**	**463**
467	479	487	491	499	503	509	**521**	**523**	541
547	557	563	**569**	**571**	577	587	593	**599**	**601**
607	613	**617**	**619**	631	**641**	**643**	647	653	**659**
661	673	677	683	691	701	709	719	727	733
739	743	751	757	761	769	773	787	797	**809**
811	**821**	**823**	**827**	**829**	839	853	**857**	**859**	863
877	**881**	**883**	887	907	911	919	929	937	941
947	953	967	971	977	983	991	997		

3.10. This is just a calculation. For example, $13(13+1) + 17 = 13(14) + 17 = 182 + 17 = 199$, which we know is prime.

3.11. Check your table in Practice Problem 3.9.

3.12.

1. $2^1 - 1 = 1$ is not prime.
2. $2^2 - 1 = 3$ is prime.
3. $2^3 - 1 = 7$ is prime.
4. $2^4 - 1 = 15$ is composite.
5. $2^5 - 1 = 31$ is prime.
6. $2^6 - 1 = 63$ is composite.
7. $2^7 - 1 = 127$ is prime.
8. $2^8 - 1 = 255$ is composite.
9. $2^9 - 1 = 511$ is composite.
10. $2^{10} - 1 = 1{,}023$ is composite.

3.13. All of the sets of three consecutive odd numbers you could make have one element divisible by 3.

3.14. $20 = 3 + 17 = 7 + 13$; $30 = 7 + 23 = 11 + 19 = 13 + 17$; $100 = 3 + 97$. Use the Sieve of Erathosthenes to find others.

Exercises

3.1. Determine which of the following numbers are prime: 1,259, 4,323, 2,391, 1,763.

3.2. Find the greatest common divisor of each of the following pairs.

(a) $(18, 27)$.

(b) $(5, -15)$.

(c) $(16, 27)$.

(d) $(1{,}000, 1{,}001)$.

(e) $(60, 72)$.

(f) $(102, 54)$.

(g) $(555 + 987{,}654{,}321 \cdot 1{,}111, 1{,}111)$.

3.3. Suppose we want to use the Sieve of Eratosthenes to get a list of all the primes up to 500. What is the largest prime we will need to check for divisibility?

3.4. Show that 1,999 is prime. Does it have a twin prime?

3.5. Write 100 as the sum of two primes in as many different ways as you can.

3.6. Goldbach's other conjecture is that every odd number greater than 5 can be written as the sum of three primes. Check this conjecture for the following numbers.
  (a) 7.
  (b) 29.
  (c) 101.
  (d) 237.

3.7. For each statement, either prove that it is true or provide an example that shows that it is not true.
  (a) If $a|b$, then $2a|4b$.
  (b) If $a|b$ and $a|c$, then $a|bc$.
  (c) If $a|c$ and $b|c$, then $ab|c$.
  (d) If $a|b$, then $a+2|b+2$.
  (e) If $a|b$, then $7a|35b$.
  (f) If $2a|4b$, then $a|b$.
  (g) If $a|b$ and $a|c$, then $a|(13b-2c)$.
  (h) If $a|b$, then $a^2|b^3$.
  (i) If $a^2|b^3$, then $a|b$.
  (j) For all integer values of a, $3|(27a-21)$.

3.8. Show that if $a \mid b$ and $c \mid d$, then $ac \mid bd$.

3.9. If n is a positive integer, what is $(0, n)$?

3.10. Decide whether each of the following statements is true. For each that is not, provide a counterexample.
  (a) If $(a, b) = 1$ and $(b, c) = 1$, then $(a, c) = 1$.
  (b) If $n|(a, b)$, then $n|a$ and $n|b$.
  (c) For all positive integer values of a, $(a, a+2) = 2$.

(d) For all positive integers a and b, $(2a, b) = 2(a, b)$.

(e) $(a + 1, b + 1) = (a, b)$.

3.11. Prove that for all positive integers a, b, and c, $(ac, bc) = (a, b)c$.

3.12. Show that $(n + 1, n) = 1$ for any positive integer n.

3.13. If p and q are twin primes, prove that $pq + 1$ is a perfect square.

3.14. Show that if p is the smallest prime factor of n, and $p > \sqrt[3]{n}$, then $\frac{n}{p}$ is prime or 1.

3.15. Show that $(2^m - 1, 2^{m+1} - 1) = 1$, and $(2^{2^n} + 1, 2^{2^{n+1}} + 1) = 1$. Think about why it might be important that these GCDs equal 1.

3.16. Let a, b, and c be any three positive integers. The greatest common divisor of a, b, and c is the largest integer d such that $d|a$, $d|b$ and $d|c$. It is denoted by (a, b, c). Prove that $(a, b, c) = ((a, b), (a, c))$.

3.17. Can you find three numbers in a row that are all composites? Can you find four? Five? Can you find a formula that will give you n composite numbers in a row?

3.18. Suppose that s is an odd positive integer, and that r is any integer.
(a) Show that for any real number x,

$$x^{rs} + 1 = (x^r + 1) \cdot (x^{(s-1)r} - x^{(s-2)r} + x^{(s-3)r} - \cdots + x^{2r} - x^r + 1).$$

Why doesn't this formula work when s is even?

(b) Use the result from part (a) to find factorizations of the polynomials $x^{20} + 1$ and $x^{24} + 1$.

(c) Use part (b) to find a nontrivial divisor of $2^{20} + 1$ and $2^{24} + 1$. (**Note:** A divisor of n is called *nontrivial* if it is not 1 or n.)

(d) Prove that if m is not a power of 2, then $2^m + 1$ is composite.

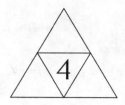

The Division and Euclidean Algorithms

4.1 The Division Algorithm

Now that we have introduced division, we should think back to when we first learned the rules for dividing one number into another. Let's start with an easy example, say, dividing 9 by 3. We may have used fractions to learn this concept, or manipulatives to divide nine things into three piles. When numbers divide evenly, this works well. If we try to divide 10 by 3, there is a slight problem because there will be 1 left over if we try to divide evenly. The 1 is called the *remainder*. Other math vocabulary words associated with the process of division, such as *quotient* and *divisor*, are helpful when we discuss the *division algorithm*.

Example. Let's divide 199 by 7 by using long division. We get that 199 is $7 \cdot 28$ with a remainder of 3.

$$
\begin{array}{r}
28 \\
7{\overline{)199}} \\
-14 \\
\hline
59 \\
-56 \\
\hline
3
\end{array}
$$

We name all these terms so that we can talk about them in the general case. The number 7 is the *divisor*, 28 is called the *quotient*, and 3 is the *remainder*. In this case, we write the equation as $199 = (7 \cdot 28) + 3$. In general, if we are dividing a number a by another number b, we get an equation

$$a = bq + r,$$

where q stands for the quotient and r represents the remainder. All of these numbers are integers.

Practice Problem 4.1. *Find q (quotient) and r (remainder) for the following numbers a and b.*

(a) $a = 200, b = 7$.

(b) $a = 35, b = 19$.

(c) $a = 72, b = 8$.

(d) $a = 12, b = 17$.

In doing Practice Problem 4.1, we notice that there is a restriction on how big and small each remainder is. First, the remainder is never a negative number. That is, $r \geq 0$. Also, the remainder is never as large as the divisor. This is because if the remainder were that large, we could increase the size of the quotient and reduce the size of the remainder so that it is less than the divisor.

Example. Let's go back to dividing 199 by 7. We could also have written $199 = 7 \cdot 27 + 10$, but this makes the remainder, 10, larger than the divisor, 7. This tells us that we did not get a large enough quotient when we divided.

These facts about division are stated in a theorem called the Division Algorithm.

Theorem 4.1 (Division Algorithm). *For any two positive integers a and b, there are unique integers q and r satisfying $a = bq + r$, where $0 \leq r < b$.*

Proof: This proof is somewhat long and tedious, but it uses the axioms we stated and the theorems we have established. Be sure to follow each step as we go.

First, let's think about the two positive integers we start with, a and b. In the first three parts of Practice Problem 4.1, $a > b$. In the fourth part, $a < b$. The theorem is also true if $a = b$. Let's do these last two cases first, since they are easy.

Case 1. $a = b$. If $a = b$, then we can write $a = b \cdot 1 + 0$. Here, since $r = 0$ and b is positive, we certainly have $0 \leq r < b$, as Theorem 4.1 requires.

Case 2. $a < b$. Since $a < b$, write $a = b \cdot 0 + a$. In this case, since $r = a$ is positive and $a < b$, we also satisfy the conditions of Theorem 4.1.

Case 3. $a > b$. This is the case in which the work will really be done; we will follow the ideas of long division. If a is a multiple of b, say $a = bk$, then the theorem is true with $q = k$ and $r = 0$. Suppose now that a is not a multiple of b. Since $a > b$, we know $a - b > 0$. Look at the numbers that result from subtracting multiples of b from a. This is the set of numbers $S = \{a - b, a - 2b, a - 3b, \cdots\}$.

Collect all the numbers from S that are positive. Write this set as $P = \{a - bk \mid k \text{ is in } \mathbb{N} \text{ and } a - bk > 0\}$. Since $a - b > 0$, we know that P is not empty. Also, since P consists of positive integers, the well-ordering axiom applies. Thus, the set P has a smallest element, called r. This smallest element r has the form

$$r = a - bq$$

for some number q in $\mathbb{N}$. We already know that $r > 0$ since r is in P. For Theorem 4.1 to be true, though, we still need to show that $r < b$.

Let's suppose instead that $r = b$ or $r > b$. First, if $r = b$, then we can rewrite a as $a = bq + b = b(q + 1)$. Therefore, a is a multiple of b, contrary to our assumption.

If instead $r > b$, then we would have $r - b > 0$ and

$$r - b = (a - bq) - b$$
$$= a - b(q + 1).$$

This means that $r - b$ would be in the set P and be smaller than r. This is a contradiction because r is the smallest number in the set P. Thus, we have shown the existence of a divisor and remainder satisfying the desired properties. We leave the proof of uniqueness to the reader. $\qquad \square$

Challenge Problem. Prove that the numbers q and r are unique. That is, assume that we can write $a = bq + r$ and at the same time we can write $a = bq_1 + r_1$, with $0 \leq r < b$ and $0 \leq r_1 < b$. Show that $q = q_1$ and $r = r_1$.

Example. Let's return to our example of dividing 7 into 199. The set P is just the numbers we get by subtracting multiples of 7 from 199. What is the set P in this particular case?

$$P = \{199 - 1 \cdot 7, 199 - 2 \cdot 7, 199 - 3 \cdot 7, \cdots, 199 - 27 \cdot 7, 199 - 28 \cdot 7\}$$
$$= \{199 - 7, 199 - 14, 199 - 21, \cdots, 199 - 189, 199 - 196\}$$
$$= \{192, 185, 178, \cdots, 10, 3\}.$$

And we know that $3 = 199 - (28 \cdot 7)$, or $199 = 7 \cdot 28 + 3$.

What the division algorithm is doing is providing a way to subtract multiples of 7 from 199 until we can't do it anymore without getting negative numbers. The numbers in the set P get smaller and smaller, and eventually we reach a smallest one. The process of long division is just the division algorithm at work, in shortcut fashion. Look back at the work for dividing 199 by 7:

$$
\begin{array}{r}
28 \\
\hline
7{\overline{)}}199 \\
-14 \\
\hline
59 \\
-56 \\
\hline
3
\end{array}
$$

How do we see the division algorithm at work here? First we do the division algorithm on 19 and get $19 = 7 \cdot 2 + 5$, or, by subtraction, $19 - 7 \cdot 2 = 5$. Next, we "bring down" the 9 and use the division algorithm to get $59 = 7 \cdot 8 + 3$. This gives $199 = 7 \cdot 28 + 3$. Writing this out in base ten notation, we see that the first use of the division algorithm gives us $19 \cdot 10 = 7 \cdot 2 \cdot 10 + 5 \cdot 10$, and the second use gives us $5 \cdot 10 + 9 = 7 \cdot 8 + 3$.

So, we have introduced the division algorithm which shows how to divide one number into another. But, what *is* an algorithm? An algorithm is a method mathematicians use to describe a way of problem solving that follows a series of steps to get to an answer. An algorithm is a process that we apply routinely, almost without thinking, which leads us through prescribed steps from a problem to a solution. Ordinarily in an algorithm, each successive step reduces the problem to a simpler one.

The words *algebra* and *algorithm* both come to us from the Arabic world of the ninth century. Commerce in that time required the use of unknown quantities in solving problems, and Al-Khwarizmi[1] wrote a

---

[1]Abu Jafar Muhammad ibn Musa al-Khwarizmi (ca. 780–850) was an astronomer who lived in Baghdad and was a member of the House of Wisdom, an association of scientists. His books on mathematics, astronomy, and geography helped to introduce

notable text about using Arabic methods to solve problems. The book was used and carried throughout the world of trade in the Mediterranean region and then spread to the rest of Europe.

The Division Algorithm with a Negative Dividend

What happens in the division algorithm if $a < 0$ and $b > 0$?

Example. Suppose $a = -31$ and $b = 7$. We have

$$-31 = 7 \cdot (-5) + 4.$$

We might have been tempted to write

$$-31 = 7 \cdot (-4) + (-3).$$

But the most important part of the division algorithm is that it requires the remainder to be nonnegative and less than the divisor b. This is what makes the proofs and applications work. So we must write $-31 = 7 \cdot (-5) + 4$ in order for the remainder to be nonnegative and less than 7.

Practice Problem 4.2. *Use the division algorithm with the following pairs of numbers.*

(a) $a = -45$, $b = 4$.

(b) $a = -17$, $b = 3$.

Challenge Problem. Prove the division algorithm for the case $a < 0$ and $b > 0$.

What happens if $b < 0$? There are two possibilities: either $a < 0$ or $a > 0$. In the first case, we can replace a and b by $-a$ and $-b$, respectively, and use the ordinary division algorithm. In the second case, we replace a and b by $-a$ and $-b$, respectively, and use the results of the challenge problem.

---

some of the Hindu ideas about mathematics and mathematical notation to Europe through commercial traders. The title of one of his texts, *Hisab al-jabr w'al-muqabala* (*Calculation by Restoration and Reduction*), is probably the origin of our words algebra and algorithm. His texts as well as other Arab books introduced the numerals we use, the use of symbols in equations, and the idea of using 0 as a placeholder.

4.2 The Euclidean Algorithm and the Greatest Common Divisor

Another algorithm allows us to find the greatest common divisor of two numbers. When we introduced the concept of the GCD of two numbers in Section 3.2, our examples were small numbers and easy to factor. But what if we wanted to find $(20{,}785, 44{,}350)$? It would not be an easy job to factor each of these numbers. Surprisingly, the Euclidean algorithm for finding the GCD is based on the division algorithm and does not involve factoring at all.

We first demonstrate the use of Euclid's method to find the greatest common divisor of two small numbers a and b. After we see how the algorithm works, we show why it does.

First, we recall Theorem 3.7(2): If a, b, and c are integers, then $(a + cb, b) = (a, b)$.

Example. Let's work with $a = 92$ and $b = 24$. We begin by using the division algorithm to write $92 = 24 \cdot 3 + 20$. Now we use Theorem 3.7(2) to see that $(92, 24) = (92 - 24 \cdot 3, 24)$, or $(92, 24) = (20, 24)$.

Another way to think about the fact that $(92, 24) = (20, 24)$ is to look at the equation we get from the division algorithm, $92 = 24 \cdot 3 + 20$. If a number divides 24 and 20, it must divide the left-hand side of the equation as well, and so it will divide 92. If we rewrite this equation as $92 - 24 \cdot 3 = 20$, we can also see that any common divisor of 24 and 92 will also divide 20.

Continuing the process of finding $(92, 24)$, we are now looking for the greatest common divisor of the two numbers 20 and 24. Use the division algorithm to write $24 = 20 \cdot 1 + 4$. A repetition of the argument we used above shows that $(20, 24) = (20, 4)$. A third use of the division algorithm gives $20 = 4 \cdot 5 + 0$, and we can conclude $(20, 4) = (4, 0) = 4$. So we have found the greatest common divisor of 92 and 24 to be 4 without ever factoring either of these numbers.

Before we state Theorem 4.2, the general case for the Euclidean algorithm, it will be useful to restate Theorem 3.8.

> **Theorem 3.8.** If c, d, q, and r are integers, and $c = dq + r$, then $(c, d) = (d, r)$.

Theorem 4.2 (The Euclidean Algorithm). *For positive integers a and b with $a > b$, use the division algorithm to write*

$$a = bq_1 + r_1, \text{ where } 0 \le r_1 < b.$$

Apply the division algorithm to b and r_1 and get

$$b = r_1 q_2 + r_2, \text{ where } 0 \le r_2 < r_1.$$

Continue this process to get

$$r_j = r_{j+1} q_{j+2} + r_{j+2}, \text{ where } 0 \le r_{j+2} < r_{j+1}$$

for $j = 0, 1, 2, \cdots, n-1$. If, in the equation

$$r_{n-1} = r_n q_{n+1} + r_{n+1},$$

we have $r_{n+1} = 0$, then $(a, b) = r_n$. That is, (a, b) is the last nonzero remainder in this process.

Proof: The steps in the division algorithm give the following set of equations:

$$
\begin{aligned}
a &= bq_1 + r_1, & &\text{where } 0 < r_1 < b, \\
b &= r_1 q_2 + r_2, & &\text{where } 0 < r_2 < r_1, \\
r_1 &= r_2 q_3 + r_3, & &\text{where } 0 < r_3 < r_2, \\
&\ \ \vdots \\
r_{j-2} &= r_{j-1} q_j + r_j, & &\text{where } 0 < r_j < r_{j-1}, \\
&\ \ \vdots \\
r_{n-2} &= r_{n-1} q_n + r_n, & &\text{where } 0 < r_n < r_{n-1}, \\
r_{n-1} &= r_n q_{n+1} + 0.
\end{aligned}
$$

We need to prove two things. The first is that we eventually get a zero remainder, and the second is that the last nonzero remainder is actually the GCD of a and b.

First, how do we know we eventually get a zero remainder? The set of remainders is the strictly decreasing sequence of nonnegative integers $r_1 > r_2 > r_3 > \cdots \ge 0$. We have a nonempty set of nonnegative remainders. By the well-ordering principle, this set has a least element, r. We claim

$r = 0$. If not, since r would be positive, we could perform the division algorithm one more time to get a smaller remainder.

Next, why is r_n the greatest common divisor of a and b? Theorem 3.8 is the key here. From the first equation in this proof, Theorem 3.8 tells us that $(a, b) = (b, r_1)$. The second equation together with Theorem 3.8 gives $(b, r_1) = (r_1, r_2)$. Continuing in this fashion, we get

$$(r_1, r_2) = (r_2, r_3) = (r_3, r_4) = \cdots = (r_{n-1}, r_n) = (r_n, 0) = r_n.$$

So, in fact, r_n is the GCD of a and b. $\qquad\qquad\qquad\qquad\qquad$ □

Example. Let's use the Euclidean algorithm to find the GCD of 128 and 154. We do the division algorithm first on these two numbers and get

$$154 = 128 \cdot 1 + 26.$$

So $(154, 128) = (128, 26)$. Now we use the two numbers 128 and 26 in the division algorithm and get

$$128 = 26 \cdot 4 + 24.$$

So $(128, 26) = (26, 24)$. The next steps give

$$26 = 24 \cdot 1 + 2,$$

so $(26, 24) = (24, 2)$, and

$$24 = 2 \cdot 12 + 0,$$

so $(24, 2) = (2, 0) = 2$.

Since 2 is the last nonzero remainder, $(154, 128) = 2$.

Practice Problem 4.3. *Use the Euclidean algorithm to find the greatest common divisor of the following pairs of numbers:*

(a) 34 and 55.

(b) 20,785 and 44,350.

How long will it take to find the GCD of two numbers a and b using the Euclidean algorithm? We have a very rough estimate in the proof. The sequence of remainders is a set of positive integers that starts with $a - bq_1$ and then decreases to 0. So the first remainder is no bigger than $a - b$. Thus, there will be no more than $a - b$ steps in the procedure.

Challenge Problem. We have an upper bound on the number of steps it will take to find the GCD of two numbers using the Euclidean algorithm. Can you find an example where the Euclidean algorithm actually takes $a - b$ steps?

The Euclidean algorithm helps to prove another easy fact about the GCD (Theorem 4.3), which leads to a nice result about divisibility.

Theorem 4.3. *Suppose b divides the product ac, and that a and b are relatively prime, that is, $(a, b) = 1$. Then b divides c.*

Proof: Since $(a, b) = 1$, the Euclidean algorithm yields the equations

$$a = bq_1 + r_1,$$
$$b = r_1 q_2 + r_2,$$
$$\vdots$$
$$r_{n-2} = r_{n-1} q_n + 1,$$

where the last remainder—that is, the GCD—is 1.

We multiply each of these steps by c to get the equations

$$ac = bq_1 c + r_1 c,$$
$$bc = r_1 q_2 c + r_2 c,$$
$$\vdots$$
$$r_{n-2} c = r_{n-1} q_n c + c.$$

Our hypothesis says ac is divisible by b. By Theorem 3.3, the term $r_1 c$ in the equation $ac = bq_1 c + r_1 c$ must also be divisible by b.

Now let's look at the next equation. Since b divides bc and b divides $q_2(r_1 c)$, b must also divide $r_2 c$. Moving through each equation, we see that each $r_j c$ is divisible by b for $j = 1, 2, \ldots, n - 1$. When we arrive at the last equation, we see that b must also divide c. □

This nice application of the Euclidean algorithm leads to Theorem 4.4.

Theorem 4.4. *If a prime p divides a product ab, then p must divide a or b.*

Proof: Suppose p divides ab but p does not divide a. Then, since p and a are relatively prime, by Theorem 4.3, p must divide b. □

We can extend this result to a product of several numbers.

Theorem 4.5. *If a prime p divides a product $a_1 \cdots a_r$ then p divides a_j for some j.*

Challenge Problem. Prove Theorem 4.5.

We also have a theorem about pairs of relatively prime numbers.

Theorem 4.6. *If $(a, b) = 1$ and $(a, c) = 1$, then $(a, bc) = 1$.*

Practice Problem 4.4. *Prove Theorem 4.6.*

4.3   The Fundamental Theorem of Arithmetic

In our first attempts at finding the greatest common divisor of two numbers (Section 3.2), we listed all the factors of both numbers. We were using small numbers and this was pretty easy. With bigger numbers, the Euclidean algorithm is a useful method of finding the GCD of two numbers. But what if we wanted to look at the factors of a single large number? From elementary arithmetic, we remember that we can do this by finding all the primes that divide the big number. Let's practice with an easy example.

Example. If we want to factor a number such as 72, we start with the smallest prime that divides it, 2. Writing $72 = 2 \cdot 36$, we then look at 36 and find a prime that divides 36. Again, we can write $36 = 2 \cdot 18$, and in turn, $18 = 2 \cdot 9$. Now 9 is no longer divisible by 2, so we try dividing by the next smallest prime, 3, and get $9 = 3 \cdot 3$. Collecting all the factors, we get $72 = 2 \cdot 2 \cdot 2 \cdot 3 \cdot 3$, or $72 = 2^3 \cdot 3^2$.

Is this the only correct way to factor 72? Could there be another way of writing 72 as a product of primes? The idea of primes as the building blocks of the integers is another of the important ideas of mathematics that seems to have become second nature to us and yet still requires a proof. This idea (Theorem 4.7) is so important that it has a name, the Fundamental Theorem of Arithmetic.

Theorem 4.7 (The Fundamental Theorem of Arithmetic). *Every positive integer greater than 1 can be written uniquely as a product of primes, with the prime factors in the product written in order of nondecreasing size.*

Proof: There are really two parts to the theorem. The first part is that we can find *at least* one way to write any positive integer as a product of primes. The second part says that there is *exactly* one way to do this; this is what the word *uniquely* means. We work on each part of the theorem separately.

The method of proof will once again be proof by contradiction. Suppose the theorem is false. Then there will be at least one positive integer x that cannot be written as the product of primes. Collect all of the integers like x into a set S. By the well-ordering axiom, since S is a nonempty set of positive integers, S has a smallest element, a. What do we know about a and the other members of S? They certainly aren't prime numbers because then each could be written as the product of a single term, namely, itself. So all the elements of our set S are composite. Since a is composite, it is the product of two smaller numbers, say, $a = b \cdot c$, where $1 < b < a$ and $1 < c < a$. Since b and c are each smaller than a, they cannot be members of the set S. This is because a is the smallest element of S. Since $b > 1$ and $c > 1$, both b and c can be written as a product of primes in the way described by the theorem. So we write $b = p_1 \cdot p_2 \cdot p_3 \cdots p_n$, and $c = q_1 \cdot q_2 \cdot q_3 \cdots q_m$, where all the p_i's and q_j's are primes. Then the product $a = bc = p_1 \cdot p_2 \cdot p_3 \cdots p_n \cdot q_1 \cdot q_2 \cdot q_3 \cdots q_m$ is a way to write a as a product of primes. This means that a is not a member of the set S, and we have a contradiction. So we have established that every positive integer can be written as a product of primes. By rearranging the prime factors in the product, we can write a with its prime factors in order of nondecreasing size.

Now we need to show that the factorization of a positive integer into primes is unique. The notation gets a little cumbersome, but we can work our way through the details in a series of easy-to-follow, logical steps.

Again, we start with the method of proof by contradiction. Suppose there is some positive integer n that has two different factorizations into primes:

$$n = p_1 \cdot p_2 \cdot p_3 \cdots p_s \text{ and } n = q_1 \cdot q_2 \cdot q_3 \cdots q_t.$$

Thus,

$$p_1 \cdot p_2 \cdot p_3 \cdots p_s = q_1 \cdot q_2 \cdot q_3 \cdots q_t,$$

where $p_1, p_2, p_3, \cdots p_s, q_1, q_2, q_3, \cdots q_t$ are all primes. Since p_1 divides the left-hand side, it also divides the right-hand side, and by Theorem 4.5, p_1 divides one of the factors on the right-hand side. Since both are prime, they must be equal. Rearrange the right-hand side so that this prime appears first. Now relabel the primes on the right-hand side so this prime is q_1. Because cancelation holds in the integers (that is, the integers are an integral domain), we can cancel p_1 and q_1. This leaves the equation

$$p_2 \cdot p_3 \cdots p_s = q_2 \cdot q_3 \cdots q_t.$$

Repeat this process until we have canceled all the primes on the left-hand side. Then there are no primes remaining on either side and we have the equation $1 = 1$. So, in fact, the original factorizations are equal, since we have matched up terms one by one and the factorizations have the primes in nondecreasing order. This proves that the prime factorization of a positive integer is unique.   □

Why We Don't Call 1 a Prime

The Fundamental Theorem of Arithmetic tells us that there is one and only one way to write a positive integer as a product of primes in nondecreasing order. Because of 1's role as the multiplicative identity, we know that if a is any number, then $1 \cdot a = a$. So what would happen to the Fundamental Theorem of Arithmetic if 1 were a prime?

Example. If we go back to writing the number 72 as a product of primes, we get $72 = 2 \cdot 2 \cdot 2 \cdot 3 \cdot 3 = 2^3 \cdot 3^2$. But we could also write $72 = 1 \cdot 2^3 \cdot 3^2$ or $72 = 1 \cdot 1 \cdot 2^3 \cdot 3^2$. This means that if 1 were a prime, there would be an infinite number of ways to write 72 as the product of primes and any number of 1s we like. If we want the factorization of a positive integer into primes to be unique, then we cannot have 1 as a prime number. In this example, notice that we can collect the prime factors that occur more than once and write them with an exponent. We can write $2 \cdot 2 \cdot 2 \cdot 3 \cdot 3 = 2^3 3^2$ and the individual prime factors still appear in nondecreasing order. In general, the Fundamental Theorem of Arithmetic says that we can write any integer $n \geq 2$ uniquely as $n = p_1^{a_1} \cdots p_k^{a_k}$, where $p_1 < p_2 < \cdots < p_k$ and the a_j's are positive integers.

Prime Factorization and the GCD

Now that we have shown that every integer has a unique prime factorization, we can use this fact to find the GCD of two numbers, although it may not be as quick as using the Euclidean algorithm. Suppose $a = p_1^{a_1} p_2^{a_2} \cdots p_n^{a_n}$ and $b = p_1^{b_1} p_2^{b_2} \cdots p_n^{b_n}$, where $p_1, p_2, \cdots, p_n$ are the primes occurring in the prime power factorization of either a or b. We include all the primes in the factorizations of both a and b in both lists, possibly with exponent 0.

Example. In finding $(12, 15)$, we would write $12 = 2^2 \cdot 3 \cdot 5^0$ and $15 = 2^0 \cdot 3 \cdot 5$.

Note that this is different from the result about unique factorization, since every prime that occurs in both factors appears here. If we write $\min\{x, y\}$ to denote the smaller of the two numbers x and y, then

$$(a, b) = p_1^{\min\{a_1, b_1\}} p_2^{\min\{a_2, b_2\}} \cdots p_n^{\min\{a_n, b_n\}},$$

since for each prime p_i, a and b share exactly $\min\{a_i, b_i\}$ factors of p_i.

Example. To find $(72, 30)$ by using prime factorizations, we write $30 = 2^1 \cdot 3^1 \cdot 5^1$ and $72 = 2^3 \cdot 3^2 \cdot 5^0$ and see that $(30, 72) = 2^1 \cdot 3^1 \cdot 5^0 = 6$.

Practice Problem Solutions and Hints

4.1.

(a) $q = 28$, $r = 4$.

(b) $q = 1$, $r = 16$.

(c) $q = 9$, $r = 0$.

(d) $q = 0$, $r = 12$.

4.2.

(a) $q = -12$, $r = 3$.

(b) $q = -6$, $r = 1$.

4.3.

(a) $55 = 34 \cdot 1 + 21$; $34 = 21 \cdot 1 + 13$; $21 = 13 \cdot 1 + 8$; $13 = 8 \cdot 1 + 5$; $8 = 5 \cdot 1 + 3$; $5 = 3 \cdot 1 + 2$; $3 = 2 \cdot 1 + 1$. $2 = 2 \cdot 1 + 0$. The number 1 is the last nonzero remainder, so $(34, 55) = 1$.

(b) $44{,}350 = 20{,}785 \cdot 2 + 2{,}780$; $20{,}785 = 2{,}780 \cdot 7 + 1{,}325$; $1{,}325 = 130 \cdot 10 + 25$; $130 = 25 \cdot 5 + 5$; $25 = 5 \cdot 5 + 0$, so $(44{,}350, 20{,}785) = 5$.

4.4. Let $d = (a, bc)$ and let p be a prime divisor of d. If $p \mid bc$, then $p \mid b$ or $p \mid c$, by Theorem 4.4. Since $(a, b) = 1$, if $p \mid a$, then $p \nmid b$. Since $(a, c) = 1$, if $p \mid a$, then $p \nmid c$. So $d = 1$.

Exercises

4.1. For each pair of numbers a and b, find the quotient q and the remainder $0 \leq r < b$ when a is divided by b.

(a) $a = 41$, $b = 13$.

(b) $a = 79$, $b = 20$.

(c) $a = 98$, $b = 99$.

(d) $a = -35$, $b = 8$.

4.2. In Case 3 of the proof of Theorem 4.1, a set P is constructed. What is P, if $a = 93$ and $b = 7$?

4.3. Use the Euclidean algorithm to find the GCDs of the following pairs of numbers.

(a) 35 and 91.

(b) 111 and 177.

(c) 881 and 997.

(d) 7,161 and 9,474.

(e) 12,390 and 63,210.

4.4. Find the GCDs of the following pairs of numbers without using the Euclidean algorithm.

(a) 1,024 and 6,216.

(b) 13,615 and 5,025.

(c) $2^1 3^3 7^2 11^3$ and $2^5 7^3 11^1 13^1 17^3$.

(d) $5^4 23^3 47^1 91^9$ and $2^4 5^2 17^8 19^3 47^3 83^2$.

4.5. Suppose that S is a nonempty subset of $\mathbb{Z}$ that is closed under subtraction.

(a) Explain why 0 must belong to S.

(b) Suppose that $S \neq \{0\}$. Show that S must contain a positive integer.

(c) Show that S is also closed under addition.

(d) Suppose that $S \neq \{0\}$, and let d be the smallest positive integer belonging to S. Prove that d divides every element of S. (Hint: If s is any element of S, apply the division algorithm with $a = s$ and $b = d$. What can be said about the remainder?)

(e) Suppose that $S \neq \{0\}$ and that d is the smallest positive element of S. Prove that S is precisely the set of all integral multiples of d.

(f) Show that S is closed under multiplication.

4.6. Let a and b be integers for which there exist integers m and n such that $am + bn = 1$. Prove that $(a, b) = 1$.

4.7. If m is a positive integer, find the greatest common divisor of $3m + 1$ and $m + 1$.

4.8. Let a and b be two integers, and let $a\mathbb{Z}$ and $b\mathbb{Z}$ denote the sets of all integer multiples of a and b, respectively. For example, if $a = 2$, then $2\mathbb{Z} = \{\dots, -4, -2, 0, 2, 4, \dots\}$; that is, the set of all even integers. Define the set $S_{(a,b)}$ as

$$S_{(a,b)} = \{x + y | x \text{ is from } a\mathbb{Z} \text{ and } y \text{ is from } b\mathbb{Z}\}.$$

In other words, $S_{(a,b)}$ consists of all integers that can be written as the sum of an element from $a\mathbb{Z}$ and an element from $b\mathbb{Z}$. (S is called the set of all *linear combinations* of a and b.)

(a) Show that $S_{(a,b)}$ is closed under subtraction.

(b) Show that a and b are elements of $S_{(a,b)}$.

(c) Show that if either a or b is nonzero, then $S_{(a,b)} \neq \{0\}$. If $a = 0$, what is $S_{(a,b)}$?

(d) Show that if d is the smallest positive element in $S_{(a,b)}$, then $d = (a, b)$. Note that this implies that the converse of Exercise 4.6 is true. Hint: Use the division algorithm, plus the fact that d is the smallest positive element of $S_{(a,b)}$, to show that the remainder of a (or b) divided by d is 0. Then, because d is an element of $S_{(a,b)}$, it can be expressed as $na + mb$ for some integers m and n. Use this to show that $(a, b) | d$, so $d = (a, b)$.

(e) Show that $S_{(a,b)} = (a, b)\mathbb{Z}$.

(f) Find $S_{(2,3)}$ and $S_{(15,24)}$.

4.9. Let a and b be two integers, and let $(a, b) = d$. According to Exercise 4.8(d), there exist integers m and n such that $ma + nb = d$. The Euclidean algorithm can be used to find one possible choice for m and n.

Consider the pair $a = 30$ and $b = 78$. Apply the Euclidean algorithm to get

$$78 = 2 \cdot 30 + 18,$$

$$30 = 1 \cdot 18 + 12,$$

$$18 = 1 \cdot 12 + 6,$$

$$12 = 2 \cdot 6 + 0.$$

Thus, $6 = (30, 78)$. Then rearrange the first three of these equations to get

$$18 = 78 - 2 \cdot 30,$$

$$12 = 30 - 1 \cdot 18,$$

$$6 = 18 - 1 \cdot 12.$$

Substitute the second equation into the third equation to get

$$6 = 18 - 1 \cdot 12 = 18 - 1 \cdot (30 - 1 \cdot 18) = 2 \cdot 18 - 1 \cdot 30.$$

Thus, 6 is expressed as a linear combination of 18 and 30. Next, substitute $78 - 2 \cdot 30$ for 18 in the above equation to get

$$6 = 2 \cdot 18 - 1 \cdot 30 = 2 \cdot (78 - 2 \cdot 30) - 1 \cdot 30 = -5 \cdot 30 + 2 \cdot 78.$$

So $m = -5$ and $n = 2$. Use this method to express (a, b) as a linear combination of a and b for the following values of a and b.

(a) $a = 91$ and $b = 35$.

(b) $a = 111$ and $b = 177$.

(c) $a = 881$ and $b = 9, 97$.

(d) $a = 7{,}161$ and $b = 9{,}474$.

(e) $a = 63{,}210$ and $b = 12{,}390$.

4.10. In the movie *Die Hard with a Vengeance,* Bruce Willis has 60 seconds to discharge a bomb. To do so, he must fill a container with exactly 4 gallons of water from a nearby fountain. He has two buckets, which hold 3 gallons and 5 gallons, respectively. First, use the Euclidean algorithm to find a solution to the equation $3m + 5n = 4$. Then, explain how Bruce Willis used n and m to disarm the bomb.

4.11. Harold has a great collection of old stamps, including lots of eight-cent stamps and lots of thirteen-cent stamps. He has a package that will cost \$3.30 to mail, and he already has figured out how to get \$3.29, using 5 thirteen-cent stamps and 33 eight-cent stamps.

(a) Describe two other ways Harold could get \$3.29 by using only eight- and thirteen-cent stamps.

(b) Use the Euclidean algorithm to find solutions to the equation $8m + 13n = 1$.

(c) Explain how Harold can use the answer to part (b) to figure out how to mail his \$3.30 package.

4.12. Prove that $(0, a) = |a|$.

4.13. If $a > 0$, show that $(ab, ac) = a(b, c)$.

4.14. Let n be any integer.

(a) Show that there is an integer q such that n can be written as $4q$, $4q + 1$, $4q + 2$, or $4q + 3$.

(b) Show that the remainder when n^2 is divided by 4 is either 0 or 1.

(c) Show that if n is odd, then the remainder when n^2 is divided by 8 is 1.

4.15. Prove that, for a, b, and c in $\mathbb{Z}$, $((a, b), c) = (a, (b, c)) = ((a, c), b)$.

4.16. Numbers a, b, and c are called *mutually relatively prime* if $((a, b), c) = 1$. As a result of Exercise 4.15, this property is symmetric in a, b, and c. Find three numbers that are mutually relatively prime, even though no pair of them is relatively prime.

Exercises 4.17–4.20 are related.

4.17. Find a number x from the set $\{1, 2, \ldots, 42\}$ that gives each of the following remainders.

(a) A remainder of 2 when divided by 7 and a remainder of 5 when divided by 6.

(b) A remainder of 0 when divided by 7 and a remainder of 5 when divided by 6.

(c) A remainder of 3 when divided by 7 and a remainder of 4 when divided by 6.

(d) A remainder of 3 when divided by 7 and a remainder of 3 when divided by 6.

4.18. Let m, n, and N be integers such that $(m, n) = 1$, $m \mid N$, and $n \mid N$.

(a) Show that $mn \mid N$. **Hint:** Use Theorem 4.3.

(b) Give a counterexample to show that mn need not divide N if the assumption that $(m, n) = 1$ is removed.

4.19. Let m, n, and M be integers, and let $(m, n) = 1$.

(a) Prove that if $(mn, M) = 1$, then $(m, M) = 1$ and $(n, M) = 1$.

(b) Prove that if $(m, M) = 1$ and $(n, M) = 1$, then $(mn, M) = 1$.

4.20. Assume that m and n are relatively prime integers, and let x and y be elements of the set $\{1, 2, \ldots, mn\}$.

(a) Show that if $mn \mid (x - y)$, then $x = y$.

(b) Show that x and y have the same remainder upon division by m if $m \mid (x - y)$, and that if $m \mid (x - y)$, then x and y have the same remainder upon division by m.

(c) Suppose that x and y have the same remainders when divided by m as when divided by n. Prove that $x = y$. **Hint:** Use Exercise 4.18.

(d) Chinese remainder theorem: Show that for each pair of numbers r_1 and r_2, with $0 \leq r_1 < m$ and $0 \leq r_2 < n$, there exists a unique number x in the set $\{1, 2, \ldots, mn\}$ such that x has remainder r_1 when divided by m and remainder r_2 when divided by n. **Hint:** How many possible pairs (r_1, r_2) are there?

4.21. Prove that if mn is a perfect square and $(m, n) = 1$, then m and n are perfect squares.

4.22. Let H be the set of positive integers that can be written in the form $4k + 1$, where $k \geq 0$ is an integer. In other words, $H = \{1, 5, 9, 13, 17, \ldots\}$.

(a) Show that H is closed under multiplication.

(b) A number h in the set H is called a *Hilbert prime* if the only way it can be written as the product of two integers in H is $h \cdot 1$ or $1 \cdot h$.

For example, 21 is not an ordinary prime, since $21 = 3 \cdot 7$, but it is a Hilbert prime since 3 and 7 are not in H. Show that every element of H can be factored into Hilbert primes.

(c) Show that the factorization of elements of H into Hilbert primes is not always unique by finding two different ways of factoring 693 into Hilbert primes.

4.23. Show that $(a, b) \le (a + b, a - b)$.

4.24. Prove that if $(a, b) = 1$ and $c | (a + b)$, then $(a, c) = (b, c) = 1$.

4.25. Consider the set $\mathbb{Z} \times \mathbb{Z}$ with the addition and multiplication defined in Exercise 1.24 and with the dictionary order defined in Exercise 2.15.

(a) Show that the existence part of the division algorithm is satisfied for $\mathbb{Z} \times \mathbb{Z}$. In other words, given (a, a') and (b, b') in $\mathbb{Z} \times \mathbb{Z}$ with $(b, b') \ne (0, 0)$, show that there exist (q, q') and (r, r') such that $(a, a') = (q, q') \cdot (b, b') + (r, r')$.

(b) Show that the uniqueness part of the division algorithm is not satisfied in $\mathbb{Z} \times \mathbb{Z}$ by finding two different quotients and remainders for $(a, a') = (33, 8)$ and $(b, b') = (5, 12)$.

4.26. Show that $(a + b, c + d) = 1$ if $|ac - bd| = 1$.

4.27. The definition of the greatest common divisor of a and b is the common divisor of a and b that is larger than every other common divisor of a and b. Show that this definition of GCD is equivalent to the following definition: The greatest common divisor of a and b is the (positive) common divisor of a and b that is a multiple of every other common divisor of a and b.

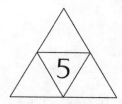

Variations on a Theme

Divisibility and its associated algorithms have applications that lead to interesting sets of numbers and some open questions in mathematics. In this chapter, we introduce several topics that directly illustrate properties of divisibility and divisors of an integer, including formulas for sum and number of divisors, perfect numbers, and the Fibonacci numbers. We also show how algorithms can be used to find least common multiples and define arithmetic in the rational numbers. We end this chapter with a discussion of Egyptian fractions.

5.1 Applications of Divisibility

Fibonacci Numbers

The properties of a well-known sequence of numbers were explored deeply in the thirteenth century by a man named Leonardo of Pisa, or Fibonacci[1] since he was the son of Bonacci. It is a sequence of numbers defined *recursively*, which just means that it tells what the first two or three terms of the sequence are, and then it gives you a rule for finding the next term of the sequence if the previous terms are known. Fibonacci used this sequence to solve a problem about rabbits multiplying.

Definition. The *Fibonacci sequence* $f_1, f_2, f_3, \cdots, f_n, \cdots$ is defined recursively with $f_1 = 1$, $f_2 = 1$, and $f_n = f_{n-1} + f_{n-2}$ for $n \geq 3$.

---

[1]Fibonacci (ca. 1180–1228), or Leonardo of Pisa, was an Italian merchant who traveled in the Arabic world of North Africa. He later wrote down what he learned in his travels and helped bring the use of Arabic notation and algorithms into Europe. His most famous book, the *Liber abbaci*, includes a problem that asks, "How many rabbits can be bred in one year from one pair?" By listing the solution to this problem as a sequence, he displayed the sequence that bears his name today.

We can list a few of the first elements of the Fibonacci sequence as

$$f_3 = f_2 + f_1 = 1 + 1 = 2,$$
$$f_4 = f_3 + f_2 = 2 + 1 = 3,$$
$$f_5 = f_4 + f_3 = 3 + 2 = 5,$$
$$f_6 = f_5 + f_4 = 5 + 3 = 8.$$

The Fibonacci numbers are a rich source of mathematical problems. Here, we show what happens if we try to find the GCD of two successive Fibonacci numbers.

Example. We use the Euclidean algorithm to find the GCD of $55 = f_{10}$ and $89 = f_{11}$:

$$89 = 1 \cdot 55 + 34,$$
$$55 = 1 \cdot 34 + 21,$$
$$34 = 1 \cdot 21 + 13,$$
$$21 = 1 \cdot 13 + 8,$$
$$13 = 1 \cdot 8 + 5,$$
$$8 = 1 \cdot 5 + 3,$$
$$5 = 1 \cdot 3 + 2,$$
$$3 = 1 \cdot 2 + 1,$$
$$2 = 1 \cdot 2 + 0,$$

so the GCD is the last nonzero remainder, or 1.

What do you notice about the successive remainders in this example?

Challenge Problem. Use your observations about the previous example to prove that for any two successive Fibonacci numbers f_n and f_{n+1}, $(f_n, f_{n+1}) = 1$.

We work with the Fibonacci numbers again in Chapter 16.

Sum and Number of Divisors

The theorem on unique factorization of integers into primes leads to some results about the divisors of a number. First, let's try to find how many positive divisors a number n has. Take a positive integer n and use the Fundamental Theorem of Arithmetic to write it as a product of primes.

$$n = p_1^{a_1} p_2^{a_2} \cdots p_m^{a_m}, \text{ where } p_1 < p_2 < \cdots < p_m.$$

To count the positive divisors of n, notice that any positive divisor d_1 of n must pair up with another positive divisor, d_2, so $n = d_1 \cdot d_2$. Together, d_1 and d_2 must contain all of the primes that appear in the factorization of n.

Example. Let's start with a power of a prime, say, $n = 32 = 2^5$. Written in prime notation, we have $32 = 2^0 \cdot 2^5 = 2^1 \cdot 2^4 = 2^2 \cdot 2^3$. The positive divisors occur in pairs: 1 and 32, 2 and 16, and 4 and 8. These six positive divisors are all the powers of 2 that are between $1 = 2^0$ and $32 = 2^5$.

Practice Problem 5.1. *Find the number of positive divisors of n for*

(a) $n = 81$.

(b) $n = 625$.

The result in Practice Problem 5.1 suggests a theorem.

Theorem 5.1. *If $n = p^m$, where p is a prime, then the number of positive divisors of n is $m + 1$.*

Proof: We write $n = p^m$. Then, if $a \mid p^m$, we must have $a = p^k$ for some $0 \le k \le m$. But also every power of p from p^0 up to p^m divides n. There are $m + 1$ of these powers, so n has $m + 1$ positive divisors. □

Next, we show a simple example of a number that has two different prime factors.

Example. Let's find the number of positive divisors of $36 = 2^2 \cdot 3^2$. Notice that all the powers of 2 that divide 36 appear as factors, and so do all the powers of 3 that divide 36, including $1 = 2^0 = 3^0$. There are some positive divisors that are not prime, but each of them contains only powers of the primes that divide 36. A positive divisor of 36 is of the form $d = 2^? \cdot 3^?$, where each ? can be 0, 1, or 2, as shown in Table 5.1.

How do we count the positive divisors carefully? Notice that in each of the positive divisors, there will be one of the three powers of 2 present $(2^0, 2^1, 2^2)$ and one of the three powers of 3 present $(3^0, 3^1, 3^2)$. This gives us $3 \cdot 3 = 9$ possibilities for positive divisors. (Notice that we do not count the 6 twice.) Thus, 36 has 9 positive divisors.

·	2^0	2^1	2^2
3^0	1	2	4
3^1	3	6	12
3^2	9	18	36

Table 5.1. Powers of primes for the positive divisors of 36.

Practice Problem 5.2. *Find the number of positive divisors of n for*

(a) $n = 72$.

(b) $n = 75$.

Now we want to generalize to a number with several prime factors. We use the symbol $\tau(n)$ to denote the number of positive divisors of n.

Theorem 5.2. *Suppose* $n = p_1^{a_1} p_2^{a_2} \cdots p_m^{a_m}$. *Then the number of positive divisors of n is* $\tau(n) = (a_1 + 1)(a_2 + 1) \cdots (a_m + 1)$.

Proof: A positive divisor of n contains some power (possibly 0) of each of the primes p_i that divide n. Possible powers for p_i are one of the integers $0, 1, 2, \cdots, a_i$. Thus, for each of the primes p_i involved in the factorization, there are $(a_i + 1)$ choices of exponent. To get the number of all possible positive divisors, we just need to multiply the numbers $(a_1 + 1)(a_2 + 1) \cdots (a_m + 1)$.   □

Practice Problem 5.3. *Find the number of positive divisors of n for*

(a) $n = 60$.

(b) $n = 240$.

Challenge Problem. Which numbers have exactly two positive divisors? Which numbers have exactly three? Which have exactly four? Which have a prime number of positive divisors?

Notice the very surprising fact that the number of positive divisors of a positive integer depends only on the exponents in the prime factorizations, and not on the actual primes themselves.

Example. The number $n = 2^3 \cdot 3^2$ has 12 positive divisors, and so does $m = 31^2 \cdot 43^3$.

Now that we know how to count the number of positive divisors of a number, let's try to answer the question of how to find the sum of the positive divisors of a number. We use the symbol $\sigma(n)$ to denote the sum of the positive divisors of n.

Example. As we did before, let's start with a number that is a power of a prime, $n = 2^5$. The positive divisors are 2^0, 2^1, 2^2, 2^3, 2^4, and 2^5. The sum of these numbers is

$$\sigma(n) = 1 + 2 + 4 + 8 + 16 + 32 = 63.$$

Theorem 5.3. *Let $n = p^a$, where p is a prime. Then $\sigma(n) = 1 + p + \ldots + p^a$.*

Proof: Clearly 1, p, p^2, ..., p^a are all the positive divisors of n. □

Theorem 5.4 presents a formula for summing powers of positive integers, including primes, that is useful here.

Theorem 5.4. *Let $k \neq 1$ be a positive integer. Then the sum*

$$1 + k + k^2 + k^3 + k^4 + k^5 + \cdots + k^a = \frac{k^{a+1} - 1}{k - 1}.$$

Proof: This proof uses an algebraic method for summing geometric series to get the result. We write the sum on the left and call it $S(k)$:

$$S(k) = 1 + k + k^2 + k^3 + k^4 + k^5 + \cdots + k^a. \qquad (5.1)$$

Now we multiply Equation (5.1) by k and get

$$\begin{aligned} k \cdot S(k) &= k(1 + k + k^2 + k^3 + k^4 + k^5 + \cdots + k^a) \\ &= k + k^2 + k^3 + k^4 + k^5 + \cdots + k^a + k^{a+1}. \end{aligned} \qquad (5.2)$$

We can subtract Equation (5.1) from Equation (5.2) to get

$$(k - 1)S(k) = k^{a+1} - 1. \qquad (5.3)$$

Now, since $k \neq 1$, we can just divide both sides of Equation (5.3) by $k - 1$ to get

$$S(k) = 1 + k + k^2 + k^3 + k^4 + k^5 + \cdots + k^a = \frac{k^{a+1} - 1}{k - 1}. \qquad □$$

Example. If we return to the previous example, we get the sum of the positive divisors of 32 to be

$$1 + 2 + 4 + 8 + 16 + 32 = \frac{2^6 - 1}{2 - 1} = 64 - 1 = 63.$$

Practice Problem 5.4. *Find the sum of the positive divisors of n for*

(a) $n = 9$.

(b) $n = 125$.

(c) $n = 49$.

Now we want to be able to sum the divisors of a more general number. Let's try one with two different prime divisors.

Example. Let's try $n = 36$. As we have seen, since $36 = 2^2 \cdot 3^2$, any positive divisor of 36 will have a power of 2 and a power of 3 in its prime factorization. If we take all the positive divisors of 36 that have a factor of $2^0 = 1$ in them, their sum will be $1 \cdot 13$. If we take all the positive divisors of 36 that have a factor of $2^1 = 2$ in them, we get $2^1 \cdot 1$, $2^1 \cdot 3$, and $2^1 \cdot 9$. The sum of these positive divisors is $2^1 \cdot 13$. If we take all the positive divisors of 36 that have a factor of $2^2 = 4$ in them, their sum will be $4 \cdot 13$. Therefore, the sum of all the positive divisors of 36 will be $(1 + 2 + 4)(13)$.

We could do the same thing with powers of 3 that divide 36. Here we see the sum of the positive divisors of 36 is $7(1 + 3 + 9)$. So $\sigma(36) = (1 + 2 + 4)(1 + 3 + 9) = 7 \cdot 13 = 91$. Table 5.2 shows this calculation.

Practice Problem 5.5. *Find the sum of the positive divisors of n for*

(a) $n = 72$.

(b) $n = 75$.

$\cdot$	2^0	2^1	2^2	**sum**
3^0	$1 \cdot 1$	$2 \cdot 1$	$4 \cdot 1$	$(1 + 2 + 4) \cdot 1$
3^1	$1 \cdot 3$	$2 \cdot 3$	$4 \cdot 3$	$(1 + 2 + 4) \cdot 3$
3^2	$1 \cdot 9$	$2 \cdot 9$	$4 \cdot 9$	$(1 + 2 + 4) \cdot 9$
sum	$1 \cdot (1 + 3 + 9)$	$2 \cdot (1 + 3 + 9)$	$4 \cdot (1 + 3 + 9)$	$(1 + 2 + 4) \cdot (1 + 3 + 9)$

Table 5.2. Summing the divisors of 36.

Again, we want to generalize to the case where n is any positive integer.

Theorem 5.5. *Suppose that $n = p_1^{a_1} p_2^{a_2} \cdots p_m^{a_m}$ is the unique factorization of n. Then the sum of the positive divisors of n is*

$$\sigma(n) = \frac{(p_1^{a_1+1} - 1)}{(p_1 - 1)} \cdot \frac{(p_2^{a_2+1} - 1)}{(p_2 - 1)} \cdots \frac{(p_m^{a_m+1} - 1)}{(p_m - 1)}$$

$$= \sigma(p_1^{a_1})\sigma(p_2^{a_2}) \cdots \sigma(p_m^{a_m}).$$

Challenge Problem. Construct a careful proof of Theorem 5.5. **Hint:** We can see that each prime power $p_i^{a_i}$ contributes a factor to the expression for the sum of the positive divisors.

Corollary 5.6. *Suppose $k = a \cdot b$, with $(a, b) = 1$. Then $\sigma(k) = \sigma(a) \cdot \sigma(b)$.*

Practice Problem 5.6. *Prove Corollary 5.6.*

Perfect Numbers

Table 5.3 is a chart of the sum of the positive divisors for the integers from 1 to 15.

Since ancient times, some numbers have been viewed as having certain mystical powers because of their special properties. One such set of numbers is that of the *perfect numbers*, those integers that are equal to half the sum of their positive divisors. From Table 5.3, we see that the number 6 is a perfect number because $\sigma(6) = 12$. Likewise, the number 28 is perfect since $\sigma(28) = 1 + 2 + 4 + 7 + 14 + 28 = 56$.

Practice Problem 5.7. *Check that there are no perfect numbers between 6 and 28.*

In ancient Greece, Euclid knew how to find all the even perfect numbers. Theorem 5.7 describes the even perfect numbers.

n	1	2	3	4	5	6	7	8	9	10	11	12	13	14	15
$\sigma(n)$	1	3	4	7	6	12	8	15	13	18	12	28	14	24	24

Table 5.3. Sum of divisors of n for $1 \leq n \leq 15$.

Theorem 5.7. *A positive integer n is an even perfect number if and only if $n = 2^{m-1}(2^m - 1)$, where m is an integer such that $m \geq 2$ and $2^m - 1$ is prime.*

Before we do the proof, let's discuss some interesting things about Theorem 5.7. This theorem appears in Euclid's *Elements*, where the proof consists of computing $\sigma(n)$. We will do the same thing here to establish part of the theorem. Remember that the numbers of the form $2^m - 1$ are called Mersenne numbers. So to find the even perfect numbers, we need to discover which Mersenne numbers are actually Mersenne primes. It has been conjectured that there are infinitely many Mersenne primes. The largest one discovered so far was found in 2008. It is the Mersenne number for $m = 43{,}112{,}609$.

As might be expected, the proof of Theorem 5.7 is fairly complicated.

Proof: When a theorem has the phrase "if and only if," there are two parts to the proof. In this theorem, we prove that if n is an even perfect number, then n has the required form. We also prove if n has that form, then it is an even perfect number. First, we show that any number of the form $n = (2^{m-1})(2^m - 1)$, where m is an integer such that $m \geq 2$ and $2^m - 1$ is prime, must be perfect.

Since $2^m - 1$ is an odd prime, we know that $(2^{m-1}, 2^m - 1) = 1$. So by Corollary 5.6,

$$\sigma(n) = \sigma(2^{m-1}) \cdot \sigma(2^m - 1).$$

By Theorem 5.3, $\sigma(2^{m-1}) = 2^m - 1$. Also, $\sigma(2^m - 1) = 2^m$ since $2^m - 1$ is prime. Therefore,

$$\sigma(n) = (2^m - 1) \cdot 2^m = 2n.$$

So n is a perfect number, and we have established the first part of the theorem.

To prove the converse, we assume n is an even perfect number and write $n = 2^r \cdot s$, where s is an odd number and $r \geq 1$. We want to show that n has the form stated in the theorem. We use $\sigma(s)$ to do this. We begin by relating $\sigma(s)$ and $\sigma(n)$. Since $(2^r, s) = 1$, by Corollary 5.6 and Theorem 5.3,

$$\sigma(n) = \sigma(2^r \cdot s) = \sigma(2^r) \cdot \sigma(s) = (2^{r+1} - 1) \cdot \sigma(s). \tag{5.4}$$

Since we are assuming n is perfect, we know that

$$\sigma(n) = 2n = 2^{r+1} \cdot s. \tag{5.5}$$

Combining Equations (5.4) and (5.5) for $\sigma(n)$, we have

$$(2^{r+1} - 1) \cdot \sigma(s) = 2^{r+1} \cdot s. \qquad (5.6)$$

Since $(2^{r+1}, 2^{r+1} - 1) = 1$, we know from Theorem 4.3 that 2^{r+1} divides $\sigma(s)$. In other words, there is an integer k such that $\sigma(s) = 2^{r+1} \cdot k$. Substituting this expression into Equation (5.6), we get

$$(2^{r+1} - 1) \cdot 2^{r+1} \cdot k = 2^{r+1} \cdot s$$

and so

$$(2^{r+1} - 1) \cdot k = s. \qquad (5.7)$$

We now know that k divides s. We also know $k \neq s$ because $r \geq 1$.

From Equation (5.7), we have

$$s + k = (2^{r+1} - 1)k + k = 2^{r+1}k = \sigma(s). \qquad (5.8)$$

Next, we claim that $k = 1$. If, instead, $k \neq 1$, then s has at least three distinct divisors, namely, $1, k$, and s, which implies that $\sigma(s) \geq 1 + k + s$, contradicting Equation (5.8). Therefore, $k = 1$ and so $s = (2^{r+1} - 1)$. Also from Equation (5.8), we now have that $\sigma(s) = s + 1$. So s must be prime, since its only positive divisors are s and 1.

We have thus shown that $n = 2^r(2^{r+1} - 1)$, where $(2^{r+1} - 1)$ is prime, as the theorem requires. □

Theorem 5.7 shows that if we want to find even perfect numbers, we need to find Mersenne primes. We give one more result (Theorem 5.8) that helps in the search for such numbers.

Theorem 5.8. *If m is a positive integer and $2^m - 1$ is prime, then m itself must be prime.*

Proof: Suppose m is a composite number and write $m = ab$, where $1 < a < m$ and $1 < b < m$. We now show how to factor $2^m - 1$. We write

$$2^m - 1 = 2^{ab} - 1 = (2^a - 1)(2^{a(b-1)} + 2^{a(b-2)} + \cdots + 2^a + 1). \qquad (5.9)$$

Both factors on the right-hand side of Equation (5.9) are greater than 1, so $2^m - 1$ is composite if m is composite. □

Several more results help determine the even perfect numbers. These may be found in the Exercises. Are there any odd perfect numbers? The answer is unknown.

5.2   More Algorithms

Rational Arithmetic and Least Common Multiples

How do we add and multiply fractions? Most of our work so far has been with the integers, so now we include the definitions of arithmetic operations in $\mathbb{Q}$. They are familiar rules.

Definition. Take two rational numbers $\frac{a}{b}$ and $\frac{c}{d}$. Then the product of the two fractions is defined by

$$\frac{a}{b} \cdot \frac{c}{d} = \frac{ac}{bd}.$$

To get the sum of two fractions we use the expressions $\frac{b}{b}$ and $\frac{d}{d}$ for the multiplicative identity 1. We then have

$$\frac{a}{b} + \frac{c}{d} = \frac{a}{b} \cdot \frac{d}{d} + \frac{c}{d} \cdot \frac{b}{b}$$

$$= \frac{ad}{bd} + \frac{bc}{bd}$$

$$= \frac{ad + bc}{bd}.$$

In both addition and multiplication, we can cancel factors common to both the numerator and denominator to get the result in lowest terms.

Example. By this algorithm, if we add $\frac{1}{14}$ and $\frac{2}{9}$, we get

$$\frac{1}{14} + \frac{2}{9} = \frac{1 \cdot 9 + 2 \cdot 14}{14 \cdot 9} = \frac{37}{126}.$$

When adding $\frac{3}{10}$ and $\frac{4}{15}$, we get

$$\frac{3}{10} + \frac{4}{15} = \frac{3 \cdot 15 + 4 \cdot 10}{150} = \frac{85}{150}.$$

In the second case, since $(10, 15) = 5$, we can cancel a factor of 5 in the answer. In the first case, the two denominators are relatively prime, and hence the sum is already reduced to lowest terms.

Challenge Problem. Given two fractions reduced to lowest terms, show that if their denominators are relatively prime, the resulting sum is already in lowest terms.

When we learn to add fractions, we usually learn to find a least common denominator before we add them. This is because sometimes the product of the two denominators is larger than necessary to add the fractions. So we look for a "least" or "lowest" common denominator.

Example. Let's add $\frac{3}{10}$ and $\frac{4}{15}$ again. A common denominator of 150 will work, but there is also a smaller number that is divisible both by 10 and 15, namely, 30, that will work as a common denominator. In fact, this is the smallest common denominator for 10 and 15. We write $\frac{3}{10} = \frac{9}{30}$ and $\frac{4}{15} = \frac{8}{30}$ and get a sum of $\frac{17}{30}$.

Definition. The *least common multiple* of two positive integers a and b is the smallest positive integer that is divisible both by a and b. The least common multiple, or LCM, is written as $[a, b]$.

Example. We just showed that $[10, 15] = 30$.

One way to figure out the LCM of two positive numbers is to make a list of the multiples of each number and look for the smallest number that appears on both lists.

Example. To find $[12, 15]$, make lists of the multiples of each number.

$$\text{Multiples of 12:} \quad 12, 24, 36, 48, 60, 72, 84, \ldots$$
$$\text{Multiples of 15:} \quad 15, 30, 45, 60, 75, \ldots$$

We notice that 60 is the first number that appears on both lists, so $[12, 15] = 60$.

An easier way is to look at the prime factorization of each number and use that to find the LCM.

Example. Let's find $[12, 15]$ again, this time using prime factorization. We write $12 = 2^2 \cdot 3$ and $15 = 3 \cdot 5$. For the LCM to be a multiple of both 12 and 15, it must be divisible by all the factors of 12 and of 15. We need both 2s from the factorization of 12, and the 5 from the factorization of 15. However, we only need one 3 in the LCM, since 3 occurs only once in both factorizations. So $[12, 15] = 2^2 \cdot 3 \cdot 5 = 60$.

Here is the rule for finding the LCM of two numbers using the prime power factorization. Let $a = p_1^{a_1} p_2^{a_2} \cdots p_n^{a_n}$ and $b = p_1^{b_1} p_2^{b_2} \cdots p_n^{b_n}$, where $p_1, p_2, \ldots, p_n$ are the primes occurring in the prime power factorizations of either a or b. Remember that we include all the primes in the factorizations of both a and b in both lists, possibly with exponent 0.

For an integer to be divisible by both a and b, we need the power of each p_i in it to be at least as large in its factorization as it is in each of a and b. Here, some new notation will be helpful. If we write $\max\{x, y\}$ to denote the larger of the two integers x and y, then we can write the LCM of a and b as

$$[a, b] = p_1^{\max\{a_1, b_1\}} p_2^{\max\{a_2, b_2\}} \cdots p_n^{\max\{a_n, b_n\}}.$$

Practice Problem 5.8. *Find the LCM of the following pairs of numbers:*

(a) $[30, 72]$.

(b) $[15, 24]$.

(c) $[100, 150]$.

Finding the LCM of two integers can be time consuming since it requires finding the prime factorizations of the integers. There is a relationship between the GCD and the LCM of two numbers that will make finding the LCM of two numbers easier once we know the GCD. Let's take a look at an example to try to discover the connection.

Example. Start with the two numbers, $30 = 2 \cdot 3 \cdot 5$ and $72 = 2^3 \cdot 3^2$. The GCD $= 2 \cdot 3 = 6$ and the LCM $= 2^3 \cdot 3^2 \cdot 5 = 360$. Notice that the product of the two numbers 72 and 30 is 2,160, the same as the product of the GCD and the LCM. Notice that each of the prime powers in 30 and 72 $(2, 3, 5, 2^3, 3^2)$ occurs in either the GCD or the LCM.

Theorem 5.9 relates the GCD and LCM.

Theorem 5.9. *If a and b are positive integers, then $[a, b] = \frac{ab}{(a,b)}$.*

Proof: Before we establish the result, we need a fact about the maximum and minimum of two numbers.

Fact. If x and y are two numbers, then $\max\{x, y\} + \min\{x, y\} = x + y$.

To see this, note that either $x \geq y$ or $x < y$. If $x \geq y$, then $\max\{x, y\} = x$ and $\min\{x, y\} = y$, so that $\max\{x, y\} + \min\{x, y\} = x + y$. If, however, $x < y$, then $\max\{x, y\} = y$ and $\min\{x, y\} = x$, so $\max\{x, y\} + \min\{x, y\} = x + y$ also.

Now that we have established this fact, we can continue with the proof of the theorem. Let $a = p_1^{a_1} p_2^{a_2} \cdots p_n^{a_n}$ and $b = p_1^{b_1} p_2^{b_2} \cdots p_n^{b_n}$, where $p_1, p_2, \ldots, p_n$ are the primes occurring in the prime power factorizations of either a or b.

Now, using the fact, we can easily establish the theorem. We know from the Euclidean algorithm in Chapter 4 that

$$(a, b) = p_1^{\min\{a_1, b_1\}} p_2^{\min\{a_2, b_2\}} \cdots p_n^{\min\{a_n, b_n\}},$$

and from the discussion above that

$$[a, b] = p_1^{\max\{a_1, b_1\}} p_2^{\max\{a_2, b_2\}} \cdots p_n^{\max\{a_n, b_n\}}.$$

To make our notation a little easier to work with, rename $\max\{a_i, b_i\} = M_i$ and $\min\{a_i, b_i\} = m_i$. The product of the GCD and LCM is then

$$
\begin{aligned}
(a, b)[a, b] &= p_1^{M_1} \cdots p_n^{M_n} \cdot p_1^{m_1} \cdots p_n^{m_n} \\
&= p_1^{(M_1 + m_1)} \cdots p_n^{(M_n + m_n)} \\
&= p_1^{(\max\{a_1, b_1\} + \min\{a_1, b_1\})} \cdots p_n^{(\max\{a_n, b_n\} + \min\{a_n, b_n\})}.
\end{aligned}
$$

From the above fact, this is just

$$
\begin{aligned}
(a, b)[a, b] &= p_1^{(a_1 + b_1)} \cdots p_n^{(a_n + b_n)} \\
&= p_1^{a_1} \cdots p_n^{a_n} \cdot p_1^{b_1} \cdots p_n^{b_n} \\
&= ab. \qquad \square
\end{aligned}
$$

This nice result allows us to find the LCM of two large numbers by first using the Euclidean algorithm to find the GCD and then using Theorem 5.9 to find the LCM. This relationship between the GCD and LCM allows us to find the lowest common denominator of two fractions when we add, since the lowest common denominator is just the LCM of the two denominators.

Egyptian Fractions

In ancient Egypt, the rule for writing fractions was to use so-called "unit fractions," or fractions that have a 1 in the numerator. The way to express a fraction $\frac{a}{b}$ was to write it as the sum of unit fractions with different denominators. The Egyptians excluded $\frac{2}{3}$ from this process; we know this from a problem on the Rhind papyrus. In this section, we assume $\frac{a}{b}$ is positive and $a < b$. Otherwise, we could rewrite our fraction as a mixed number with an integer plus a fraction $\frac{a}{b}$ with $a < b$.

Examples. Let's practice trying to write some fractions as the sum of unit fractions.

1. Let's begin with an easy one, say, $\frac{3}{5}$. How should we go about writing $\frac{3}{5}$ as the sum of unit fractions? We know $\frac{3}{5}$ is bigger than $\frac{1}{2}$, and since $\frac{1}{2}$ is the biggest unit fraction, we will try to use that as part of the sum. Think about what else we need to add to $\frac{1}{2}$ to get $\frac{3}{5}$. We are really solving the equation

$$x = \frac{3}{5} - \frac{1}{2}$$

and then writing x as a sum of unit fractions. Here, elementary arithmetic tells us that $\frac{3}{5} - \frac{1}{2} = \frac{1}{10}$, which is itself a unit fraction. So we can write $\frac{3}{5} = \frac{1}{2} + \frac{1}{10}$, which is the sum of two unit fractions.

   Always remember that denominators in unit fractions must be different. Otherwise, you could write $\frac{3}{5} = \frac{1}{5} + \frac{1}{5} + \frac{1}{5}$, but in Egyptian fractions, repeated denominators are not allowed.

2. Let's try a harder one, say, $\frac{5}{7}$. Again, $\frac{1}{2}$ is the largest unit fraction less than $\frac{5}{7}$, so we write $\frac{5}{7} - \frac{1}{2} = \frac{3}{14}$, and then we try to write $\frac{3}{14}$ as a sum of unit fractions. Now we want to find the largest unit fraction that is less than $\frac{3}{14}$ and then subtract it from $\frac{3}{14}$ in order to continue toward a solution. A little rational arithmetic tells us that we want to find n so that $\frac{1}{n}$ is the biggest unit fraction less than $\frac{3}{14}$. This involves solving the inequalities

$$\frac{1}{n} < \frac{3}{14} \text{ and } \frac{3}{14} < \frac{1}{n-1},$$

or

$$14 < 3n \text{ and } 3(n-1) < 14.$$

Together, these give

$$14 < 3n < 17.$$

The n that makes both of these inequalities true is 5. So the unit fraction $\frac{1}{5}$ is the largest one less than $\frac{3}{7}$, and we will use it to continue toward a solution of the problem. So far, we have

$$\frac{5}{7} = \frac{1}{2} + \frac{3}{14}$$

$$= \frac{1}{2} + \frac{1}{5} + \left(\frac{3}{14} - \frac{1}{5} \right),$$

and $\left(\frac{3}{14} - \frac{1}{5} \right) = \frac{1}{70}$.

A solution to writing $\frac{5}{7}$ as a sum of unit fractions is

$$\frac{5}{7} = \frac{1}{2} + \frac{1}{5} + \frac{1}{70}.$$

Trying to write fractions the Egyptian way leads to three questions. First, how do we know this process will always work? Second, if it works, is there only one way to make it work? And finally, can we estimate how many steps it will take for a particular fraction? We can develop an algorithm for writing Egyptian fractions that will lead to answers to these questions.

Practice Problem 5.9. *Write the following fractions in the Egyptian method of using sums of unit fractions with no repeated denominators. Try to do this in more than one way.*

(a) $\frac{4}{9}$.

(b) $\frac{5}{12}$.

(c) $\frac{7}{8}$.

(d) $\frac{2}{3}$ *(despite the Egyptians!).*

Let's develop an algorithm for writing Egyptian fractions. Start with a fraction $\frac{a}{b}$, with $1 < a < b$. Find the smallest integer n_1 so that

$$\frac{1}{n_1} < \frac{a}{b} < \frac{1}{n_1 - 1} \tag{5.10}$$

Notice that $n_1 > 1$ because $a < b$. Then subtract $\frac{1}{n_1}$ from $\frac{a}{b}$ to give $\frac{a}{b} - \frac{1}{n_1} = \frac{an_1 - b}{bn_1}$. How big is this numerator? From Equation (5.10), we know

$$a(n_1 - 1) < b, \quad \text{or}$$
$$an_1 - a < b, \quad \text{or}$$
$$an_1 - b < a.$$

So the new numerator is smaller than a. If $an_1 - b = 1$ we are done; otherwise $an_1 - b > 1$.

Now, as we saw in the examples, at the next step we want to find the smallest integer n_2 that satisfies the inequalities $\frac{1}{n_2} < \frac{an_1 - b}{bn_1} < \frac{1}{n_2 - 1}$. Notice that the denominator of the new fraction is greater than b. So the numerator has decreased by at least 1 and the denominator has increased.

Continuing in this process, at each stage the numerator is at least 1 less than the previous numerator, and the denominator is greater than the previous denominator. We started with the numerator $a > 1$, and now have found a collection of positive integers that will eventually end in 1 because of the well-ordering principle. This answers two of our earlier questions at once. We can use Egyptian fractions to express any positive fraction $\frac{a}{b}$ with $a < b$ as a sum of unit fractions with different denominators. This algorithm also tells us it will take no more than $a - 1$ steps, and therefore a unit fractions, to write any fraction by using Egyptian fractions.

Notice that this algorithm doesn't necessarily get us to a final answer in the fewest number of steps.

Example. We can write $\frac{9}{20} = \frac{1}{4} + \frac{1}{5}$. But by using the algorithm, we would get $\frac{9}{20} = \frac{1}{3} + \frac{1}{10} + \frac{1}{60}$.

Example. A fraction that takes the maximum number of steps is $\frac{4}{25}$. You should try working the algorithm to see that the sum is correct.

$$\frac{4}{25} = \frac{1}{7} + \frac{1}{59} + \frac{1}{5{,}163} + \frac{1}{53{,}307{,}975}.$$

We have shown just a few applications of divisibility. There are many others that occur; they are expanded in the Exercises that follow.

Practice Problem Solutions and Hints

5.1.

(a) The number $n = 81 = 3^4$ has five positive divisors: 1, 3, 9, 27, and 81.

(b) The number $n = 625 = 5^4$ has five positive divisors: 1, 5, 25, 125, and 625.

5.2.

(a) The number $n = 72 = 2^3 \cdot 3^2$ has 12 positive divisors:

·	2^0	2^1	2^2	2^3
3^0	1	2	4	8
3^1	3	6	12	24
3^2	9	18	36	72

(b) The number $n = 75 = 3 \cdot 5^2$ has six positive divisors:

$\cdot$	5^0	5^1	5^2
3^0	1	5	25
3^1	3	15	75

5.3.

(a) The number $n = 60 = 2^2 \cdot 3 \cdot 5$ has $3 \cdot 2 \cdot 2 = 12$ positive divisors: 1, 2, 3, 5, 2^2, $2 \cdot 3$, $2 \cdot 5$, $3 \cdot 5$, $2^2 \cdot 3$, $2^2 \cdot 5$, $2 \cdot 3 \cdot 5$, and $2^2 \cdot 3 \cdot 5$.

(b) The number $n = 240 = 2^4 \cdot 3 \cdot 5$ has $5 \cdot 2 \cdot 2 = 20$ positive divisors.

5.4.

(a) The number $n = 9 = 3^2$ has a positive divisor sum of $1 + 3 + 9 = 13$, or $\frac{3^3 - 1}{3 - 1} = \frac{27 - 1}{2} = 13$.

(b) The number $n = 125 = 5^3$ has a positive divisor sum of $\frac{5^4 - 1}{5 - 1} = \frac{624}{4} = 131$.

(c) The number $n = 49 = 7^2$ has a positive divisor sum of $\frac{7^3 - 1}{7 - 1} = \frac{342}{6} = 57$.

5.5.

(a) The number $n = 72 = 2^3 \cdot 3^2$ has a positive divisor sum of $\left(\frac{2^4 - 1}{2 - 1}\right)\left(\frac{3^3 - 1}{3 - 1}\right) = 15 \cdot 13 = 195$.

(b) The number $n = 75 = 3 \cdot 5^2$ has a positive divisor sum of $\left(\frac{3^2 - 1}{3 - 1}\right)\left(\frac{5^3 - 1}{5 - 1}\right) = 4 \cdot 31 = 124$.

5.6. Let $a = p_1^{a_1} \cdots p_m^{a_m}$ and $b = q_1^{b_1} \cdots q_n^{b_n}$. Because $(a, b) = 1$, we can write k as the product $p_1^{a_1} \cdots p_m^{a_m} \cdots q_1^{b_1} \cdot q_n^{b_n}$, where $p_i \neq q_j$ for any i or j. By Theorem 5.5, $\sigma(a) = \sigma(p_1^{a_1}) \cdots \sigma(p_m^{a_m})$ and $\sigma(b) = \sigma(q_1^{b_1}) \cdots \sigma(q_n^{b_n})$, so $\sigma(ab) = \sigma(a)\sigma(b)$.

5.7. The following chart lists the sum of the positive divisors for 16 to 27.

n	16	17	18	19	20	21	22	23	24	25	26	27
$\sigma(n)$	31	18	39	20	42	32	36	24	60	31	42	40

5.8.

(a) $[30, 72] = 360$.

(b) $[15, 24] = 120$.

(c) $[100, 150] = 300$.

5.9.
  (a) $\frac{4}{9} = \frac{1}{3} + \frac{1}{9} = \frac{1}{4} + \frac{1}{6} + \frac{1}{36}$.
  (b) $\frac{5}{12} = \frac{1}{3} + \frac{1}{12} = \frac{1}{4} + \frac{1}{6}$.
  (c) $\frac{7}{8} = \frac{1}{2} + \frac{1}{4} + \frac{1}{8}$.
  (d) $\frac{2}{3} = \frac{1}{2} + \frac{1}{6} = \frac{1}{3} + \frac{1}{4} + \frac{1}{12}$.

Exercises

5.1. Find $(21, 55)$, $(55, 144)$, and $(89, 233)$.

5.2. What do you think (f_n, f_{n+2}) is for a positive integer n? Try to prove your conjecture.

5.3. Prove the following for all positive integers n.
  (a) $f_{n+3} = 2 \cdot f_{n+1} + f_n$.
  (b) $f_{n+4} = 3 \cdot f_{n+1} + 2 \cdot f_n$.
  (c) $f_{n+5} = 5 \cdot f_{n+1} + 3 \cdot f_n$.
  (d) For any integer m, $f_{n+m} = f_m \cdot f_{n+1} + f_{m-1} \cdot f_n$.

5.4. Find $\tau(n)$ and $\sigma(n)$ for the following values of n:
  (a) 51.
  (b) 234.
  (c) 272.
  (d) 180.
  (e) 9,261.

5.5. How many numbers between 0 and 100 inclusive have exactly one positive divisor? How many have exactly two positive divisors? Three? Four? Five?

5.6. What is the greatest number of positive divisors that any number between 0 and 100 has?

5.7. Show that $\tau(n)$ is odd if and only if n is a perfect square.

5.8. At Mathematica High School, there are 1,000 students and 1,000 numbered lockers. On the opening day, the students line up facing the numbered lockers, and each student is assigned the number of the locker in front of them. All the students are instructed to open their numbered lockers. Then Student 1 is told to close them all (phew!). Student 2 then opens all the even-numbered lockers. Student 3 switches each locker whose number is divisible by 3, meaning locker 3 gets opened, 6 gets closed, 9 gets opened, and so on. Each student switches all lockers with numbers divisible by the student's number. When student 1,000 is finished, which locker doors will be closed?

5.9. Show that if $\tau(n)$ is odd, then $\sigma(n)$ is odd. Give an example to show that if $\sigma(n)$ is odd, $\tau(n)$ can be even.

5.10. Show that $\tau(n) \leq 2\sqrt{n}$ for every positive integer n. **Hint:** Use Exercise 2.10.

5.11. Show that $\sigma(n) \leq n \cdot \tau(n)$ for every positive integer n. Conclude, using the previous problem, that $\sigma(n) \leq 2n\sqrt{n}$.

5.12. When is $\sigma(n)$ prime?

  (a) Find all primes p such that $\sigma(p)$ is also prime.

  (b) Show that if $\sigma(n)$ is prime, then $n = p^r$ for some prime number p and some positive integer r.

  (c) Suppose that $n = 2^r$. Show that $\sigma(n)$ is prime if and only if $2 \cdot n - 1$ is a Mersenne prime.

  (d) For which integers n is $\tau(n)$ a prime number?

5.13. Suppose that $\sigma(a) = 91$. What must a be?

5.14. Suppose that $\sigma(b) = 124$. What are the two numbers that b could be?

5.15. Suppose that $\sigma(c) = 7,905$. What are all the possible values for c? How do you know you have found all of them?

5.16. Suppose that $(m, n) = 1$. Prove that $\tau(mn) = \tau(m) \cdot \tau(n)$ and $\sigma(mn) = \sigma(m) \cdot \sigma(n)$. Give a counterexample that shows that these relationships are not necessarily true when m and n are not relatively prime.

5.17. Remember that an integer n is *perfect* if $\sigma(n) = 2n$. An integer n is *deficient* if $\sigma(n) < 2n$, and it is *abundant* if $\sigma(n) > 2n$. Determine whether the following integers are deficient or abundant.

(a) 45.

(b) 24.

(c) 40.

(d) 945.

(e) p^k, where p is prime and k is a positive integer.

5.18. Suppose m is an abundant number and n is relatively prime to m. Prove that mn is abundant.

5.19. Find all pairs a and b for which $(a, b) = 28$ and $[a, b] = 1{,}512$.

5.20. Prove that $[a, b] = ab$ if and only if a and b are relatively prime.

5.21. Suppose that $[a, b] = ma$ and $[a, b] = nb$. Prove that m and n are relatively prime.

5.22. Let a and b be any integers, and let $a\mathbb{Z}$ and $b\mathbb{Z}$ denote the set of all integer multiples of a and b, respectively (see Exercise 4.8). Let $S = a\mathbb{Z} \cap b\mathbb{Z}$ be the intersection of these two sets; that is, $S = \{x \mid x$ is in $a\mathbb{Z}$ and $b\mathbb{Z}\}$.

(a) Show that S is the set of all common multiples of a and b.

(b) Let m be the smallest nonnegative element of S. Show that m is the LCM of a and b.

(c) Show that S is closed under subtraction. Conclude that $S = [a, b]\mathbb{Z}$.

(d) Let a and b be any integers. Prove that every common multiple of a and b is divisible by the least common multiple of a and b.

5.23. Write the following numbers as the sums of unit fractions.

(a) $\frac{3}{7}$.

(b) $\frac{4}{25}$.

(c) $\frac{5}{19}$.

(d) $\frac{6}{13}$.

5.24. The Egyptian algorithm gives a method for writing a fraction as a sum of unit fractions, where each term is as big as possible. Some fractions, however, can be written as a sum of unit fractions in more than one way. For example, $\frac{5}{12} = \frac{1}{3} + \frac{1}{12} = \frac{1}{4} + \frac{1}{6}$. Write the following fractions as sums of unit fractions in at least two different ways.

(a) $\frac{7}{36}$.

(b) $\frac{63}{100}$.

5.25. Suppose that n is a positive integer, and that $a_1 < \ldots < a_r$ are positive integers such that

$$\frac{1}{n} = \frac{1}{a_1} + \ldots + \frac{1}{a_r}.$$

(a) Prove that $a_1, \ldots, a_r$ are all greater than n.

(b) Prove that any fraction of the form $\frac{1}{n}$ can be written as the sum of at least two distinct unit fractions.

(c) Prove that $\frac{m}{n}$ can be written as a sum of distinct unit fractions in infinitely many ways. **Hint:** Start with one such expression, and then decompose the unit fraction with the highest denominator into a sum of distinct unit fractions. Why does the resulting decomposition of the original fraction $\frac{m}{n}$ still contain distinct unit fractions?

5.26. Assume that the Goldbach conjecture is true, that is, that we can write every even integer greater than 2 as the sum of two primes. Prove that for every even number N, there exist prime numbers p_1 and p_2 such that $\sigma(p_1) + \sigma(p_2) = N$.

5.27. Two different integers m and n are said to be *amicable* if $\sigma(m) - m = n$ and $\sigma(n) - n = m$. In other words, the sum of the divisors of m, not including m itself, is n, and vice versa.

(a) Show that 220 and 284 are amicable.

(b) *Thabit's rule:* Let $s \geq 2$ be an integer, and suppose that $p = 3 \cdot 2^{s-1} - 1$, $q = 3 \cdot 2^s - 1$, and $r = 9 \cdot 2^{2s-1} - 1$ are all prime. Prove that $2^s pq$ and $2^s r$ are amicable.

5.28. Show that $(f_m, f_n) = f_{(m,n)}$.

5.29. Let n be an even perfect number. Show that

$$\sum_{d|n} \frac{1}{d} = 2.$$

5.30. Each of the following sets has the property that the minimum number of unit fractions required to write each fraction is the same for all fractions in a set. (You should check this.) What property do they have in common that allows them to be written this way? Can you prove that any number with that property can be written as the sum of that number of unit fractions?

(a) $\{\frac{2}{5}, \frac{2}{7}, \frac{2}{9}, \frac{2}{121}, \frac{2}{347}\}$.

(b) $\{\frac{2}{3}, \frac{3}{5}, \frac{4}{7}, \frac{17}{33}\}$.

(c) $\{\frac{3}{7}, \frac{3}{13}, \frac{3}{19}, \frac{3}{37}, \frac{3}{601}\}$.

(d) $\{\frac{4}{25}, \frac{4}{49}, \frac{4}{241}, \frac{4}{24,001}\}$.

5.31. An equation of the form

$$a_1 + \cfrac{1}{a_2 + \cfrac{1}{a_3 + \cfrac{1}{a_4 + \dots + \frac{1}{a_n}}}}$$

is known as a *continued fraction*. Any fraction can be rewritten as a continued fraction. For example, we can write $\frac{376}{100}$ as

$$3 + \cfrac{1}{1 + \cfrac{1}{3 + \frac{1}{8}}}.$$

Rewrite the following fractions as continued fractions.

(a) $\frac{104}{15}$.

(b) $\frac{87}{36}$.

(c) $\frac{89}{55}$.

(d) $\frac{13,615}{5,025}$.

5.32. Prove that every fraction can be written as a continued fraction.

5.33. What is the maximum number of steps it can take to write $\frac{m}{n}$ as a continued fraction? Can you find a fraction that takes this many steps?

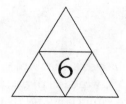

Congruences and Groups

6.1 Congruences and Arithmetic of Residue Classes

Now that we have worked carefully and extensively with integers, we want to make up a new system for which the rules of arithmetic may be a little different. This new system is based on the ideas we used in Chapter 1 when we did Even-Odd arithmetic and one's digit arithmetic.

Definition. Start with a positive integer $m \geq 2$. Take two integers a and b. We say that *a is congruent to b modulo m* if m divides $a - b$. We write this as $a \equiv b \pmod{m}$.

Let's rewrite the definition in terms of our earlier notion of divisibility. For m to divide $a - b$, we must have an integer k so that $a - b = km$. Written another way, for a and b to be congruent modulo m requires an integer k that makes the equation $a = km + b$ true.

Examples. What numbers are congruent modulo 4?

1. Let's start with $m = 4$. Take two integers, say, $a = 13$ and $b = 9$. Are 13 and 9 congruent modulo 4? Subtracting, we get $13 - 9 = 4$, which is divisible by 4. So 13 and 9 are congruent modulo 4.

2. Let's try $a = 13$ and $b = 37$. Subtracting, we get $13 - 37 = -24$ and -24 is divisible by 4. So 13 and 37 are congruent modulo 4.

3. Now let's try $a = 13$ and $b = 6$. This time, $13 - 6 = 7$, which is not divisible by 4. So 13 and 6 are not congruent modulo 4.

Which integers are congruent to 13 modulo 4? We need to find integers b whose difference from 13 is divisible by 4. The number 9 works, since $13 - 9 = 4$. The number 5 also works, since $13 - 5 = 8$.

Practice Problem 6.1. *Make a list of positive integers that are congruent to 13 modulo 4.*

Are there any negative integers that are congruent to 13 modulo 4? Try -3. Then $13 - (-3) = 16$, which is divisible by 4. There is also a set of negative integers that are congruent to 13 modulo 4, namely, the set $\{-3, -7, -11, \ldots\}$.

Practice Problem 6.2. *Make a list of integers congruent to 1 modulo 4. Compare this list to the list of numbers congruent to 13 modulo 4.*

Let's try another number. What numbers will be congruent to 12 modulo 4? The set $\{\ldots, -12, -8, -4, 0, 4, 8, 12, \ldots\}$ contains the numbers congruent to 12 modulo 4.

Practice Problem 6.3. *Find all numbers congruent to 11 modulo 4. Find all numbers congruent to 10 modulo 4.*

If we look for numbers that are congruent to 9 modulo 4, we find the set to be $\{\ldots, -11, -7, -3, 1, 5, 9, 13, \ldots\}$, which is the same set of numbers that are congruent to 13 modulo 4.

Let's prove some simple properties of congruences that are easy consequences of divisibility facts.

Theorem 6.1. *Let m be a positive integer greater than or equal to 2. Congruences modulo m satisfy the following three properties:*

1. $a \equiv a \pmod{m}$.

2. If a and b are integers satisfying $a \equiv b \pmod{m}$, then $b \equiv a \pmod{m}$.

3. If $a, b,$ and c are integers satisfying $a \equiv b \pmod{m}$ and $b \equiv c \pmod{m}$, then $a \equiv c \pmod{m}$.

Proof:

1. For a to be congruent to itself modulo m, we need $a - a = 0$ to be divisible by m. But 0 is divisible by any integer, so this statement is true.

2. Since $a \equiv b \pmod{m}$, $m \mid a - b$. By Theorem 3.4, $m \mid -1(a - b)$; that is, $m \mid b - a$. So $b \equiv a \pmod{m}$.

3. Since $a \equiv b \pmod{m}$, $m \mid a - b$. Since $b \equiv c \pmod{m}$, $m \mid b - c$. So by Theorem 3.2, $m \mid (a - b) + (b - c) = a - c$. So $a \equiv c \pmod{m}$.   □

Let's return to working in the congruence system modulo 4. Notice that the integers are divided neatly into four different sets that together include all the integers. No two of these sets intersect, and each set contains only those numbers congruent to each other mod 4. We want to select one number from each set and do some arithmetic in the mod 4 system. How should we select such a representative? Let's look at the four sets and see if we can figure out a logical choice.

$$A = \{\ldots, -12, -8, -4, 0, 4, 8, 12, \ldots\},$$
$$B = \{\ldots, -11, -7, -3, 1, 5, 9, 13, \ldots\},$$
$$C = \{\ldots, -10, -6, -2, 2, 6, 10, 14, \ldots\},$$
$$D = \{\ldots, -9, -5, -1, 3, 7, 11, 15, \ldots\}.$$

What do the integers in set A have in common? If we divide any of them by 4, we get a remainder of 0. The integers in set B share a similar property, since any integer in set B leaves a remainder of 1 when we divide it by 4. For example, $9 = 4 \cdot 2 + 1$ and $-11 = 4(-3) + 1$. For set C, the common remainder is 2, and for set D, the common remainder is 3. By the division algorithm, if a is any integer and we divide a by m, we will get $a = mq + r$, where $0 \leq r < m$. If m is 4, then the possible values for r are $0, 1, 2$, and 3. We will select as our representative from each set the common remainders of $0, 1, 2$, or 3 when we are working modulo 4.

Practice Problem 6.4. *Choose a different number for m, say, $m = 5$. Divide the integers into sets of numbers that are congruent modulo 5, and select a set of representatives for this collection, as was done when m was 4.*

Definition. The set of numbers that correspond to the same remainder r when divided by m forms a *residue class modulo m*. For the remainder of the chapter, we assume that $m \geq 2$.

Example. When we worked with $m = 4$, sets A, B, C, and D each formed a residue class modulo 4.

Now we want to do arithmetic in congruence systems. First, we have to show that when we add, subtract, and multiply, congruence arithmetic behaves in a way we would expect.

Theorem 6.2. *Start with a positive integer $m \geq 2$ and two integers a and b satisfying $a \equiv b \pmod{m}$. If c is any integer, then*

1. $a + c \equiv b + c \pmod{m}$.

2. $a - c \equiv b - c \pmod{m}$.

3. $ac \equiv bc \pmod{m}$.

Proof:

1. We need to show $(a + c) - (b + c)$ is divisible by m. But $(a + c) - (b + c) = a - b$, which is divisible by m since we assumed that $a \equiv b \pmod{m}$.

2. The proof is similar. The reader should be able to supply the details.

3. Here, we need $ac - bc$ to be divisible by m. But $ac - bc = (a - b)c$. Since $a - b$ is divisible by m, so is $(a - b)c$. $\qquad \square$

This important theorem tells us that if we want to add, subtract, or multiply in the congruence arithmetic system, we can take any element we like from a residue class to perform the operation. The answer will be the same no matter which element we choose as a representative. Therefore, to make life simple, we always use the set of remainders as the set of representatives for doing arithmetic in congruences.

Definition. The set of remainders $\{0, 1, 2, 3, \ldots, m - 1\}$ is called a *residue system modulo m*. Every integer is congruent modulo m to exactly one member of this set. Each remainder represents the residue class in which it is contained. We call these remainders *residues*.

Example. For $m = 8$, a residue system is the set $\{0, 1, 2, 3, 4, 5, 6, 7\}$. The number 1 represents the residue class $\{\ldots, -15, -7, 1, 9, 17, \ldots\}$.

We left division, or cancelation, out of Theorem 6.2. Let's think about why we did this. Remember that cancelation did not work in the system of one's digit arithmetic. What happens when we divide both sides of a congruence equation by an integer?

Example. We know that $10 \equiv 6 \pmod{4}$. That is, $5 \cdot 2 \equiv 3 \cdot 2 \pmod{4}$. But we cannot cancel the factor of 2, since 5 is not congruent to 3 modulo 4. This shows that cancelation doesn't always work with congruences.

Challenge Problem. Figure out when it is possible to cancel a common factor from a congruence equation and still have a valid congruence equation.

We are preparing to define addition and multiplication for residue classes. These operations will satisfy most of the Rules of Arithmetic.

Definition. Suppose m is a positive integer greater than or equal to 2. Suppose A and B are residue classes modulo m, and a is an element of A and b is an element of B. Define the sum $A + B$ to be the residue class containing the integer $a + b$. Define the product AB to be the residue class containing the element ab.

Example. Let's show how these operations work modulo 4. Remember that $A = \{\ldots, -8, -4, 0, 4, 8, 12, \ldots\}$ and $B = \{\ldots, -7, -3, 1, 5, 9, 13, \ldots\}$. Take one element from A and one from B, say, $a = -12$ and $b = 5$, and add them. Then $a + b = -12 + 5 = -7$, which is congruent to 1 modulo 4. In fact, if we choose any element from set A and any element from set B, their sum is in set B, so, in this case, $A + B = B$. Multiplication works the same way. If we take $a = 8$ and $b = -11$, $ab = -88$, which is congruent to 0 modulo 4. If we choose any element from A and any element from B, their product will be in the set A, so $AB = A$.

This seems very natural, but one problem arises in many parts of mathematics when we use sets in definitions. That is, we must show that these operations yield the same answer no matter which a and b are picked from the residue classes A and B. This property says that the operations are *well-defined*. In fact, this is what Theorem 6.3 tells us.

Theorem 6.3. *If $a \equiv b \pmod{m}$ and $c \equiv d \pmod{m}$, then $a + c \equiv b + d \pmod{m}$ and $ac \equiv bd \pmod{m}$.*

Proof: We know that $a + c \equiv b + c \pmod{m}$ and $b + c \equiv b + d \pmod{m}$ by Theorem 6.2(1). Therefore, by Theorem 6.1(3), $a + c \equiv b + d \pmod{m}$. The proof of the second statement is similar.   ☐

Now let's do some arithmetic with the residue system $\{0, 1, 2, 3\}$ in modulo 4. We start with addition. The first few facts are easy:

$$0 + 0 \equiv 0 \pmod{4},$$
$$0 + 1 \equiv 1 \pmod{4},$$
$$0 + 2 \equiv 2 \pmod{4},$$
$$0 + 3 \equiv 3 \pmod{4}.$$

If we add 0 to any of the residues, we get the same residue back. This is a familiar property of the integer 0. In fact, set A, containing the element 0, is the identity for addition of residue classes. We use residue 0 to represent set A.

Now let's see what happens when we add 1 to each of the residues. Again, the first few are easy:

$$0 + 1 \equiv 1 \pmod 4,$$
$$1 + 1 \equiv 2 \pmod 4,$$
$$2 + 1 \equiv 3 \pmod 4.$$

But what is $3 + 1$ modulo 4? In ordinary arithmetic, we would get 4 as an answer. In congruence arithmetic modulo 4, we look for the integer representing the residue class of the number 4, which is 0. So $3 + 1 \equiv 0 \pmod 4$.

Let's create the table for addition of residue classes modulo 4 (Table 6.1).

What do we notice about the table? We can see addition is commutative because of symmetry in the table. Does every number have an additive inverse? A quick check of Table 6.1 shows that every number does have an additive inverse.

Now let's do the multiplication table for congruence arithmetic modulo 4 (Table 6.2).

Again, it is easy to check that 1 still acts as the multiplicative identity in this number system. Multiplication is commutative since Table 6.2 is symmetric across the diagonal. Do the nonzero elements have multiplicative inverses? The answer is that some do and some don't. The numbers 1 and 3 are inverses for themselves, but 2 does not have a multiplicative inverse in this system. This reminds us of the situation in one's digit arithmetic.

+ (mod 4)	0	1	2	3
0	0	1	2	3
1	1	2	3	0
2	2	3	0	1
3	3	0	1	2

Table 6.1. Addition modulo 4.

· (mod 4)	0	1	2	3
0	0	0	0	0
1	0	1	2	3
2	0	2	0	2
3	0	3	2	1

Table 6.2. Multiplication modulo 4.

We want to see which of the Rules of Arithmetic work in congruence arithmetic. First we prove a theorem that shows addition and multiplication of residue classes modulo m are commutative and associative. We want to show that if A, B, and C are any residue classes modulo m, then $A + B = B + A$, $AB = BA$, $(A + B) + C = A + (B + C)$ and $(AB)C = A(BC)$.

Theorem 6.4. *Suppose A, B, and C are residue classes modulo m. Then,*

1. $A + B = B + A$ *and* $AB = BA$.

2. $(A + B) + C = A + (B + C)$ *and* $(AB)C = A(BC)$.

Proof:

1. Take a in A, and b in B. Then the integer $a + b$ defines the class $A + B$. In $\mathbb{Z}$, by Axiom A1, $a + b = b + a$. But $b + a$ defines the class $B + A$. So $A + B = B + A$.

2. Similarly, the integer ab defines the class AB. In $\mathbb{Z}$, by Axiom M1, $ab = ba$. But ba defines the class BA. So $AB = BA$.   □

Practice Problem 6.5. *Show that addition and multiplication of residue classes are also associative.*

Next we consider Axioms A3 and M3. In modulo 4 arithmetic, we observed that the residue class of 0 acts as an identity for addition of residue classes modulo 4 and the residue class of 1 acts as the identity for multiplication of residue classes modulo 4. In fact, this is the case for residue classes modulo m for any m.

Challenge Problem. Let m be a positive integer greater than or equal to 2. Show the residue class of 0 is the identity for addition of residue classes modulo m. Show that the residue class of 1 is the identity for multiplication of residue classes modulo m.

We look now at Axiom A4, the additive inverse. In addition modulo 4, we saw that the additive inverse of the residue class A containing the residue a was the residue class containing $-a$, which we can call $-A$. Again this is true modulo m for any positive integer m, since 0 is in the residue class $A + -A$. What residue will represent $-A$? If a is a nonzero residue, the element $-a$ is not a residue modulo m. However, the number $m - a$ is in the same residue class as $-a$, and $m - a$ is a residue. So $m - a$ represents $-A$, and the element $m - a$ is the additive inverse of a in the residue system.

Example. For $m = 4$, the additive inverse of 3 is $4 - 3 = 1$. That is, $-3 = 1$.

From the examples, we know Axiom M4 may not hold in all these systems. So let's look next at the arithmetic tables for a particular value of m. Let's create tables in the residue classes modulo 7. Instead of writing A, B, C, D, E, F, and G to represent the residue classes, we use the residue system $\{0, 1, 2, 3, 4, 5, 6\}$. We use congruence arithmetic to determine the entries of Tables 6.3 and 6.4. The tables are partially filled in; supply the missing numbers before we continue.

In Table 6.3, 0 is the additive identity and every element has an additive inverse. For example, the additive inverse of 3 is $7 - 3 = 4$, so that $-3 = 4$ in this system.

In Table 6.4, 1 is the multiplicative identity and every nonzero number has a multiplicative inverse. For example, the multiplicative inverse of 5 is 3. That is, $5^{-1} = 3$ in this system.

Practice Problem 6.6. *Write the multiplicative inverses for all the nonzero elements in the modulo 7 table (Table 6.4).*

By now you may have recognized that we have seen arithmetic systems like these before. When we did the Even-Odd arithmetic system, we were really working in congruences modulo 2. So to bring the notation for the Even-Odd system into compliance with congruence arithmetic, we just need to recognize that "Even" is the same as 0 and "Odd" is the same as 1. Table 6.5 shows the two tables in each system for comparison.

Moreover, the one's digit arithmetic we worked through so carefully is really what happens when we work modulo 10. If we think about it,

+ (mod 7)	0	1	2	3	4	5	6
0	0	1	2	3	4	5	6
1	1	2	3	4	5	6	0
2	2	3	4	5	6	0	1
3	3						
4	4						
5	5						
6	6						

Table 6.3. Addition modulo 7.

· (mod 7)	0	1	2	3	4	5	6
0	0	0	0	0	0	0	0
1	0	1	2	3	4	5	6
2	0	2	4	6	1	3	5
3	0						
4	0						
5	0						
6	0						

Table 6.4. Multiplication modulo 7.

working modulo 10 means taking the remainder when a number is divided by 10, which is the same as looking at the one's digit.

⊕	Even	Odd
Even	Even	Odd
Odd	Odd	Even

+ (mod 2)	0	1
0	0	1
1	1	0

⊙	Even	Odd
Even	Even	Even
Odd	Even	Odd

· (mod 2)	0	1
0	0	0
1	0	1

Table 6.5. Comparison of Even-Odd and modulo 2 arithmetics.

6.2   Groups and Other Structures

It is time to gather some of these concepts and name the structures that are similar. Mathematicians like to categorize and name things so that when we see them again, we can recognize them for what they are and also so that we can prove theorems that tell us something about all of the examples. The definitions in this section describe the formal algebraic structures that include as examples some of the systems we are using.

The simplest notion is that of a group, which we first mentioned in Chapter 1.

Definition. A *group* is a set of elements G closed under a binary operation $*$ that satisfies the following three axioms:

Axiom I. The operation $*$ is associative.

Axiom II. There is an element e in G that acts as the identity for the group. That is, if g is any element of the set G, then $e * g = g$ and $g * e = g$.

Axiom III. For every element g in the set G, there is an associated element g^{-1} that acts as the inverse for the element g. That is, $g^{-1} * g = e$ and $g * g^{-1} = e$.

Examples. We already have seen many examples of groups, such as the following four.

1. The set $\mathbb{Z}$ of integers together with the operation $+$ of addition forms a group. Addition is associative, the number 0 acts as the identity and if a is an integer, then $-a$ is its additive inverse.

2. The set of nonzero rational numbers together with the operation of multiplication, which we call $\mathbb{Q}^{\times}$, forms a group. Here, multiplication is associative, the number 1 acts as the multiplicative identity, and if $\frac{a}{b}$ is a nonzero rational number, then $\frac{b}{a}$ acts as its multiplicative inverse, since a is not zero.

3. The residue classes modulo 7 together with the operation of addition form a group. Again, addition is associative, 0 is the identity, and every element has an additive inverse.

4. The set of nonzero residue classes modulo 7 together with the operation of multiplication forms a group. Multiplication is associative, 1 is the identity, and every element has a multiplicative inverse.

Practice Problem 6.7. *How many other groups can you think of? Which sets of residue classes will be groups if the operation is addition? Which sets of nonzero residue classes will be groups if the operation is multiplication?*

Definition. If the group operation is commutative, we call the group a *commutative group*. We also call such a group *abelian*, after the Norwegian mathematician Abel.[1]

Example. All of the groups we have worked with so far are commutative, or abelian, groups.

---

[1]Niels Henrik Abel (1802–1829) was a famous Norwegian mathematician. He read the works of the other mathematicians of his time, including Euler, and decided to study mathematics. He spent his time studying how to find roots of certain kinds of equations. He was unable to find a position at a university, so he published important results in brief texts he paid for himself. He died of tuberculosis when he was only 28.

· (mod 10)	1	3	7	9
1	1	3	7	9
3	3	9	1	7
7	7	1	9	3
9	9	7	3	1

Table 6.6. Multiplication of elements of $\mathbb{Z}_{10}$ with multiplicative inverses.

Notation. If we are working with the integers modulo some integer $m \geq 2$, we use the symbol $\mathbb{Z}_m$ for the residue system $\{0, 1, 2, \ldots, m-1\}$. Remember that each integer here represents the entire residue class in which it is contained. Since we have used two different operations with these systems, we have to be careful to indicate whether the operation is addition or multiplication modulo m. Most of the time, the operation we are using will be clear from the context. In the discussion above, we showed that axioms A1, A2, A3, and A4 hold in the system of residues modulo m under addition, so we state this result here as Theorem 6.5:

Theorem 6.5. *For any integer $m \geq 2$, $\mathbb{Z}_m$ with the operation of addition is an abelian group.*

Unfortunately, the same kind of theorem is not true when the operation is multiplication on the nonzero residue classes modulo m. We get a group for some values of m, and we don't for others. A careful check of Table 6.4 for $\mathbb{Z}_7$ shows that the identity and all nonzero multiplicative inverses are present. Since the set of nonzero elements in $\mathbb{Z}_7$ is closed under multiplication, these elements form a group with the operation of multiplication.

However, when we look at $\mathbb{Z}_{10}$ (one's digit arithmetic) and think about the operation of multiplication, not all of the nonzero residues have multiplicative inverses. In fact, only four elements, namely, 1, 3, 7, and 9, have multiplicative inverses. Table 6.6 is the multiplication table in $\mathbb{Z}_{10}$ for these elements.

Let's check the group axioms within the set $\{1, 3, 7, 9\}$. First, the set is closed under the binary operation of multiplication, since the only elements that appear in Table 6.6 are the four numbers 1, 3, 7, and 9. The operation is still associative, inherited from this property in $\mathbb{Z}_{10}$. Is there an identity? The element 1 acts like an identity, as it did in all of $\mathbb{Z}_{10}$. Every element has a multiplicative inverse in the set, since that is how we chose these numbers. Therefore, this is a group.

For a positive integer m, we want to give a name to the nonzero elements in $\mathbb{Z}_m$ that have multiplicative inverses.

Definition. If a is a nonzero element in $\mathbb{Z}_m$ and a has a multiplicative inverse, we say a is a *unit* in $\mathbb{Z}_m$. The units of $\mathbb{Z}_m$ are the set of nonzero elements in $\mathbb{Z}_m$ that have multiplicative inverses, denoted by $\mathbf{U}(m)$.

Example. In $\mathbb{Z}_7$, the units are all the nonzero elements, so $\mathbf{U}(7) = \{1, 2, 3, 4, 5, 6\}$. Table 6.7 is the multiplication table for the set $\mathbf{U}(7)$. It is clear from Table 6.7 that the set is closed under multiplication, that 1 acts as the identity, and that every element has a multiplicative inverse. Since associativity of multiplication is inherited from the original group $\mathbb{Z}_7$, (see Practice Problem 6.5), we know that $\mathbf{U}(7)$ is a group.

How does an element of the residue system $\{0, 1, 2, \ldots, m-1\}$ become an element of $\mathbf{U}(m)$?

Theorem 6.6. *Suppose m is a positive integer greater than or equal to 2. An element a in the residue system $\{0, 1, 2, \ldots, m-1\}$ is an element of $\mathbf{U}(m)$ if and only if $(a, m) = 1$.*

Proof: Suppose first that $(a, m) = 1$. Consider the set of numbers $\{a \cdot 0, a \cdot 1, \ldots, a \cdot (m-1)\}$. We want to show this set contains a representative from each of the m different residue classes mod m. Do any of these numbers represent the same residue class as any others? Suppose $a \cdot j \equiv a \cdot k \pmod{m}$ and $j < k$. Then m divides $a \cdot k - a \cdot j = a \cdot (k - j)$. Since $(a, m) = 1$, this means that m divides $k - j$, by Theorem 4.3. This implies that $k - j = 0$, since $k - j < m$. So $j = k$. So the list $\{a \cdot 0, a \cdot 1, \ldots, a \cdot (m-1)\}$ does contain a representative from each of the m different residue classes. One of them must be congruent to 1

· (mod 7)	1	2	3	4	5	6
1	1	2	3	4	5	6
2	2	4	6	1	3	5
3	3	6	2	5	1	4
4	4	1	5	2	6	3
5	5	3	1	6	4	2
6	6	5	4	3	2	1

Table 6.7. Multiplication on $\mathbf{U}(7)$.

modulo m, so a has a multiplicative inverse modulo m. Therefore, a is in $\mathbf{U}(m)$.

Now suppose that a has a multiplicative inverse modulo m. Say there is an element k so that $ak \equiv 1 \pmod{m}$, and $a < m$ and $k < m$. Then there is a number c so that $ak = cm + 1$, or $ak - cm = 1$. Set $d = (a, m)$. Then d divides the left-hand side of this equation, so d divides 1. Therefore $d = 1$, and a and m are relatively prime.   □

For a positive integer m, there is a special function that counts the number of integers that are less than m and relatively prime to m. It is called the *Euler-ϕ function* and is written as $\phi(m)$. This function tells us how many elements are in the set $\mathbf{U}(m)$ for any m. Table 6.8 shows the values for $\phi(m)$ for $1 \le m \le 15$.

m	1	2	3	4	5	6	7	8	9	10	11	12	13	14	15
$\phi(\mathbf{m})$	0	1	2	2	4	2	6	4	6	4	10	4	12	6	8

Table 6.8. Values of $\phi(m)$ for $1 \le m \le 15$.

Example. Consider the set of units in $\mathbb{Z}_8$. The set $\mathbf{U}(8) = \{1, 3, 5, 7\}$. Table 6.9 is the multiplication table for $\mathbf{U}(8)$. Again, the set of units in $\mathbb{Z}_8$ forms a group (check the axioms).

· (mod 8)	1	3	5	7
1	1	3	5	7
3	3	1	7	5
5	5	7	1	3
7	7	5	3	1

Table 6.9. Multiplication on $\mathbf{U}(8)$.

Theorem 6.7. *The set $\mathbf{U}(m)$ is closed under the operation of multiplication.*

Proof: Suppose a and b are both elements of $\mathbf{U}(m)$. Then $(a, m) = 1$ and $(b, m) = 1$. But then $(ab, m) = 1$, by Theorem 4.6.   □

Corollary 6.8. $\mathbf{U}(m)$ *forms a group under multiplication.*

Cyclic Groups

Think about the group $\mathbb{Z}_{10}$ under addition. If we start with the element 1, and add it to itself, we get the number 2, which is also in $\mathbb{Z}_{10}$. If we add 1 to itself three times we get the number 3, which is also in $\mathbb{Z}_{10}$. In fact, if we add 1 to itself enough times, we can get any element of $\mathbb{Z}_{10}$ we want. We call a number such as 1 in $\mathbb{Z}_{10}$ a *generator* of $\mathbb{Z}_{10}$ because any element in $\mathbb{Z}_{10}$ can be written as the sum of 1s.

Definition. Suppose the binary operation on a group G is addition. If G has an element a so that every element of G can be written as a sum of a's and $-a$'s, then we call G a *cyclic group generated by a*.

Example. If $G = \mathbb{Z}_5$ under addition, the number 2 generates G since

$$0 \equiv 2 + 2 + 2 + 2 + 2 \pmod{5},$$
$$1 \equiv 2 + 2 + 2 \pmod{5},$$
$$2 \equiv 2 \pmod{5},$$
$$3 \equiv 2 + 2 + 2 + 2 \pmod{5},$$
$$4 \equiv 2 + 2 \pmod{5}.$$

So every element of G can be written as a sum of 2s.

Practice Problem 6.8. *For $G = \mathbb{Z}_8$ under addition, show that the element $a = 5$ generates G by writing every element of G as the sum of 5s. Show that if $a = 2$, then a does not generate G.*

Definition. If the group operation is multiplication instead of addition, then a *generator* of G will be an element a, so that we can write every element of G as a product of a's and a^{-1}'s.

Example. Let G be the group $\mathbf{U}(7)$ under multiplication. Then the element 3 generates G since

$$1 \equiv 3^6 \pmod{7},$$
$$2 \equiv 3^2 \pmod{7},$$
$$3 \equiv 3^1 \pmod{7},$$
$$4 \equiv 3^4 \pmod{7},$$
$$5 \equiv 3^5 \pmod{7},$$
$$6 \equiv 3^3 \pmod{7}.$$

Thus, every element of G can be written as a product of 3s.

Practice Problem 6.9. *In the group* $\mathbf{U}(5)$*, show that the element* $a = 2$ *generates the group by writing every element of* G *as a product of* $2s$*.*

Practice Problem 6.10. *Show that the group* $\mathbf{U}(8)$ *is not a cyclic group.*

Definition. In a group G, if x is an element of G, we define the *order* of x to be the smallest positive integer n such that $x^n = e$, where $x^n = x \cdot x \cdots \cdot x$ (n times). In this case, x is said to be of order n. If we write G additively, then the order of x is the smallest positive integer such that $nx = e$, where $nx = x + x + \cdots + x$ (n times). If such an n does not exist, we say that x has *infinite order*.

Examples. What are the orders of the indicated elements of the following groups?

1. In $G = \mathbb{Z}_5$ under addition, the element 3 has order 5, since $3 + 3 + 3 + 3 + 3 = 0$ and no smaller number of 3s works.

2. In the group $\mathbf{U}(5)$ under multiplication, 3 has order 4 since $3 \cdot 3 \cdot 3 \cdot 3 = 1$ and no smaller number of 3s works.

3. In $\mathbb{Z}$ under addition, the element 2 has infinite order, since there is no n such that $2 + \cdots + 2n$ times $= 0$.

Definition. The *order* of G, written $|G|$, is the number of elements in G. If G has n elements for some natural number n, we say G has *finite order* equal to n. If G has an infinite number of elements, then G is called an *infinite group*.

Examples. What are the orders of the following groups?

1. The order of $\mathbb{Z}_n$ under addition is n.

2. The order of $\mathbf{U}(8)$ under multiplication is 4.

3. $\mathbb{Z}$ under addition is an infinite group.

Practice Problem 6.11. *Show that if* G *is a group and* a *is an element of* G*, then the order of* a *is less than or equal to the order of* G*.*

If x generates a cyclic group G, and $|G| = n$, then the order of x is also n.

Rings

In Chapter 2, we discussed the fact that mathematicians have a term for
a set with not just one, but two different binary operations. The following
is a reminder of the definition of a commutative ring.

Definition. A *commutative ring* is a set R closed under two binary operations, $+$ and $*$, that satisfy Axioms A1–A4, M1–M2, and D of the Rules
of Arithmetic.

Axioms A1–A4 tell us that the elements of the ring R form a commutative group with the binary operation $+$. The definition of a commutative
ring requires multiplication to be associative and commutative. Notice
also that there is no requirement for a multiplicative identity to exist.
Of course, without a multiplicative identity, there are no multiplicative
inverses!

Examples. The following are commutative rings:

1. The set $\mathbb{Z}$ of integers under ordinary addition and multiplication is
   a commutative ring with multiplicative identity 1. There are only
   two elements with multiplicative inverses, namely, 1 and -1.

2. The set $\mathbb{Z}_m = \{0, 1, 2, 3, \ldots, m - 1\}$ under addition and multiplication modulo m is a commutative ring. There is a multiplicative
   identity, 1, but multiplicative inverses don't always exist. As we
   have seen, the elements with multiplicative inverses in $\mathbb{Z}_m$ are those
   relatively prime to m.

3. The set $2\mathbb{Z} = \{\ldots, -4, -2, 0, 2, 4, \ldots\}$ is a commutative ring with
   no multiplicative identity.

Zero Divisors and Fields

In arithmetic in $\mathbb{Z}$, we know that if the product of two numbers is 0, one of
the original numbers must be 0. This is a consequence of the cancelation
property discussed in Chapter 2. We have stated this as an axiom for the
integers. However, in some of the rings we have seen, the product of two
elements may be zero, despite the fact that neither is zero.

Example. Consider $\mathbb{Z}_{10}$, or one's digit arithmetic. Here, $2 \cdot 5 = 0$, but
certainly neither 2 nor 5 is 0 in this ring.

Recall that a nonzero element a in a commutative ring R is called a *zero divisor* if there is a nonzero element b in R such that $ab = 0$.

Practice Problem 6.12. *List the zero divisors in $\mathbb{Z}_m$ for each value of m.*

(a) $m = 4$.

(b) $m = 6$.

(c) $m = 8$.

Also recall from Chapter 2 that a commutative ring R that has a multiplicative identity is called an integral domain if the cancelation axiom is satisfied. That is, if a, b, and c are in the ring, and $ab = ac$ with $a \neq 0$, then $b = c$.

Examples. The following are integral domains:

1. The ring of integers, $\mathbb{Z}$.

2. The ring $\mathbb{Z}_7$.

There is a natural relationship between a commutative ring being an integral domain and having no zero divisors. In fact, they are exactly the same, as Theorem 6.9 states.

Theorem 6.9. *A commutative ring R is an integral domain if and only if it has no zero divisors.*

Proof: Suppose first that R has no zero divisors. We want to show that cancelation holds in R. Suppose a, b, and c are in the ring. Now, $ab = ac$ implies that $ab - ac = 0$ or $a(b - c) = 0$. If $a \neq 0$ and there are no zero divisors, then we must have $b - c = 0$, or $b = c$, so cancelation holds.

Now we assume cancelation holds and show that there are no zero divisors. Suppose the product $ab = 0$. If $a \neq 0$, then we can write $ab = a \cdot 0$, and cancel the a's to get $b = 0$.   □

There is one final system to define, which is a special kind of integral domain.

Definition. A commutative ring that has a multiplicative identity is called a *field* if every nonzero element is a unit.

Examples. The following sets are fields.

1. The rational numbers $\mathbb{Q}$ form a field. Every nonzero element of $\mathbb{Q}$ has a multiplicative inverse.

2. The real numbers $\mathbb{R}$ form a field.

3. The set $\mathbb{Z}_m$ is a field if and only if m is a prime. We have already seen that $\mathbb{Z}_m$ is a commutative ring that has a multiplicative identity. From our work with $\mathbf{U}(m)$, we know that a nonzero element a of $\mathbb{Z}_m$ has a multiplicative inverse if and only if $(a, m) = 1$. Thus, the only way for every nonzero element of $\mathbb{Z}_m$ to have a multiplicative inverse is for m to be prime.

Practice Problem Solutions and Hints

6.1. A few numbers on this list are 1, 5, 9, 13.

6.2. This list is the same as the list in Practice Problem 6.1.

6.3. The set $\{\ldots, -9, -5, -1, 3, 7, 11, 15, \ldots\}$ contains the numbers that are congruent to 11 modulo 4. The set $\{\ldots, -10, -6, -2, 2, 6, 10, 14, \ldots\}$ contains the numbers that are congruent to 10 modulo 4.

6.4. If $m = 5$, we get the following sets.

$$A = \{\ldots - 10, -5, 0, 5, 10, 15, \ldots\},$$
$$B = \{\ldots - 9, -4, 1, 6, 11, 16, \ldots\},$$
$$C = \{\ldots - 8, -3, 2, 7, 12, 17, \ldots\},$$
$$D = \{\ldots - 7, -2, 3, 8, 13, 18, \ldots\},$$
$$E = \{\ldots - 6, -1, 4, 9, 14, 19, \ldots\}.$$

A set of representatives could be the remainders when dividing by 5, namely, 0, 1, 2, 3, and 4.

6.5. For associativity of addition, take $a \in A$, $b \in B$, and $c \in C$. We want to show $(A + B) + C = A + (B + C)$. The number $a + b$ defines the class $(A + B)$, and $(a + b) + c$ defines the class $(A + B) + C$. In $\mathbb{Z}$, by Axiom A2, $(a + b) + c = a + (b + c)$. But $b + c$ defines $B + C$, so $a + (b + c)$ defines $A + (B + C)$. Multiplicative associativity is shown similarly.

6.6.

a	1	2	3	4	5	6
a^{-1}	1	4	5	2	3	6

6.7. If the operation is addition, and m is a positive integer, any set of residue classes modulo m will form a group. If the operation is multiplication, and p is prime, the set of nonzero residue classes modulo p will form a group. There are many other examples.

6.8. $0 \equiv 5+5+5+5+5+5+5+5 \pmod 8$,
$\quad 1 \equiv 5+5+5+5+5 \pmod 8$,
$\quad 2 \equiv 5+5 \pmod 8$,
$\quad 3 \equiv 5+5+5+5+5+5+5 \pmod 8$,
$\quad 4 \equiv 5+5+5+5 \pmod 8$,
$\quad 5 \equiv 5 \pmod 8$,
$\quad 6 \equiv 5+5+5+5+5+5 \pmod 8$,
$\quad 7 \equiv 5+5+5 \pmod 8$.

A sum of 2s is always an even number modulo 8, so 2 cannot generate $\mathbb{Z}_8$ because every element of G is not included. See the list below.

$$2 \equiv 2 \pmod 8,$$
$$2+2 \equiv 4 \pmod 8,$$
$$2+2+2 \equiv 6 \pmod 8,$$
$$2+2+2+2 \equiv 0 \pmod 8,$$
$$2+2+2+2+2 \equiv 2 \pmod 8,$$
$$2+2+2+2+2+2 \equiv 4 \pmod 8.$$

$$\vdots$$

6.9. $2^1 \equiv 2 \pmod 5$,
$\quad 2^2 \equiv 4 \pmod 5$,
$\quad 2^3 \equiv 3 \pmod 5$,
$\quad 2^4 \equiv 1 \pmod 5$.

6.10. In $\mathbf{U}(8)$, $3^2 = 1$, $5^2 = 1$, and $7^2 = 1$, so no element generates the group.

6.11. In G, $a^0, a^1, \ldots, a^m$ are all distinct if a has order m. So for a in G, G must have order $\geq m$ to contain all these elements.

6.12.

(a) Zero divisors in $\mathbb{Z}_4$ include just the number 2.

(b) Zero divisors in $\mathbb{Z}_6$ are 2, 3, and 4.

(c) Zero divisors in $\mathbb{Z}_8$ are 2, 4, and 6.

Exercises

6.1. For each of the following problems, find five integers n (at least one of them negative) that satisfy the given congruence equation.

(a) $n \equiv 3 \pmod{7}$.

(b) $n \equiv 1 \pmod{10}$.

(c) $n \equiv 8 \pmod{12}$.

(d) $n \equiv 5 \pmod{13}$.

6.2. Which of the following pairs of integers are congruent modulo 5? Modulo 6?

(a) 4 and 13.

(b) 5 and 1,005.

(c) −2 and 77.

(d) 14 and 1,414.

(e) −1,000 and −100.

6.3. Find all positive integers $m \geq 2$ for which the following are true.

(a) $-17 \equiv 10 \pmod{m}$.

(b) $77 \equiv 19 \pmod{m}$.

(c) $-29 \equiv 16 \pmod{m}$.

(d) $117 \equiv 54 \pmod{m}$.

6.4. How many positive integers $m \geq 2$ are there for which $27 \equiv 387 \pmod{m}$?

6.5. Find values of x that make the following equations true.

(a) $2x \equiv 8 \pmod{13}$.

(b) $2x \equiv 8 \pmod{14}$.

(c) $2x \equiv 9 \pmod{13}$.

(d) $2x \equiv 9 \pmod{14}$.

6.6. Find the multiplicative and additive inverses, modulo 13, for the following numbers: 2, 3, 4, 5, 12.

6.7. Find the residue class of the following factorials.

(a) $6!$ $\pmod 7$.

(b) $10!$ $\pmod{11}$.

(c) $12!$ $\pmod{13}$.

(d) $16!$ $\pmod{17}$.

6.8. Make a multiplication table for $\mathbf{U}(9)$. Is this a cyclic group? If so, find all the generators.

6.9. Find all the generators of $\mathbb{Z}_{20}$. Find the elements of $\mathbf{U}(20)$. How do these sets compare? Can you explain this?

6.10. Explain why, for any positive integer n, there are exactly n residue classes modulo n.

6.11. Prove that if $a \equiv b \pmod n$ and k is a positive integer, then $a^k \equiv b^k \pmod n$.

6.12. Find all integers n such that $n \equiv n^2 \pmod{10}$.

6.13. Let r and m be integers such that $1 \leq r < m$. Show that if r is in $\mathbf{U}(m)$, then so is $m - r$.

6.14. If G is a group, prove that e is the only element of G having order 1.

6.15. Let G be a group with the binary operation $*$ and identity e. Let g be any element of G. Prove the following for any m and n in $\mathbb{Z}$.

(a) $g^m * g^n = g^{m+n}$.

(b) $(g^m)^n = g^{mn}$.

(c) If G is an abelian group, then $g^n * h^n = (g * h)^n$.

(d) If G is a group and $g^2 * h^2 = (g * h)^2$ for all $g, h \in G$, then G is abelian.

6.16. Suppose that g is in a group G and that g has order n. Prove that the order of g^{-1} is n as well.

6.17. Let g be an element of a group with identity e, and suppose that the order of g is n. Let m be a positive integer. Prove that $g^m = e$ if and only if $n|m$. **Hint:** Use the division algorithm.

6.18. Let G be a finite abelian group with binary operation $*$ and identity e. Assume that $|G| = N$. This problem outlines a proof that the order of any element g of G is a divisor of N.

(a) Make a list of the elements of G: $g_1, g_2, \ldots, g_N$. Select one of these, say, g_i. Prove that no two elements of the set

$$\{g_i * g_1, g_i * g_2, \ldots, g_i * g_N\}$$

are equal.

(b) Prove that $G = \{g_i * g_1, g_i * g_2, \ldots, g_i * g_N\}$.

(c) Prove that $g_1 * g_2 * \cdots * g_N = (g_i * g_1) * (g_i * g_2) * \cdots * (g_i * g_N)$.

(d) Prove that $(g_i)^N = e$ and use Exercise 6.17 to deduce that the order of g_i is a divisor of N.

6.19. Find the order of the element 2 in each of the following groups with the indicated operation. If the group is finite, verify that the order of 2 is a divisor of the order of the group.

(a) $(\mathbb{Z}_5, +)$.

(b) $(\mathbf{U}(5), \cdot)$.

(c) $(\mathbf{U}(7), \cdot)$.

(d) $(\mathbb{Z}, +)$.

(e) $(\mathbf{U}(13), \cdot)$.

(f) $(\mathbb{Z}_{12}, +)$.

(g) $(\mathbf{U}(15), \cdot)$.

6.20. Let G be an abelian group with binary operation $*$. Let g be an element of G having finite order n.

(a) Suppose that k and l are two integers with $1 \le k \le l \le n$ such that $g^k = g^l$. Prove that $k = l$.

(b) Show that no two elements in the set

$$\{g, g^2, g^3, \ldots, g^n\}$$

are equal.

(c) Assume that G is a finite abelian group having N elements. Prove that G is cyclic if and only if G contains an element having order N.

(d) Suppose that G has a prime number of elements. Prove that G must be cyclic.

(e) Give an example of a cyclic group that does not have a prime number of elements.

6.21. Find the order of all the elements of the cyclic group $(\mathbb{Z}_{12}, +)$. Find all the generators of $(\mathbb{Z}_{12}, +)$. What common property do these generators possess?

6.22. A 2×2 matrix (over $\mathbb{Q}$) is an array of the form $\left(\begin{smallmatrix} a & b \\ c & d \end{smallmatrix}\right)$, where $a, b, c,$ and d are in $\mathbb{Q}$.

Denote the collection of all such matrices as $M_2(\mathbb{Q})$. If $A = \left(\begin{smallmatrix} a & b \\ c & d \end{smallmatrix}\right)$ and $A' = \left(\begin{smallmatrix} a' & b' \\ c' & d' \end{smallmatrix}\right)$, define the product

$$A \cdot A' = \begin{pmatrix} a & b \\ c & d \end{pmatrix} \cdot \begin{pmatrix} a' & b' \\ c' & d' \end{pmatrix} = \begin{pmatrix} aa' + bc' & ab' + bd' \\ ca' + dc' & cb' + dd' \end{pmatrix}.$$

(a) Show that this multiplication is a binary operation on $M_2(\mathbb{Q})$.

(b) Show that this multiplication is associative on $M_2(\mathbb{Q})$.

(c) If $I = \left(\begin{smallmatrix} 1 & 0 \\ 0 & 1 \end{smallmatrix}\right)$, show that I is an identity for this multiplication.

(d) If $A = \left(\begin{smallmatrix} a & b \\ c & d \end{smallmatrix}\right)$ and $ad - bc \neq 0$, show that

$$A^{-1} = \begin{pmatrix} \frac{d}{ad-bc} & \frac{-b}{ad-bc} \\ \frac{-c}{ad-bc} & \frac{a}{ad-bc} \end{pmatrix}$$

is the multiplicative inverse for A; that is, $A \cdot A^{-1} = A^{-1} \cdot A = I$.

(e) Let $M_2(\mathbb{Q})^\times = \{\left(\begin{smallmatrix} a & b \\ c & d \end{smallmatrix}\right)$ in $M_2(\mathbb{Q})$ satisfying $ad - bc \neq 0\}$. Find matrices A and B in $M_2(\mathbb{Q})^\times$ satisfying $AB \neq BA$.

(f) Conclude that $M_2(\mathbb{Q})^\times$ is a nonabelian group.

6.23. In Exercise 6.22, replace $\mathbb{Q}$ by the group $\mathbb{Z}_2$. So $M_2(\mathbb{Z}_2)^\times = \{\left(\begin{smallmatrix} a & b \\ c & d \end{smallmatrix}\right)$ such that a, b, c, and d are in $\mathbb{Z}_2$ and satisfy $ad - bc \neq 0\}$. Show that $M_2(\mathbb{Z}_2)^\times$ is a nonabelian group of order six.

Exercises 6.24–6.27 provide a numerical formula for the Euler function $\phi(n)$.

6.24. Suppose that $(m, n) = 1$, $a \equiv b \pmod{m}$, and $a \equiv b \pmod{n}$.
 (a) Prove that $a \equiv b \pmod{mn}$.
 (b) Give a counterexample to show that this is not necessarily true if m and n are not relatively prime.

6.25. Let p be a prime number, and let r and n be positive integers.
 (a) Prove that $(n, p^r) = 1$ if and only if $(n, p) = 1$.
 (b) Show that exactly p^{r-1} of the integers $1, 2, 3, \ldots, p^r$ are not relatively prime to p. Deduce that $\phi(p^r) = p^r - p^{r-1} = p^r(1 - \frac{1}{p})$.
 (c) How many elements are in the set $\mathbf{U}(32)$? In the set $\mathbf{U}(81)$? In the set $\mathbf{U}(125)$?

6.26. Assume m and n are integers with $(m, n) = 1$.
 (a) Let r_1 be an element of $\mathbf{U}(m)$ and r_2 be an element of $\mathbf{U}(n)$. Prove that there is a unique element r of $\mathbf{U}(mn)$ such that $r \equiv r_1 \pmod{m}$ and $r \equiv r_2 \pmod{n}$. **Hint:** See Exercise 4.20.
 (b) Let a be an integer such that $1 \leq a \leq mn$, and suppose that $a \equiv r_1 \pmod{m}$ and $a \equiv r_2 \pmod{n}$. Prove that $(a, mn) = 1$ if and only if $(a, r_1) = (a, r_2) = 1$. **Hint:** Theorem 3.8 tells us that $(a, m) = (r_1, m)$ and $(a, n) = (r_2, n)$; then use Exercise 4.19.
 (c) Prove that $\phi(mn) = \phi(m) \cdot \phi(n)$.

6.27. Suppose n is a positive integer with prime factorization $p_1{}^{r_1} p_2{}^{r_2} \cdots p_s{}^{r_s}$. Derive the following formula for $\phi(n)$:

$$\phi(n) = n \left(1 - \frac{1}{p_1}\right) \left(1 - \frac{1}{p_2}\right) \cdots \left(1 - \frac{1}{p_s}\right).$$

6.28. Find the number of elements in $\mathbf{U}(100)$, $\mathbf{U}(120)$, and $\mathbf{U}(360)$.

6.29. What is the relationship between units and zero divisors in an integral domain?

6.30. Cancelation in groups: Let G be a group and suppose a, b, and c are in G. Show that if $ab = ac$, then $b = c$.

6.31. Solutions of equations in groups: Let G be a group and suppose a and b are in G. Show that there exist elements x and y in G so that $xa = b$ and $ay = b$.

For Exercises 6.32–6.37, we define a *subgroup* of a group as follows.

Definition. Let G be a group with binary operation $*$. A nonempty subset H of G is a *subgroup* of G if H is also a group with the same binary operation $*$. Thus,

1. If a and b are in H, then $a * b$ is in H.

2. If e is the identity on G, then e is in H.

3. If a is in H, then a^{-1} is in H.

6.32.
   (a) For any group G, show that $\{e\}$ and G are subgroups of G.

   (b) Find all four subgroups of the group $\mathbb{Z}_6$ under addition.

6.33. If H is a subgroup of group G, why must the binary operation $*$ be associative on H?

6.34. Let G be a group and H a nonempty subset of G. Suppose that for any two elements a and b in H, the element $a * b^{-1}$ is also in H. Prove that H is a subgroup of G.

6.35. Let n be a positive integer.
   (a) Determine all the subgroups of the group $\mathbb{Z}_n$ under addition.

   (b) Suppose G is a cyclic group of order n. Suppose that H is a subgroup of G and $|H| = m$. Show that H must be a cyclic group and that m divides n.

6.36. Let $G = M_2(\mathbb{Q})^\times$ as in Exercise 6.22. Set

$$H = \left\{ A = \begin{pmatrix} a & b \\ c & d \end{pmatrix} \text{ in } M_2(\mathbb{Q})^\times \text{ satisfying } (ad - bc) = 1 \right\}.$$

(a) Show H is a subgroup of G.

(b) Is there a similar subgroup H if $G = M_2(\mathbb{Z}_2)^\times$?

6.37. Which of these are rings? Why or why not?

(a) $\mathbb{Q}$.

(b) $3\mathbb{Z} = \{\ldots, -6, -3, 0, 3, 6, \ldots\}$.

(c) $3\mathbb{Z} + 1 = \{\ldots, -5, -2, 1, 4, 7, \ldots\}$.

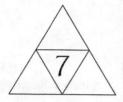

Applications of Congruences

In this chapter, we illustrate how congruences are used in a number of applications. Some of these applications show the usefulness of congruence in calculations, such as tests for divisibility or determining the day of the week. Other applications show how congruences can be used to encode information via check digits.

7.1 Divisibility Tests

One of the most interesting applications of congruences is in establishing divisibility tests for integers. These are tests that students learn to quickly determine divisibility of a large number by a smaller one, usually by a prime. Most elementary students can describe how to determine whether a number is divisible by 3, but have never learned why this test works. In fact, these divisibility tests aren't tricks at all, but consequences of facts about congruences.

Divisibility by Powers of 2

The easiest test is determining whether a number is divisible by 2. This is simply a matter of observing whether the number is even or odd. That is, we need to know whether the number is congruent to 0 modulo 2 or is congruent to 1 modulo 2. Thus, we just need to check the one's digit to see if it is divisible by 2 (or is congruent to 0 modulo 2) to see whether the number itself is divisible by 2.

We can extend this result to determine whether a number is divisible by powers of 2, such as 4, 8, 16, and so on, and see why our methods work. Notice that when we check for divisibility by 2, we are checking the digit in the one's place (10^0) for its residue class modulo 2. Divisibility by powers of 2 will require that we check up to the ten's place (10^1) for divisibility by 2^2, up to the hundred's place (10^2) for divisibility by 2^3, and so on, when checking the number for divisibility by $2^2, 2^3$, or higher power of 2.

To prove why the divisibility tests work, we need an easy way of writing what we mean by a digit in a particular place in decimal notation. In the decimal expansion of an integer n, the numeral in the 10^m place will be labeled x_m. Then in the decimal expansion, $n = x_m 10^m + x_{m-1} 10^{m-1} + \cdots + x_2 10^2 + x_1 10^1 + x_0 10^0$, where each x_j is a digit between 0 and 9, and $x_m \neq 0$ unless $n = 0$, in which case we write $x_0 = 0$.

Example. If $n = 35{,}627$, we have

$$x_0 = 7, \text{ which represents } 7 \cdot 10^0,$$
$$x_1 = 2, \text{ which represents } 2 \cdot 10^1,$$
$$x_2 = 6, \text{ which represents } 6 \cdot 10^2,$$
$$x_3 = 5, \text{ which represents } 5 \cdot 10^3,$$
$$x_4 = 3, \text{ which represents } 3 \cdot 10^4.$$

This notation should be familiar.

Since $10 \equiv 0 \pmod 2$, then $10^k \equiv 0 \pmod{2^k}$ for all positive integers k. This fact comes from Exercise 3.8, which says that if $a \mid b$ and $c \mid d$, then $ac \mid bd$. So we have $n \equiv x_0 \pmod 2$.

So, to check for divisibility by 2, we need only check the digit x_0 for its residue class modulo 2. Tests for divisibility by powers of 2 depend on residue classes of powers of 10 modulo powers of 2. For higher powers of 2, we have

$n \equiv x_1 x_0 \pmod 4$,

where $x_1 x_0$ represents the two-digit number $10x_1 + x_0$;

$n \equiv x_2 x_1 x_0 \pmod 8$,

where $x_2 x_1 x_0$ represents the three-digit number $100x_2 + 10x_1 + x_0$;

$$\vdots$$

$n \equiv x_{k-1} x_{k-2} \cdots x_2 x_1 x_0 \pmod{2^k}$,

which has a similar interpretation.

Consequently, n is divisible by 2^k if the k-digit number $x_{k-1} \cdots x_0$ is divisible by 2^k.

Example. Is 2,376 divisible by 4? As we explained above, $2 \cdot 1,000$ and $3 \cdot 100$ are divisible by 4, so all we need to check is the 76. Since $76 = 4 \cdot 19$, we conclude that $4 \mid 2,376$.

Example. Let $n = 1,723,408$. What is the highest power of 2 that divides n? We have $2 \mid n$ since $2 \mid 8$; $4 \mid n$ since $4 \mid 08$; $8 \mid n$ since $8 \mid 408$; and $16 \mid n$ since $16 \mid 3,408$; but $32 \nmid n$ since $32 \nmid 23,408$.

Divisibility by Powers of 5

Because we are working in base 10, divisibility tests for powers of 5 are similar. As we observed for 2, we have the congruences $10^k \equiv 0 \pmod{5^k}$, since 5 divides each power of 10 in the product 10^k. So again, we need to check only the last k digits of a number n to test for divisibility by 5^k.

Example. Let $n = 352,250$. What is the highest power of 5 that divides n? We have that $5 \mid n$, since $5 \mid 0$; $25 \mid n$, since $25 \mid 50$; and $125 \mid n$, since $125 \mid 250$; but $625 \nmid n$, since $625 \nmid 2,250$.

Practice Problem 7.1. *Check the following numbers to find the highest power of 2 and the highest power of 5 that divides each.*

(a) $n = 468,750$.

(b) $n = 60,000$.

(c) $n = 4,480$.

Divisibility by 3 and 9

It may seem somewhat mysterious why the tests for divisibility by 3 and 9 work, but they are also consequences of simple facts about congruences. Notice that the congruences $10 \equiv 1 \pmod 3$ and $10 \equiv 1 \pmod 9$ both hold. So if we multiply to get powers of 10, we will have that both $10^k \equiv 1 \pmod 3$ and $10^k \equiv 1 \pmod 9$.

Using our earlier decimal notation, we write

$$n = x_m 10^m + x_{m-1} 10^{m-1} + \cdots + x_2 10^2 + x_1 10^1 + x_0 10^0.$$

Then

$$n \equiv x_m + x_{m-1} + \cdots + x_2 + x_1 + x_0 \pmod 3$$

and

$$n \equiv x_m + x_{m-1} + \cdots + x_2 + x_1 + x_0 \pmod 9$$

by Theorem 6.3 for adding and multiplying congruences. This just says that to find out whether a number n is divisible by 3 or by 9, we need to check whether the sum of the digits is divisible by 3 or by 9.

Example. Let $n = 874,326$. The sum of the digits of n is $8 + 7 + 4 + 3 + 2 + 6 = 30$, which is congruent to 0 modulo 3 and to 3 modulo 9. So n is divisible by 3 but not by 9.

Practice Problem 7.2. *Check the following numbers to find whether* 3 *or* 9 *divides each.*

(a) $n = 234,648$.

(b) $n = 525$.

(c) $n = 8,019,346,725$.

The congruence of the sum of the digits of a number modulo 9 also forms the basis for the process of "casting out nines." This procedure asks what the remainder is when we divide a number by 9. The congruence

$$n \equiv x_m + x_{m-1} + \cdots + x_2 + x_1 + x_0 \pmod 9$$

shows that the residue for n modulo 9 will be the same as the residue of the sum of the digits modulo 9.

Example. For $n = 87,523$, $n \equiv 8 + 7 + 5 + 2 + 3 \equiv 5 \pmod 9$, so the remainder when we divide n by 9 will be 5.

Divisibility by 11

In this case, the fact that $10 \equiv -1 \pmod{11}$ is the basis for testing for divisibility by 11. Multiplying congruences, we have $10^2 \equiv 1 \pmod{11}$, $10^3 \equiv -1 \pmod{11}$, and so on. Thus, we have

$$n = x_m 10^m + x_{m-1} 10^{m-1} + \cdots + x_1 10^1 + x_0 10^0$$
$$\equiv x_m (-1)^m + x_{m-1}(-1)^{m-1} + \cdots + x_1 (-1)^1 + x_0 (-1)^0 \pmod{11}.$$

In simple terms, this is an alternating sum of the digits. We start with the one's digit and alternately subtract and add the remaining digits. If the resulting sum is divisible by 11, then n is also divisible by 11. Furthermore, as with "casting out nines," this alternating sum tells us the remainder when n is divided by 11.

Example. Let $n = 38,564$. The sum $4 - 6 + 5 - 8 + 3 = -2 \equiv 9 \pmod{11}$, so n is not divisible by 11. In fact, n leaves a remainder of 9 when divided by 11.

Divisibility by 7, 11, and 13

There is a clever way of testing for divisibility for the three primes 7, 11, and 13 all at once. The test uses two facts: $7 \cdot 11 \cdot 13 = 1,001$ and $10^3 \equiv -1 \pmod{1,001}$. This test, like the test for divisibility by 11, uses a particular alternating sum of digits. Because of the congruence $10^3 \equiv -1 \pmod{1,001}$, we group the digits together in threes, and then alternate the signs of these sums. We thus have

$$n = x_m 10^m + x_{m-1} 10^{m-1} + \cdots + x_2 10^2 + x_1 10^1 + x_0 10^0.$$
$$n = (x_0 + 10x_1 + 100x_2) + 1,000(x_3 + 10x_4 + 100x_5)$$
$$+ 1,000^2(x_6 + 10x_7 + 100x_8) + \cdots$$
$$\equiv (x_0 + 10x_1 + 100x_2) + (-1)(x_3 + 10x_4 + 100x_5)$$
$$+ (1)(x_6 + 10x_7 + 100x_8) + \cdots \pmod{1,001}$$
$$\equiv (x_2 x_1 x_0) - (x_5 x_4 x_3) + (x_8 x_7 x_6) - \cdots \pmod{1,001}.$$

Remember that $x_j x_k x_l$ stands for the number $100x_j + 10x_k + x_l$.

Thus, n is congruent, modulo 1,001, to the integer we get by successively adding and subtracting the three-digit integers formed by taking blocks of three decimal digits from n, beginning with the right-most digit. Since 7, 11, and 13 are divisors of 1,001, n will be divisible by each of these integers if this alternating sum and difference of three-digit blocks is divisible by 7, 11, or 13.

Example. Suppose $n = 15,331,680$. The alternating sum of three-digit blocks is $680 - 331 + 15 = 364$. Since $7 \mid 364$ and $13 \mid 364$, n is divisible by 7 and 13. However, $11 \nmid 364$, so n is not divisible by 11.

7.2 Days of the Week

What day of the week were you born? If you don't remember (or even if you do) in this section, we show you a way to figure out the day of the week of any given date by using congruence arithmetic and the calendar.

Our present-day calendar has some quirks built into it, so that the number of days in a year correspond as closely as possible to the time it actually takes Earth to complete one full revolution around the sun. Early attempts at this resulted in a calendar that had exactly 365 days. Julius Caesar changed this to a new calendar, called the Julian calendar, with $365\frac{1}{4}$ days. To account for this extra $\frac{1}{4}$ day, he put in a leap year every fourth year. This calendar still was not accurate enough to reflect the true length of a year, which is approximately 365.2422 days.

Practice Problem 7.3.

(a) *Using a calendar with* 365 *days, how far off will the calendar be after* 100 *years? After* 1,000 *years?*

(b) *Using a calendar with* $365\frac{1}{4}$ *days, how far off will the calendar be after* 100 *years? After* 1,000 *years?*

In 1582, Pope Gregory created a new calendar, called the Gregorian calendar, that had leap years in every year divisible by 4, except for those divisible by 100, which would be leap years only when divisible also by 400. Thus, the year 1900 was not a leap year, but the year 2000 was. The error in this calendar is now only 0.0003 days per year; so a future correction will be needed again, but not for about 3,000 more years.

Practice Problem 7.4. *How far off will the Gregorian calendar be after* 3,000 *years?*

When we do congruence arithmetic with dates, we ave to take into account whether a particular year is or is not a leap year.

Another quirk is the number of days in each month, which is usually 30 or 31, but is either 28 or 29 in February. When we do congruence arithmetic with dates, we also have to take into account how many days are in each month.

Most of the calculating we do with this problem is modulo 7, since there are seven days in a week. We assign a number 0 through 6 to each day of the week, starting with Sunday = 0 and going through Saturday = 6. Then, when we do our congruence work and come up with an answer of 3, we know that answer will correspond to Wednesday.

Calculating from the First Date of Any Year

In order to figure out the day of the week of a particular date, we must first have a reference date from which to calculate. We begin by knowing that January 1, 2000, was a Saturday. Since Saturday is represented by 6 modulo 7, we assign January 1, 2000, the number 6.

Example. Let's figure out what day of the week Valentine's Day was in 1999. Our strategy is first to figure out the day of the week of January 1, 1999, then add enough days to get to February 14, 1999.

Since 1999 was not a leap year, it had 365 days. We know that 365 is congruent to 1 modulo 7, so to get the day of the week of January 1 that year, we subtract 1 from the 6 that represents January 1, 2000, and get 5. Since 5 represents Friday, we know that January 1, 1999, was a Friday.

Next, we want to count forward to get to February 1, 1999. Since January has 31 days, and 31 is congruent to 3 modulo 7, we add 3 to 5 modulo 7 and get 1. Since 1 represents Monday, now we know that February 1 was a Monday in 1999. If we add 13 to 1 to get February 14, we get 14, which is congruent to 0 modulo 7. So Valentine's Day fell on a Sunday in 1999.

Practice Problem 7.5. *Use congruence arithmetic to show that Independence Day, July 4, was a Sunday in 1999.*

How to Find the Day of the Week

Now that we know how to get to any day of the year from January 1 in that same year, let's figure out how to get the day of the week of January 1 in any given year.

For example, let's find the day of the week of January 1, 1933. There are 67 years between 1933 and 2000. Our task would be easier if every year were the same length, but we have to account for some leap years. To account for leap years, we have a useful calculational tool.

Definition. The *greatest integer* in a real number x, denoted by $[x]$, is the largest integer less than or equal to x. That is, $[x]$ is the integer satisfying the inequality $[x] \leq x < [x] + 1$. The function $[x]$ is called the *greatest integer function*.

Example. From the definition, we have $[\frac{7}{2}] = 3$, $[\frac{-7}{2}] = -4$, $[5] = 5$, and $[0] = 0$.

Practice Problem 7.6. *Suppose k is an integer. Show that $[x+k] = [x]+k$.*

There is a relationship between the division algorithm and the greatest integer function. If we divide a by b, by the division algorithm we can write $a = bq + r$, for $0 \leq r < b$. Think about the greatest integer for the quotient $\frac{a}{b}$. It will be q. Thus, $[\frac{a}{b}] = q$.

Let's return to the problem of finding the day of the week of January 1, 1933. We need to subtract elapsed years, including the extra day in each leap year since 1933. Since every fourth year is a leap year, there will be $[\frac{67}{4}] = 16 \equiv 2 \pmod 7$ leap years, each of which adds one day to the total number of days between January 1, 1933, and January 1, 2000.

Remembering that 365 is congruent to 1 modulo 7, we need to subtract $67 \cdot 1$ modulo 7 to account for the elapsed years, and then subtract another 16 modulo 7 to account for the leap years.

$$365 \cdot 67 \equiv 1 \cdot 67 \pmod 7 \equiv 4 \pmod 7,$$
$$[\tfrac{67}{4}] = 16 \equiv 2 \pmod 7.$$

So

$$365 \cdot 67 + [\tfrac{67}{4}] \equiv 4 + 2 \pmod 7 \equiv 6 \pmod 7.$$

Since 67 is congruent to 4 modulo 7, and 16 is congruent to 2 modulo 7, we just subtract 6 from 6 (representing Saturday, January 1, 2000) to get 0, or Sunday as the day of the week of January 1, 1933.

Practice Problem 7.7. *Find the day of the week of January 29, 1933.*

Practice Problem 7.8. *Find the day of the week of Independence Day (July 4), 1953.*

This algorithm works to find the day of the week of any date in the century between January 1, 1901, and January 1, 2000. During this period, if we want to find the day of the week of January 1 in year N, we use the formula

$$6 - [2000 - N] \cdot 365 - [\tfrac{2000-N}{4}] \pmod 7.$$

However, before January 1, 1901, and after January 1, 2000, some century years are leap years and others are not.

Practice Problem 7.9. *Use the greatest integer function to modify the formula to compute the day of the week of January 1 in any year after 1583 and before 4926.*

From this, we can compute any day of the week of that year. Some dates to try are included in the Exercises.

7.3   Check Digits

One of the more commonly used applications of congruences is finding errors in strings of digits. First, let's see how congruences are used to find errors in strings of 0s and 1s, as used in computer data, and then we show how congruences are used to detect errors in decimal digits used in identifications of products, books, and other information of this type.

In computer work, data are in the form of 0s and 1s, originally used to indicate whether a certain switch was on or off. A message sent via computer consists of a string of digits that include only 0s and 1s.

Example. The numbers $a = 000110100$ and $b = 110010100$ could be computer messages.

We want to have a simple way of checking to be sure that the message has been received correctly. One way to do this is to insert an additional digit at the end of the message, called a *check digit*, to the original message string in such a way that errors may be detected. The simplest way to do this is to compute the residue of the sum of the digits in the message modulo 2, and then insert this residue as an additional digit. This kind of a check digit is called a *parity check digit*.

Example. In $a = 000110100$, the sum of the digits is 3, which is congruent to 1 modulo 2. Thus, message a with a parity check digit inserted would be $a' = 0001101001$. In $b = 110010100$, the sum of the digits is 4, which is congruent to 0 modulo 2, so the new message will be $b' = 1100101000$.

How is this useful? If we receive a message $c = 110110001$, we can tell from the discrepancy between the parity check digit and the sum of the remaining digits that there has been an error in transmitting the data.

Practice Problem 7.10. *Tell whether the following messages are valid.*

(a) 1101100111.

(b) 1010101010.

(c) 1111111111.

This method of adding a parity check digit will help detect a single error. However, if more than one error is made, the parity check digit may not catch them. There are more complicated schemes for adding check digits to strings of numbers so that errors may be detected and even corrected. Two of the more common check digit schemes are those used to write ISBN numbers on books and UPC numbers on products.

ISBNs and UPC Numbers

Almost all books published today are assigned a number by their publishing company. This number is called the International Standard Book Number (ISBN). These numbers encode certain information for the publisher. For example, some of the digits tell in what language the book is written; others may be a code for the publishing company. The check digit in an ISBN is designed to detect the most common errors made when ISBNs are copied, which are using a wrong digit or transposing two digits.

Here is how to compute the ISBN check digit. We write the ten-digit ISBN as $x_1x_2x_3x_4x_5x_6x_7x_8x_9x_{10}$, where x_{10} is the check digit. The first nine numbers are in the set $\{0, 1, 2, \cdots, 9\}$. To get x_{10}, we compute a weighted sum mod 11.

$$x_{10} \equiv x_1 + 2x_2 + 3x_3 + 4x_4 + 5x_5 + 6x_6 + 7x_7 + 8x_8 + 9x_9 \pmod{11}.$$

Example. 0-865-68164-3 is a valid ISBN since $1 \cdot 0 + 2 \cdot 8 + 3 \cdot 6 + 4 \cdot 5 + 5 \cdot 6 + 6 \cdot 8 + 7 \cdot 1 + 8 \cdot 6 + 9 \cdot 4 = 223 \equiv 3 \pmod{11}$.

What happens if the check digit turns out to be 10? The clever solution to this problem is to use the Roman numeral X for 10.

Example. 0-387-90462-X is a valid ISBN since $1 \cdot 0 + 2 \cdot 3 + 3 \cdot 8 + 4 \cdot 7 + 5 \cdot 9 + 6 \cdot 0 + 7 \cdot 4 + 8 \cdot 6 + 9 \cdot 2 = 197 \equiv 10 \pmod{11}$.

Another way to write the weighted sum is

$$x_1 + 2x_2 + 3x_3 + 4x_4 + 5x_5 + 6x_6 + 7x_7 + 8x_8 + 9x_9 + 10x_{10} \equiv 0 \pmod{11}.$$

This equation helps to explain how the check digit detects certain errors.

Suppose that a single error has been made in copying the ISBN. Then one of the digits will be written incorrectly. For some i, instead of x_i, we will have another number, $x_i + k$, where k is a number between -9 and 9, and $k \neq 0$. The term in the incorrect weighted sum will now be $i(x_i + k) = ix_i + ik$. Since the original weighted sum is congruent to 0 modulo 11, this new sum will instead be congruent to ik modulo 11. But since $11 \nmid i$ and $11 \nmid k$, $11 \nmid ik$, so ik is not congruent to 0 modulo 11. Therefore, we know that the copied ISBN must have an error. Notice that this works for any i between 0 and 10, since 11 is prime.

Suppose now that two unequal digits of the original ISBN are transposed. Then there will be two different integers j and k where the digits

x_j and x_k are in the wrong place. In the new ISBN's weighted sum, there will be terms jx_k and kx_j. How will the weighted sum of the new ISBN differ from the original? The terms jx_j and kx_k will be missing, and instead the terms jx_k and kx_j will be added. Look at this difference, $(jx_k - jx_j) + (kx_j - kx_k) = (j - k)(x_k - x_j)$. Since x_j and x_k were different, $11 \nmid (x_k - x_j)$. Also, $11 \nmid (k - j)$. So since this sum is not congruent to 0 modulo 11, we can detect an interchange of two different digits.

Challenge Problem. Can you devise an error in an ISBN that will not be detected by the check digit?

The check digit here is a useful tool for the people who sell books, but it really offers no information for the buyer. The Universal Product Code (UPC) is similar; it is the 12-digit number printed on many items bought in stores. The last number is a check digit, found by using a formula created to detect errors.

We write the UPC number $x_1x_2x_3x_4x_5x_6x_7x_8x_9x_{10}x_{11}x_{12}$ so that x_{12} is the check digit. The congruence that is satisfied by the UPC number is

$$x_{12} \equiv -(3x_1 + x_2 + 3x_3 + x_4 + 3x_5 + x_6 + 3x_7 + x_8 + 3x_9 + x_{10} + 3x_{11}) \pmod{10}$$

Example. The number 7 15793 00126 3 is a valid UPC number on a product. To see that this is a valid number, we do the computation to see that $3 \equiv -(21 + 1 + 15 + 7 + 27 + 3 + 0 + 0 + 3 + 2 + 18) \pmod{10}$.

Challenge Problem. Show that the weighted sum for UPC numbers can detect single errors and transposition errors, provided the transposed digits a and b are adjacent and $|a - b| \neq 5$.

Practice Problem Solutions and Hints

7.1.

   (a) $2 \mid 468{,}750$ and $5^4 \mid 468{,}750$.

   (b) $2^5 \mid 60{,}000$ and $5^4 \mid 60{,}000$.

   (c) Our test tells us only that $2^4 \mid 4{,}480$. In fact, $2^7 \mid 4{,}480$. Also $5 \mid 4{,}480$.

7.2.

   (a) The sum of the digits of 234,648 is 27. So both 3 and 9 divide n.

(b) The sum of the digits of 525 is 12. So 3 divides n but 9 does not.

(c) The sum of the digits of 8,019,346,725 is 45. So both 3 and 9 divide n.

7.3.

(a) The calendar will be off by about 24 days after 100 years, and about 242 days after 1,000 years.

(b) The calendar will be off by about 1 day after 100 years, and about 8 days after 1,000 years.

7.4. After 3,000 years, the calendar will be off by about a day.

7.5. January 1, 1999, is assigned 5. From January 1 to July 1 is $31 + 28 + 31 + 30 + 31 + 30 = 181$ days, congruent to 6 modulo 7. Since $5 + 6 \equiv 4$ modulo 7, July 1, 1999, was a Thursday. We add 3 more days to get to July 4, 1999, a Sunday.

7.6. Let $[x] = m$, so m is an integer,

$$m \leq x < m + 1.$$

By Axiom O3,

$$m + k \leq x + k < m + k + 1.$$

So $m + k = [x] + k$ is the greatest integer less than or equal to $x + k$. Thus, $[x + k] = [x] + k$.

7.7. January 1, 1933, was a Sunday. We add 28 days, which is congruent to 0 modulo 7. So January 29, 1933, was a Sunday.

7.8. The number of elapsed years from 2000 to 1953 is 47. To account for leap years, we must also subtract $[\frac{47}{4}] = 11 \equiv 4 \pmod 7$.

$$365 \cdot 47 \equiv 1 \cdot 47 \pmod 7 \equiv 5 \pmod 7,$$

$$[\tfrac{47}{4}] = 11 \equiv 4 \pmod 7.$$

We want to subtract $5 + 4 = 9 \equiv 2 \pmod 7$. So,

$$6 - 2 \equiv 4 \pmod 7,$$

so January 1, 1953, was a Thursday. To get to July 4, 1953, we add

$$31 + 28 + 31 + 30 + 31 + 30 + 3 \equiv 2 \pmod 7.$$

Since $2 + 4 = 6$, July 4, 1953, was a Saturday.

7.9. Hint: Use the greatest integer function to account for elapsed century leap years. In the 1500s, use $[\frac{1500}{4}]$.

7.10.

(a) The message 110110011 has digit sum congruent to 0 modulo 2. Since the check digit is 1, this message isn't valid.

(b) Invalid.

(c) Valid.

Exercises

7.1. Find the highest power of 2 and the highest power of 5 that divides the following numbers. Are the original numbers divisible by 3, 7, 9, or 11?

(a) 85,140.

(b) 111,650.

(c) 1,268,904.

(d) 24,248,484.

7.2. Compute the following:

(a) $[\frac{7}{3}]$.

(b) $[\pi]$.

(c) $[-\pi]$.

(d) $[5]$.

7.3. Find the day of the week of the following dates:

(a) September 18, 1981.

(b) November 5, 1938.

(c) October 31, 1966.

(d) March 26, 1915.

7.4. The great astrologer Zoltronimus has predicted the world will end on Monday, July 17, 6666. Prove him wrong.

7.5. Let x be a real number. Prove that there exists a unique real number y, with $0 \leq y < 1$, such that $x = [x] + y$.

7.6. Prove that if x and y are any two real numbers, then $[x+y] \geq [x]+[y]$. Find conditions on x and y for which $[x+y] = [x]+[y]$.

7.7. Show that $[a] + [-a] = 0$ if and only if a is an integer. What is $[a] + [-a]$ equal to if a is not an integer?

7.8. What are the least and greatest number of Friday the 13ths that can occur in one year?

7.9. Find the ISBN for this book (on the back of the title page). Verify that it has the correct check digit.

7.10. Which of these are valid ISBNs?
  (a) 0-7963-4010-2.

  (b) 1-2895-2215-7.

  (c) 0-2075-3098-6.

  (d) 2-3070-8651-8.

7.11. Compute the missing digits in the following ISBNs. **Hint:** First make a chart of the inverses of all the elements in $\mathbf{U}(11)$.
  (a) 0-69?-05919-7.

  (b) 0-12-163251-?.

  (c) 0-387-90?18-9.

  (d) 3-540-9637?-8.

  (e) 0-?3-487174-X.

7.12. Compute the missing digit in the following UPC numbers.
  (a) 0 74865 26491 ?.

  (b) 0 43100 0?512 9.

  (c) ? 81515 99300 2.

7.13. Check digits are also used by banks. One way a bank ID number might be made is to use the form $x_1x_2x_3x_4x_5x_6x_7x_8x_9$ with

$$x_9 \equiv 7x_1 + 3x_2 + 9x_3 + 7x_4 + 3x_5 + 9x_6 + 7x_7 + 3x_8 \pmod{10}$$

Is 123456789 a valid bank ID number?

7.14. Which errors can this type of the bank ID number check digit can catch?

7.15. Some credit card numbers are 16 digits long, say,

$$x_1x_2x_3x_4x_5x_6x_7x_8x_9x_{10}x_{11}x_{12}x_{13}x_{14}x_{15}x_{16}.$$

The last digit is a check digit that satisfies the congruence

$$x_{16} \equiv -(2x_1 + x_2 + 2x_3 + \cdots + x_{14} + 2x_{15} + N) \pmod{10},$$

where N is the number of digits in the odd positions that are 5 or higher. Find an actual 16-digit credit card number and check that it satisfies this congruence.

7.16. If you start with a valid credit card number and change one digit, is the new number ever valid? If so, when? If not, prove it.

7.17. What happens if, starting with a valid credit card number, you switch two consecutive digits? Can the new number still be a valid credit card number? If so, when? If not, prove it.

7.18. Let n be a positive integer.

(a) Let p be a prime number and r a positive integer. Show that $[\frac{n}{p^r}]$ is equal to the number of positive integers less than or equal to n that are divisible by p^r.

(b) Show that $[\frac{n}{p^k}] = 0$ for sufficiently large k.

(c) Show that the highest power of p dividing $n!$ is p^k, where

$$k = \left[\frac{n}{p}\right] + \left[\frac{n}{p^2}\right] + \left[\frac{n}{p^3}\right] + \cdots$$

(d) Compute the powers of 2 and 3 occurring in the prime factorization of 50!. Then find the complete prime factorization of 50!.

7.19. How many zeroes are at the end of the expanded form of 1,000,000!?

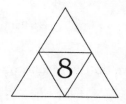

Rational Numbers
and Real Numbers

We have focused our attention mostly on properties of integers: divisibility, primes, and congruences. Before we leave the topic of number theory, we want also to explore the properties of rational and irrational numbers. To begin, we explore the relationship between fractions and decimals. Then we consider the set $\mathbb{Q}$ itself and ask how large this set is. Finally, we move on to the rest of the real numbers, or the irrationals, and ask how many of them there are.

8.1   Fractions to Decimals

Early in the text, we defined a rational number as the quotient of two integers, a and b, with $b \neq 0$. It will be useful to be able to write fractions as decimals. To do this we divide b into the number a, written as the decimal $a.00\ldots$, and look at the remainders. In this section, we assume $0 < a < b$.

Example. Let $a = 1$ and $b = 4$. In the first step in the familiar process of long division, we get

$$
\begin{array}{r}
0.2 \\
4\overline{)1.0} \\
-8 \\
\hline
2
\end{array}
$$

This is just an application of the division algorithm. Notice that the remainder is 2, and when we do the second division, we get

$$
\begin{array}{r}
0.25 \\
4\overline{)1.00} \\
-\underline{8} \\
20 \\
-\underline{20} \\
0
\end{array}
$$

and since the remainder is 0, we have the equation that $\frac{1}{4} = 0.25$.

Example. Suppose $a = 4$ and $b = 11$. Then the division becomes

$$
\begin{array}{r}
0.363 \\
11\overline{)4.00} \\
-\underline{33} \\
70 \\
-\underline{66} \\
40 \\
-\underline{33} \\
7
\end{array}
$$

At this stage, we notice that the remainder of 7 is the same as the remainder at the first step. This process continues indefinitely, and so we get a repeating decimal $0.363636\ldots$ for the representation of $\frac{4}{11}$. We write a bar over the part of the decimal that repeats, as in $0.\overline{36}$, to indicate the repetition.

Decimals that represent rational numbers will be one of two types. Either the decimal will terminate, as $0.25 = \frac{1}{4}$ did, or the decimal will repeat, as $0.\overline{36} = \frac{4}{11}$ did. How do we know that one of these two must occur? The clue is in the work we did with $\frac{4}{11}$. In the division algorithm, when we divide by an integer b, the only remainders will be numbers between 0 and b. So as we do the long division, either we will eventually get a remainder of 0, or some remainder will repeat, as in $\frac{4}{11}$. Then the decimal expansion will repeat in a pattern of length at most $b - 1$. This repetition must take place in at most $b - 1$ steps.

Example. Suppose $a = 1$ and $b = 7$. The long division yields

$$
\begin{array}{r}
0.142857 \\
7\overline{)1.0} \\
-7 \\
\hline
30 \\
-28 \\
\hline
20 \\
-14 \\
\hline
60 \\
-56 \\
\hline
40 \\
-35 \\
\hline
50 \\
-49 \\
\hline
1
\end{array}
$$

Take a look at the remainders. They are in the pattern 3, 2, 6, 4, 5, and 1, which are all the possible nonzero remainders when we are dividing by $b = 7$. We have hit all the remainders. When we get back to 1, the process repeats itself because the initial step also was dividing 7 into 1.0. So the decimal for $\frac{1}{7}$ has a repeating block of $6 = 7 - 1$ numbers.

Thus, it follows from the division algorithm that we can write any rational number as a terminating or repeating decimal.

Definition. The *period* of a repeating decimal is the number of digits in the repeating part.

Examples. Examples of repeating decimals include:

1. For $\frac{1}{3} = 0.\overline{3}$, the period is 1.

2. For $\frac{1}{7} = 0.\overline{142857}$, the period is 6.

3. For $0.05\overline{21}$, the period is 2.

We can consider the decimal 0.25 as a repeating decimal if we write it as $0.25000\ldots = 0.25\overline{0}$. The repeating part is 0 and the period is 1.

Even though the period of a fraction with denominator b must be no more than $b - 1$, many fractions have shorter periods. The period for $\frac{1}{3}$ is 1, not 2. The period of $\frac{1}{11}$ is 2, not 10.

Practice Problem 8.1. *Divide out $\frac{4}{13}$ and show that the period is 6.*

Challenge Problem. Let p be a prime less than 100. When does the fraction $\frac{1}{p}$ have period $p-1$? Note that, for general p, this is an unsolved problem.

To summarize, we can write any rational number as a repeating decimal.

8.2 Decimals to Fractions

If we can write every rational number as a terminating or repeating decimal, can we also write either of these types of decimals as a fraction? For a terminating decimal, the answer is clear. Just represent the decimal with a denominator that is the appropriate power of 10.

Example. If our decimal is 0.789456, we write it in fraction form as $\frac{789,456}{1,000,000}$. Then we can reduce to lowest terms and get $\frac{43,091}{62,500}$.

What about repeating decimals? The idea here is to use an algorithm to convert this decimal into fraction form. Suppose x is a repeating decimal. Look at the period of x. If the period is n, then we will multiply x by 10^n and subtract x.

Example. Convert $x = 0.28\overline{145}$ to a fraction. The period is 3, so we multiply this equation by 10^3 and get $1,000x = 281.45\overline{145}$. Now we subtract $\overline{x}$:

$$
\begin{array}{r}
1,000x = 281.45\overline{145} \\
-x = -0.28\overline{145} \\
\hline
999x = 281.17
\end{array}
$$

Now we solve for x, and get $x = \frac{28,117}{99,900}$.

Practice Problem 8.2. *Write the following repeated decimals as fractions.*

(a) $x = 0.\overline{56}$.

(b) $x = 0.145\overline{34}$.

This simple algorithm for writing repeating decimals as fractions leads to an interesting fact, namely, that $0.9999\ldots = 1$.

If we just follow the algorithm outlined above, we have $x = 0.\overline{9}$ and so $10x = 9.\overline{9}$. Subtracting, we get

$$
\begin{array}{rl}
10x = & 9.\overline{9} \\
-x = & -0.\overline{9} \\
\hline
9x = & 9
\end{array}
$$

which shows that $x = 1$.

This statement makes many students uncomfortable. Part of the problem in grasping this fact has to do with what it means to have an infinitely repeating decimal. Here the idea of infinity plays with our conception of what we think *should* be true.

To sum up, we can write any terminating or repeating decimal in fraction form. Along with the previous discussion, this shows there is a one-to-one correspondence between rational numbers and repeating decimals. The only possible problem in this correspondence is with those decimals that repeat in all nines and therefore have two decimal representations. In those cases, we round up and use the terminating decimal version for the one-to-one correspondence. For example, we write the number 1 as $1.\overline{0}$ rather than $0.\overline{9}$.

8.3   Infinity

So far, we have been doing work with numbers throughout the text, and now we turn to a concept of something that is not a number: infinity. Students usually describe infinity as something that's too big to be counted. Actually, that's not a bad idea to start with. As usual, this is a definition we want to make precise. So we start with a formal definition of finite, and then follow with a discussion of infinity.

What is a finite set? We have the idea that we can count the things in a finite set and come up with some natural number n that tells how many elements are in the set. We can label a finite set by this number n and use n to describe the size of the finite set. Students do this very early in school, when the classic question, "If you have three apples and someone gives you two more, how many apples do you have in all?" helps students learn the operation of addition. Finite sets come in all sizes, and two of them are of equal size if we can count them and get the same number n to describe how many elements are in each set. In other words, we could "match up" the elements of the two sets, and they would pair up exactly. This is the idea of one-to-one correspondence. Observe that a one-to-one correspondence is simply a function from one set to another that is both one-to-one and onto.

Definition. Start with two sets A and B. Sets A and B are said to be in *one-to-one correspondence* if there is a function $f : A \rightarrow B$ that is one-to-one and onto. We then say that A and B have the same *cardinal number*.

Example. If $A = \{a, b, c, d, \ldots, y, z\}$ and $B = \{1, 2, 3, \ldots, 25, 26\}$, there is a one-to-one correspondence between A and B with the pairing $1 \rightarrow a$, $2 \rightarrow b$, and so on. Each set is of size $n = 26$.

Practice Problem 8.3. *Find another one-to-one correspondence between the two sets $A = \{a, b, c, d, \ldots, y, z\}$ and $B = \{1, 2, 3, \ldots, 25, 26\}$.*

Definition. A *finite set with n elements* is one that can be put into one-to-one correspondence with the set of natural numbers $\{1, 2, 3, \ldots, n\}$.

This idea of being able to count the elements of a finite set can be expanded to infinite sets very naturally. The first numbers we learn are the counting numbers, or $\mathbb{N}$. How many of them are there? Certainly the numbers never stop. This is an example of an infinite set. Infinity is a slippery concept, and we begin to see the trouble right away in answering this seemingly innocent question.

Question. Which is a bigger set: the set $\mathbb{N}$ of all natural numbers or the set $E = \{2, 4, 6, 8, \cdots\}$ of even natural numbers?

At first, most students think that the set of all the natural numbers must be bigger. Their reasons may include that the even numbers are a proper subset of the natural numbers and therefore must be of smaller size, maybe only half the size of $\mathbb{N}$. Let's take a closer look at that idea, remembering the concept of one-to-one correspondence. Is there a way to put the natural numbers into one-to-one correspondence with the even natural numbers? There is a very simple way to do it, namely, pair up the natural number n with the even number $2n$:

$$1 \rightarrow 2, 2 \rightarrow 4, 3 \rightarrow 6, \ldots$$

That is, write $f(n) = 2n$ for this correspondence. Also, the even natural number $2m$ pairs up with the natural number m:

$$2 \rightarrow 1, 4 \rightarrow 2, 6 \rightarrow 3, \ldots$$

Here, for even natural numbers m, we write $g(m) = m/2$.

This is a perfectly good correspondence, since every natural number is paired with precisely one even natural number and vice versa. This seems to be a paradox! The way out of this paradox is to realize that infinite sets do not follow the same rules that finite sets do. We need to make a definition of size that describes infinite sets.

Definition. We call a set *infinite* if it can be put into one-to-one correspondence with a proper subset of itself.

Practice Problem 8.4. *Find a one-to-one correspondence between $\mathbb{N}$ and the set $O = \{1, 3, 5, 7, \ldots\}$ of odd natural numbers.*

We have shown that the two sets O and $\mathbb{N}$ and the two sets E and $\mathbb{N}$ have the same size. We want to label the size of $\mathbb{N}$ with a special number.

Definition. We label the size of the set of natural numbers $\mathbb{N}$ with the symbol aleph naught, or $\aleph_0$, and say that $\mathbb{N}$ is *countably infinite*, or *countable*.

Any set that can be put into one-to-one correspondence with the natural numbers $\mathbb{N}$ is also called countable and is said to be of size $\aleph_0$. We don't want the word countable to be confusing. Of course, we can count the elements in a finite set; however, we reserve the word countable to describe infinite sets.

Example. The set of all powers of 2, $\{1, 2, 4, 8, 16, \ldots\}$, is a countable set.

Practice Problem 8.5. *Show that the set $\mathbb{Z}$ of integers is countable.*

Let's think about other infinite sets and see whether they are countable. The next logical place to look is at $\mathbb{Q}$, the rational numbers, or fractions.

8.4   Rational Numbers

How big is the set $\mathbb{Q}$? We will use a clever method to put the set of positive rational numbers in one-to-one correspondence with $\mathbb{N}$. We can write every positive rational number as a fraction, so let's do that in a systematic way. Begin with the rational numbers that can be expressed with denominator 1. Then make a list of all those rational numbers that can be written with a denominator of 2, eliminating those that have been listed already in the first row. Do the same with fractions with denominator 3, eliminating those that have appeared already in the first two rows, and continue in this way. The rows will look like this:

$$\frac{1}{1}, \frac{2}{1}, \frac{3}{1}, \frac{4}{1}, \frac{5}{1}, \frac{6}{1}, \cdots$$

$$\frac{1}{2}, \frac{3}{2}, \frac{5}{2}, \frac{7}{2}, \frac{9}{2}, \frac{11}{2}, \cdots$$

$$\frac{1}{3}, \frac{2}{3}, \frac{4}{3}, \frac{5}{3}, \frac{7}{3}, \frac{8}{3}, \cdots$$

$$\frac{1}{4}, \frac{3}{4}, \frac{5}{4}, \frac{7}{4}, \frac{9}{4}, \frac{11}{4}, \cdots$$

$$\frac{1}{5}, \frac{2}{5}, \frac{3}{5}, \frac{4}{5}, \frac{6}{5}, \frac{7}{5}, \cdots$$

Now we use a diagonal method for counting these numbers so that each positive rational number is counted exactly once.

$$
\begin{array}{cccccc}
\frac{1}{1} & \frac{2}{1} \rightarrow \frac{3}{1} & \frac{4}{1} \rightarrow \frac{5}{1} & \frac{6}{1} \rightarrow \cdots \\
\frac{1}{2} & \frac{3}{2} & \frac{5}{2} & \frac{7}{2} & \frac{9}{2} & \frac{11}{2} & \cdots \\
\frac{1}{3} & \frac{2}{3} & \frac{4}{3} & \frac{5}{3} & \frac{7}{3} & \frac{8}{3} & \cdots \\
\frac{1}{4} & \frac{3}{4} & \frac{5}{4} & \frac{7}{4} & \frac{9}{4} & \frac{11}{4} & \cdots \\
\frac{1}{5} & \frac{2}{5} & \frac{3}{5} & \frac{4}{5} & \frac{6}{5} & \frac{7}{5} & \cdots \\
\vdots & \vdots & \vdots & \vdots & \vdots & \vdots
\end{array}
$$

This shows a one-to-one correspondence between the positive rational numbers and the natural numbers as follows:

$$\frac{1}{1} \rightarrow 1, \frac{1}{2} \rightarrow 2, \frac{2}{1} \rightarrow 3,$$

and so on through the grid. Thus, the positive rational numbers are a countable set. It is easy to see that this method gives a one-to-one correspondence, but it is difficult to label it.

Challenge Problem. What rational numbers correspond to $n = 17$, $n = 43$, or $n = 1,001$ in this correspondence? What natural numbers correspond to the rational numbers $r = \frac{4}{7}$ or $r = \frac{18}{11}$ in this correspondence?

Practice Problem 8.6. *Explain why all of $\mathbb{Q}$ (including the negative rational numbers) is a countable set.*

Example. The following is another way to see that $\mathbb{Q}$ is countable. This is an explicit one-to-one correspondence between the positive rational numbers and the natural numbers: Suppose $\frac{a}{b}$ is a positive rational number with $(a, b) = 1$. Use the Fundamental Theorem of Arithmetic to write $a = p_1^{\alpha_1} p_2^{\alpha_2} \ldots p_h^{\alpha_h}$ and $b = q_1^{\beta_1} q_2^{\beta_2} \ldots q_k^{\beta_k}$. Let $f(1)$ correspond to 1, and otherwise define

$$
f(a/b) = \begin{cases}
q_1^{2\beta_1-1} q_2^{2\beta_2-1} \cdots q_k^{2\beta_k-1} & \text{if } a = 1 \text{ and } b \neq 1, \\
p_1^{2\alpha_1} p_2^{2\alpha_2} \cdots p_h^{2\alpha_h} & \text{if } a \neq 1 \text{ and } b = 1, \\
p_1^{2\alpha_1} p_2^{2\alpha_2} \cdots p_h^{2\alpha_h} q_1^{2\beta_1-1} q_2^{2\beta_2-1} \cdots q_k^{2\beta_k-1} & \text{if } a \neq 1 \text{ and } b \neq 1.
\end{cases}
$$

Then the correspondence given by f is one-to-one from the positive rational numbers to the natural numbers. The proof that this is a one-to-one correspondence is outlined in Exercise 8.15.

8.5   Irrational Numbers

The irrational numbers are those that cannot be expressed as the quotient of two integers. Therefore, irrational numbers are those numbers whose decimal expansions do not repeat. It is hard to think about numbers that we cannot write a certain way. So how *do* we write them?

Example. The number $0.101001000100001000001\ldots$ has a pattern that does not repeat, so it is an irrational number.

Many students, when asked about irrational numbers have two favorites that they "know" are irrational: $\sqrt{2}$ and π. How do we know that these are irrational numbers? It is not enough to know that π is "about"

3.14... and it doesn't appear to repeat. We can prove that $\sqrt{2}$ is irrational, but the proof that π is irrational takes a little more machinery than we have developed.

Theorem 8.1. *The number $\sqrt{2}$ is irrational.*

Proof: We do the proof by contradiction and assume that $\sqrt{2}$ is rational. That means there are positive integers a and b, $b \neq 0$, where $\sqrt{2} = \frac{a}{b}$. We can assume $(a, b) = 1$, since otherwise we can cancel common factors to put the fraction $\frac{a}{b}$ in lowest terms. Since $\sqrt{2} = \frac{a}{b}$, we can multiply both sides by b and get the equation

$$b\sqrt{2} = a.$$

Now square both sides of this equation and get

$$2b^2 = a^2. \tag{8.1}$$

Since 2 divides the left-hand side of the equation, it also divides a^2, and therefore must divide a by Theorem 4.5. So there is an integer m satisfying $a = 2m$. Then $a^2 = (2m)^2 = 4m^2$. We substitute the value $4m^2$ for a^2 in Equation (8.1), and we get

$$2b^2 = 4m^2,$$
$$b^2 = 2m^2.$$

This means that 2 divides b^2, and so divides b. But if $2 \mid a$ and $2 \mid b$, that is a contradiction to $(a, b) = 1$. Therefore, $\sqrt{2}$ must be irrational. $\square$

Practice Problem 8.7. *Prove that $\sqrt{3}$ is irrational.*

Now we want to use some of the divisibility theorems to get an even stronger result. First we prove a helping theorem, sometimes called a lemma.

Lemma 8.2. *If x is an irrational number and k is any nonzero integer, then kx is also an irrational number.*

Proof: Suppose x is irrational but kx is rational. Then we can write $kx = \frac{a}{b}$, where a and b are integers, $b \neq 0$. Dividing by k, we get $x = \frac{a}{kb}$, which is a rational number. This is a contradiction, so kx is irrational. $\square$

Theorem 8.3. *Suppose n is a positive integer that is not a perfect square. Then $\sqrt{n}$ is irrational.*

Proof: This proof uses the same idea as the proof that $\sqrt{2}$ is irrational. First, since n is not a perfect square, we can write $n = m^2 \cdot p_1 \cdot p_2 \cdots p_k$, where m is a positive integer and $p_1, p_2, \ldots, p_k$ are distinct primes.

Let $h = p_1 \cdot p_2 \cdots p_k$. To show that n is irrational, it is enough to show that $\sqrt{h}$ is irrational, since $\sqrt{n} = m\sqrt{h}$. This is because, by the lemma, if $\sqrt{h}$ is irrational, so is $m\sqrt{h} = \sqrt{n}$.

We proceed by contradiction. Assume that $\sqrt{h}$ is rational; that is, there are integers a and b with $(a, b) = 1$, and $\sqrt{h} = \frac{a}{b}$. We then have

$$\frac{a}{b} = \sqrt{h}, \text{ or}$$
$$a = b\sqrt{h}.$$

As we did in the proof for $\sqrt{2}$, we square both sides of this equation to get

$$a^2 = b^2 h. \tag{8.2}$$

Since p_1 divides h, p_1 also divides a^2, so p_1 divides a. Write $p_1 r = a$. Then $a^2 = p_1^2 r^2$. Substitute this for a^2 in Equation (8.2). We then get

$$p_1^2 r^2 = b^2 h.$$

Because p_1^2 does not divide h, p_1 must divide b. But $(a, b) = 1$, so this is a contradiction. Therefore $\sqrt{n}$ is irrational. □

8.6   How Many Real Numbers?

Now we would like to count the real numbers. Let's work with a subset of the real numbers, those that are in the interval between 0 and 1. We write this as $(0, 1)$. We show that the set of real numbers in this interval is infinite, and its size is larger than $\aleph_0$.

Every real number in $(0, 1)$ has a decimal expansion of the form $0.x_1 x_2 x_3 x_4 \ldots$, where each of the digits x_i is in the set $\{0, 1, 2, 3, 4, 5, 6, 7, 8, 9\}$. Suppose that the set of these numbers is countably infinite. We could then list these numbers as

$$r_1 = 0.x_{11}x_{12}x_{13}x_{14}x_{15}x_{16}\cdots$$
$$r_2 = 0.x_{21}x_{22}x_{23}x_{24}x_{25}x_{26}\cdots$$
$$r_3 = 0.x_{31}x_{32}x_{33}x_{34}x_{35}x_{36}\cdots$$
$$r_4 = 0.x_{41}x_{42}x_{43}x_{44}x_{45}x_{46}\cdots$$
$$\vdots$$
$$r_n = 0.x_{n1}x_{n2}x_{n3}x_{n4}x_{n5}x_{n6}\cdots$$
$$\vdots$$

and this list would give a one-to-one correspondence between the real numbers in $(0,1)$ and the natural numbers. We don't want any duplicates here. Remember, for example, that we proved $.3\overline{9}$ is the same as $.4\overline{0}$, so we include just one of these on the list.

If all the real numbers in the interval $(0,1)$ were in this list, then the set would be countable. If we can show there exists a real number in $(0,1)$ not on this list, then the set would not be countable, so we would have to produce a real number y in $(0,1)$ that is not in this list.

Because we want y in the interval $(0,1)$, y will have a decimal expansion of the form $0.y_1y_2y_3y_4y_5y_6\cdots$ with each digit y_i in the set $\{0,1,2,3,4,$ $5,6,7,8,9\}$. We first select a digit for y_1 by looking at the first digit of the first number r_1 on the list. If x_{11} is 3, then we will make $y_1 = 2$. If x_{11} is not 3, then we will make $y_1 = 3$. Thus, we know that our new number y is not the same as r_1, since they differ in the first digit.

Now we select a digit for y_2 in the same manner, this time by looking at the second digit of r_2, which is x_{22}. If $x_{22} = 3$, we make $y_2 = 2$; otherwise, we make $y_2 = 3$. We know our new number y is not the same as r_2, since they differ in the second digit.

We keep up this process, choosing the nth digit of y to be different from r_n's nth digit, x_{nn}.

Thus, this new number y is not on the list, which was supposed to be a list of all the numbers in the interval $(0,1)$. It is therefore impossible to list all the numbers in $(0,1)$ in one-to-one correspondence with the natural numbers. This means that the set of real numbers is not countable and that there are more real numbers than natural numbers. We could carry out the above process on any list proposed to be a list of all real numbers. This means that there cannot be a one-to-one correspondence between the real numbers and the natural numbers.

Since the set of real numbers $\mathbb{R}$ is larger than $\mathbb{N}$, we need another symbol to describe its size.

Definition. The set of real numbers is *uncountably infinite*. The symbol $\mathfrak{c}$ stands for the size of the set of real numbers, and we write $\mathfrak{c} > \aleph_0$.

Practice Problem Solutions and Hints

8.1. $\frac{4}{13} = 0.\overline{307692}$.

8.2.
  (a) $x = \frac{56}{99}$.
  (b) $x = \frac{14,389}{99,000}$.

8.3. Another correspondence would be: $1 \to z, 2 \to y, 3 \to x$, etc. There are many other pairings.

8.4. You could use the correspondence: $1 \to 1, 2 \to 3, 3 \to 5$, or more generally, $n \to 2n - 1$.

8.5. A possible correspondence is $0 \to 1, 1 \to 2, -1 \to 3, 2 \to 4, -2 \to 5, 3 \to 6, -3 \to 7$, or, in general, $0 \to 1$ and for each positive integer n, $n \to 2n$ and $-n \to 2n + 1$.

8.6. Use the natural numbers to count positive rational numbers, the negative integers to count negative rational numbers, and 0 to count 0. We know $\mathbb{Z}$ is countable, so $\mathbb{Q}$ is too.

8.7. Say $\sqrt{3} = \frac{m}{n}$, $n \neq 0$ and $(m, n) = 1$. Squaring both sides, we get $3 = \frac{m^2}{n^2}$, or $3n^2 = m^2$. Since $3 \mid m^2$, we have $3 \mid m$ by Theorem 4.6. So $m = 3k$ for some integer k and $m^2 = 9k^2$. By substitution, $3n^2 = 9k^2$ or $n^2 = 3k^2$. We have $3 \mid n^2$, so $3 \mid n$. This is a contradiction since $(m, n) = 1$. So $\sqrt{3}$ is irrational.

Exercises

8.1. Show that the following sets are infinite.

  (a) The set of all integers that are palindromes (that is, they are the same number when read left to right or right to left).

(b) The set of all integers that have a 3 in their base 10 representation.

(c) The set of all integers whose only digits are 1 and 7.

(d) The set of all possible words of any length, made up of any letters of the English alphabet.

8.2. Convert the following fractions into decimals. If the decimal does not terminate, find its period.

(a) $\frac{13}{8}$.

(b) $\frac{1}{75}$.

(c) $\frac{137}{1,250}$.

(d) $\frac{15}{14}$.

(e) $\frac{1}{17}$.

8.3. Convert the following repeating decimals into fractions:

(a) $0.\overline{3967}$.

(b) $0.0\overline{25641}$.

(c) $0.034\overline{59101}$.

(d) $12.14\overline{1214}$.

8.4. Tell whether the result of the following operations on the given type of number must be rational, must be irrational, or could be either.

(a) The sum of two rational numbers.

(b) The sum of two irrational numbers.

(c) The sum of a rational number and an irrational number.

(d) The product of two rational numbers.

(e) The product of two irrational numbers.

(f) The product of a rational number and an irrational number.

8.5. Suppose x is a real number and assume that x^n is irrational for some positive integer n. Prove that x must be irrational as well.

8.6. Let $x = \sqrt{2} + \sqrt{3}$. Use the result of Exercise 8.5 to show that x is irrational.

8.7. Let m and n be two positive integers. When is $\sqrt{m} + \sqrt{n}$ rational?

8.8. Prove that $\sqrt[3]{2}$ is irrational.

8.9. Suppose m and n are positive integers and $(m, n) = 1$. Show that the decimal representation of $\frac{m}{n}$ terminates if and only if there is some positive integer r such that $10^r(\frac{m}{n})$ is an integer. Deduce that a fraction in lowest terms yields a terminating decimal if and only if 2 and 5 are the only possible prime divisors of its denominator.

8.10. Suppose $m > 0$ and $n > 1$ are integers, with $(m, n) = 1$ and $(10, n) = 1$. Let α be the period of $\frac{m}{n}$.

(a) Explain why α is the smallest positive integer for which $10^\alpha(\frac{m}{n}) - \frac{m}{n} = (10^\alpha - 1)\frac{m}{n}$ is a terminating decimal.

(b) Show that α is the smallest positive integer such that n divides $10^\alpha - 1$.

(c) Show that in the group $\mathbf{U(n)}$ (defined in Section 6.2), the order of 10 is α.

8.11. Find the prime factorization of $10^r - 1$ for $r = 1, 2, 3, 4, 5,$ and 6.

8.12. Use the results of Exercise 8.10(b) to find all primes p such that the following are true.

(a) $\frac{1}{p}$ has period 1.

(b) $\frac{1}{p}$ has period 2.

(c) $\frac{1}{p}$ has period 3.

(d) $\frac{1}{p}$ has period 4.

(e) $\frac{1}{p}$ has period 5.

(f) $\frac{1}{p}$ has period 6.

8.13. Prove that there are only finitely many primes p for which $\frac{1}{p}$ has a given period α.

8.14. Suppose m and n are positive integers with $(m, n) = 1$. Assume also that $(m, 10) = (n, 10) = 1$. Suppose the period of $\frac{1}{m}$ is α_m and the period of $\frac{1}{n}$ is α_n. Prove that the period of $\frac{1}{mn}$ is the LCM of α_m and α_n.

8.15. Prove the correspondence given by f in Section 8.4 is one-to-one by answering the following questions.

(a) Why is $f(a/b)$ in $\mathbb{N}$?

(b) What is $f(2/3)$? What is $f(1/2)$?

(c) Which rational number corresponds to the natural number 72 in this correspondence? Which rational number corresponds to the natural number 3?

(d) Show that, given a natural number k, you can find a unique rational number a/b so that $f(a/b) = k$ by writing k as a product of primes that groups odd powers and even powers separately, and showing how to get a and b from this grouping.

8.16. This exercise creates a new set of numbers, the *complex numbers*, which extend the real numbers $\mathbb{R}$. The real number system does not allow square roots of negative numbers. Mathematicians have created a new number system in which $\sqrt{-1}$ makes sense. Let $\mathbb{C} = \{(a, b)$ such that a and b are in $\mathbb{R}\}$. Define addition and multiplication in $\mathbb{C}$ by $(a, b) + (c, d) = (a + b, c + d)$, $(a, b) \cdot (c, d) = (ac - bd, ad + bc)$. Exercise 1.25 can be extended to $\mathbb{C}$ to show that these operations make $\mathbb{C}$ a commutative ring with additive identity $(0, 0)$ and multiplicative identity $(1, 0)$.

(a) Suppose $(a, b) \neq (0, 0)$. Show that $(a/(a^2 + b^2), -b/(a^2 + b^2))$ is a multiplicative inverse of (a, b). Conclude that $\mathbb{C}$ is a field.

(b) Show that $(0, 1)^2 = (-1, 0)$ and that $(0, -1)^2 = (-1, 0)$.

(c) The real numbers $\mathbb{R}$ can be identified as a subset of the complex numbers $\mathbb{C}$ by letting the real number a correspond to the complex number $(a, 0)$. Show that under this correspondence the real numbers are a subfield of $\mathbb{C}$. (Subfields, like subgroups, are exactly what you think they are.)

(d) Under the correspondence in part (c), can be written 1 as $(1, 0)$ and -1 as $(-1, 0)$. Set $i = (0, 1)$. Show that $i^2 = -1$ and that $(-i)^2 = -1$.

(e) Show that every complex number can be written in the form $a + bi$, where a and b are real numbers. Compute the sum and product of complex numbers using this notation.

(f) Prove that the complex numbers are uncountable, but the Gaussian integers (see Exercise 1.25) are countable.

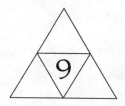

Introduction to
Geometry and Symmetry

In the first half of this book, we worked with numbers and number theory. Another topic that intrigues mathematicians is geometry and symmetry. Our approach to geometry is mainly through the study of symmetry and patterns, which provides a connection to number theory. Students should find this quite different from the Euclidean geometry typically studied in school. Here, we focus first on polygons and their construction with compass and straightedge. We continue with the study of the symmetries of polygons and introduce symmetry groups. These groups will provide the student with a direct link to the groups studied in the number theory sections of the book. Some of the groups will be familiar; in particular, the rotation groups of the polygons will have the same structure as $\mathbb{Z}_n$. However, some of the groups will be nonabelian and provide us with interesting links to new topics such as permutations. Moving into three dimensions, we study polyhedra and their symmetry groups. As we did in number theory, we also study an aspect of geometry that connects us to the infinite, namely, infinitely repeating patterns in two and three dimensions. We end the book with a discussion of two direct connections linking number theory and geometry, that is, Fibonacci numbers and the golden ratio, and the topic of constructible numbers.

Most of us are familiar with different kinds of symmetry and can name everyday objects that have pleasing symmetry. In the Triangle Game in Chapter 0, we used the fact that the equilateral triangle can be flipped or rotated to give six solutions to the game every time we found one set of numbers that worked. These two motions show that an equilateral triangle has two kinds of symmetry.

We can rotate the triangle 120° about a center point and the triangle occupies the same space it did originally. This kind of symmetry is *rota-*

Figure 9.1. Objects with rotational symmetry.

tional symmetry. Many objects have rotational symmetry; the ying-yang symbol and a pinwheel have rotational symmetry (Figure 9.1).

An equilateral triangle also allows us to "flip" it over an altitude, and it occupies the same space it did originally. This is symmetry across a line, or *reflectional symmetry.* A butterfly has reflectional symmetry, and so does a valentine heart. Some of the letters of the alphabet and some numerals we use have symmetry.

Practice Problem 9.1. *Which of the capital letters of the alphabet have rotational or reflectional symmetry? Do any letters have both?*

There is also symmetry in other types of patterns. Patterns that extend indefinitely can have these symmetries as well as other types of symmetries. For example, a tiled floor usually has a pattern of squares that repeats in rows. We call this kind of pattern a *tessellation.* A tessellation uses various shapes, or tiles, to completely cover the plane so there are no gaps or overlaps. Mathematicians usually use the word tessellation to refer to patterns that cover the whole plane; we also use the word tessellation to describe patterns that cover a part of the plane but could be extended to cover the whole plane. Thus, we call the patterns formed by square tiles on a floor, bricks making up a wall, or the honeycomb pattern of beehives tessellations. These patterns also have symmetries. Some of them have rotational symmetry. For example, consider the indefinitely repeating pattern described by the square tiles on a floor. If we rotate the whole pattern 90° around the center of any tile or around any vertex, we will have the same shapes in the same places as the original (see Figure 9.2).

However, for the infinite pattern described by the bricks in Figure 9.3, a 90° rotation won't do because then the bricks are standing on their edges, perpendicular to their original positions.

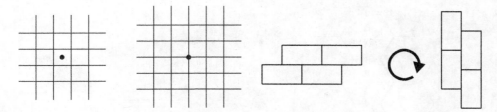

Figure 9.2. Centers of rotation for an infinite grid.

Figure 9.3. Rotation of bricks by 90°.

Practice Problem 9.2. *Is there a point in this brick tessellation for which rotation by 90° is not a symmetry, but rotation by 180° is a symmetry?*

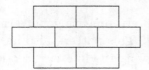

Consider the infinitely repeating pattern described by a honeycomb (Figure 9.4). This pattern uses hexagons as tiles, so a 60° rotation around the center of any tile is a symmetry of this pattern.

Some of these patterns also have reflectional symmetry. If you imagine a line through the middle of one row of square tiles and you flip the whole design over the line, the pattern will occupy its original place (see Figure 9.5). We can do the same thing with the honeycomb, using a vertical line through the center of any hexagon (see Figure 9.6).

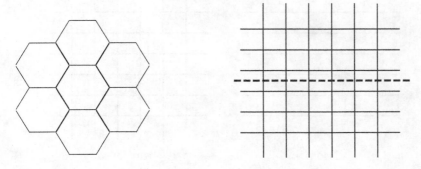

Figure 9.4. Honeycomb pattern.   **Figure 9.5.** Reflection of an infinite grid.

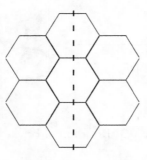

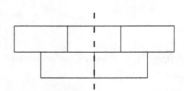

Figure 9.6. Reflection of the honeycomb.   **Figure 9.7.** Reflection of bricks.

Practice Problem 9.3. *What other lines through the patterns of squares or hexagons have reflectional symmetry?*

Consider also the infinitely repeating pattern described by the way that bricks are usually laid. This pattern also has reflectional symmetry, if we are careful about drawing the line of reflection (see Figure 9.7).

These infinitely repeating patterns also have symmetries that the single shapes, such as the equilateral triangle or pinwheel, do not. We can shift the entire pattern one or more tiles to the right or left or up or down, and the pattern will look exactly the same (see Figure 9.8). This kind of symmetry is called *translational symmetry*.

Practice Problem 9.4. *Look around your home or school or neighborhood to find repeating patterns that could be extended infinitely, and describe the symmetries of these infinite patterns.*

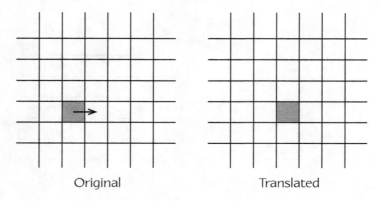

Original   Translated

Figure 9.8. Translation of an infinite grid.

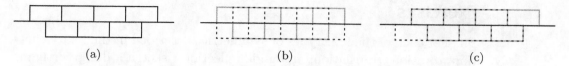

Figure 9.9. Glide reflection.

Another kind of symmetry that infinitely repeating patterns can have is called *glide reflection*. Glide reflection combines two of the symmetries we have seen already, although it is neither of these types of symmetries. An example of a pattern with glide reflection as a symmetry is the infinite pattern described by two rows of a brick wall.

If we reflect Figure 9.9(a) over the line between the two rows, we see that the resulting pattern (Figure 9.9(b)) is not the same as the original. However, if we now translate the pattern half a brick to the left (Figure 9.9(c)), we get the original pattern. Thus, a glide reflection is a combination of a translation and a reflection.

By studying symmetries, we can better understand the geometric figures that are the building blocks of our world. As we saw in number theory, and are seeing now in geometry, there are similarities and differences between finite patterns and infinite patterns. In this part of the book, we study geometry and symmetry as well as their connections to number theory.

Practice Problem Solutions and Hints

9.1. The letters that have horizontal reflectional symmetry are B, C, D, E, H, I, O and X. The letters that have vertical reflectional symmetry are A, H, I, M, O, T, U, V, W, X, and Y. The letters that have 180° rotational symmetry are H, I, N, O, S, X, and Z. The letters with all three are H, I, O and X. (Depending on how you draw it, K can also have horizontal symmetry.)

9.2. The center of the middle brick.

9.3. Diagonal lines through the center of the squares, at 45°, and through the center of the hexagons, at 60°, have reflectional symmetry.

9.4. Look, for example, at floor or wall tile patterns, or at wallpaper or fabric prints.

Exercises

9.1. For each of the four types of symmetries described in the text (reflection, rotation, translation, and glide reflection), sketch a design or name a common object that has that symmetry type and none of the others.

9.2. For each of the following, name the types of symmetries of the object.

  (a) A single snowflake.

  (b) The recycling symbol.

  (c) The Canadian flag.

  (d) Footprints in the sand.

  (e) A row of smiling faces that continues forever in both directions.

  (f) A straight line of railroad track.

9.3. Sketch patterns that have each of the following combinations of symmetry types.

  (a) All four symmetries described in this section.

  (b) Exactly three of the four types of symmetries. Which one can it lack?

  (c) Exactly two of the four types of symmetries.

9.4. How many different lines of reflectional symmetry does each of the following figures have?

  (a) An equilateral triangle.

  (b) A square.

  (c) A rectangle.

  (d) A circle.

9.5. Elementary school children are taught to use the symmetry of a heart to cut out a heart more easily, by folding the paper in half. Using this idea, how many times should you fold a piece of paper to cut out a rhombus? A snowflake? A paper doll? Try these.

9.6. How could you use folding to simplify the process of cutting out the following shape? Try it!

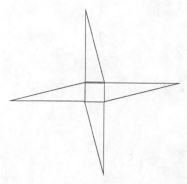

9.7. Suppose an object in the plane has both vertical and horizonal symmetry. Show that it must also have rotational symmetry and that the center of rotation will be the intersection of the two lines of symmetry.

9.8. What kinds of symmetry does a tessellation using parallelograms, as shown below, have?

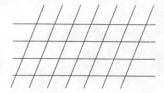

9.9. Can you write the letter Q so that it has some type of symmetry? What kind of symmetry does it have? What about K?

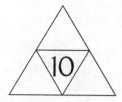

Polygons and
Their Construction

10.1 Polygons and Their Angles

In order to understand some of the deeper mathematical properties of symmetry, we begin in this chapter with a study of some familiar two-dimensional geometric figures. As usual, we want to make a precise definition so that the mathematical description is unambiguous.

Definition. A *polygon* is a finite collection of line segments in a plane, each meeting exactly two others, one at each endpoint, such that no proper subset of the line segments has this same property. The line segments of a polygon are called its sides, or *edges*. We call the endpoints or corners of a polygon its *vertices*. We also require that no three successive vertices be collinear. If a polygon has n sides, we call it an *n-gon*.

Examples. A triangle is a three-sided polygon. There are many terms that describe different kinds of triangles. There are even more names for polygons with four sides, which are called quadrilaterals or 4-gons.

We can continue adding more sides to make different polygons. A 5-gon is more commonly called a pentagon, and a 6-gon is a hexagon. These names are from the Latin and Greek languages. Table 10.1 lists the names of polygons for different values of n.

Triangles

We have already seen an *equilateral triangle*, in which all three sides are equal and all three angles are equal (Figure 10.1).

There are several other names for triangles, each describing a different shape. In an *isosceles triangle*, two of the three sides are equal and the angles opposite the equal sides are equal (Figure 10.2).

Figure 10.1. Equilateral triangle.

Figure 10.2. Isosceles triangle.

Number	Name
3	Triangle
4	Quadrilateral
5	Pentagon
6	Hexagon
7	Heptagon
8	Octagon
9	Nonagon
10	Decagon
11	Undecagon
12	Dodecagon
n	n-gon

Table 10.1. Names of polygons.

Figure 10.3. Scalene triangle.

A *scalene triangle* has three unequal sides and therefore three unequal angles (Figure 10.3).

Angles greater than 0° and less than 90° are called *acute angles*. Angles greater than 90° and less than 180° are called *obtuse angles*. A 90° angle is a *right angle*.

A *right triangle* contains a right angle (Figure 10.4) and an *obtuse triangle* contains an obtuse angle (in both cases, only one such angle is possible). A triangle is called an *acute triangle* if all three angles are acute.

Since the names of triangles all involve the sizes of the angles, we want to know how many degrees there are in all three angles of a triangle.

Figure 10.4. Right triangle.

Fact. The sum of the degrees in the interior angles of a triangle is 180°.

The simple fact above is a consequence of one of the axioms of Euclidean geometry (see the Appendix). We use this result to establish other facts about polygons.

Practice Problem 10.1. *Determine the relationships among the different types of triangles. For example, can a right triangle be an isosceles triangle? Can a scalene triangle be a right triangle?*

Quadrilaterals

Figure 10.5. Square.

A *square* is a quadrilateral with four right angles and four equal sides (Figure 10.5).

A *rectangle* is a quadrilateral with four right angles (Figure 10.6).

A *rhombus* is a "slanted square" in which all four sides are equal (Figure 10.7).

A *parallelogram* is a quadrilateral with opposite sides parallel and of equal length (Figure 10.8).

A *trapezoid* is a quadrilateral with one pair of parallel sides (Figure 10.9).

A *kite* is a quadrilateral in which each side is equal in length to one of its adjacent sides (Figure 10.10).

There is also a quadrilateral that, unlike all the others so far, has an interior angle of more than 180° (Figure 10.11). There is a special name (nonconvex) to distinguish this kind of polygon from the previous ones.

Definition. A *convex* polygon has the property that if you connect any two vertices of the polygon, the line segment connecting the vertices will lie entirely within the polygon. Any polygon that does not have this property is called *nonconvex*.

Practice Problem 10.2. *Determine the relationships among the different types of quadrilaterals. For example, is a rhombus a kite? Is a parallelogram a trapezoid? Is the following figure a trapezoid?*

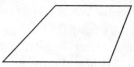

Definition. A *regular polygon* is one in which all the sides are of equal length and all the interior angles are equal.

Example. Regular 3-gons are precisely equilateral triangles. Regular quadrilaterals are squares.

We know that there are 180° in the sum of the interior angles of any triangle. How many degrees are there in the sum of the interior angles of a quadrilateral? The answer is easy to figure out for the rectangle and square, since they each have four right angles, so the sum is 360°. Is this true for the other quadrilaterals? In fact, it is true. If you draw a line connecting opposite vertices in any of the convex quadrilaterals, the result is two triangles. Since each triangle has 180° as its interior angle sum, each contributes 180° to the quadrilateral's interior angle sum of

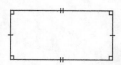

Figure 10.6. Rectangle.

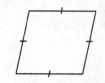

Figure 10.7. Rhombus.

Figure 10.8. Parallelogram.

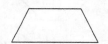

Figure 10.9. Trapezoid.

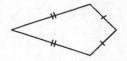

Figure 10.10. Kite.

Figure 10.11. Something different.

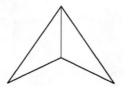

Figure 10.12. Triangulation of a nonconvex quadrilateral.

360°. This same fact is true of nonconvex quadrilaterals; you just have to be careful to connect the right vertices to get the two triangles (see Figure 10.12).

n-gons

Practice Problem 10.3. *Draw a pentagon and make a conjecture about the sum of its interior angles.*

We want to prove our first theorem about polygons.

Theorem 10.1. *The sum of the interior angles of a convex n-gon is $180(n-2)°$.*

Figure 10.13. Our favorite convex *n*-gon.

Proof: Draw any convex *n*-gon. We demonstrate the proof here by using an octagon (Figure 10.13), our favorite convex *n*-gon.

We choose any vertex and draw lines that connect that vertex to the $n-3$ nonadjacent vertices. When we draw the first line, we break the interior of the polygon into two regions. Each additional line adds one more interior region. So, after $n-3$ steps, we have $n-2$ regions, each of which is a triangle (see Figure 10.14).

Figure 10.14. Triangulated octagon.

Since each of the $n-2$ triangles contributes 180° to the sum, the total number of degrees in all these triangles is $180(n-2)°$.   □

Figure 10.15. Triangulated octagon (second approach).

Here is a different approach to the proof of the theorem. Again, draw any convex *n*-gon; we still use the octagon. Put a point in the interior of the *n*-gon and connect this new point to all *n* vertices of the polygon (Figure 10.15). We can do this because the polygon is convex (see Exercise 10.4).

This creates *n* triangles. Their interior angle sum is $180n°$. However, they all have an angle in the center of the polygon. The sum of these should not count toward the sum of the polygon's interior angles, so we need to subtract all those central angles. This is easy since they completely fill a circle, and we know that a circle has $360°$. So the sum of the polygon's interior angles is $180n° - 360° = 180(n-2)°$.

This same conclusion is also true of nonconvex polygons. The sum of the interior angles of a nonconvex n-gon will still be $180(n-2)°$. This is much harder to prove.

Challenge Problem. Try to prove that the sum of the interior angles of a nonconvex n-gon is also $180(n-2)°$. First, try several examples by dividing the interior of the polygon into triangles using only the vertices of the polygon as vertices of your triangles.

We have one more result from Theorem 10.1.

Corollary 10.2. *The number of degrees in each interior angle of a regular n-gon is $\frac{180(n-2)}{n}°$.*

Proof: We know the sum of the degrees in all the interior angles is $180(n-2)°$. Since a regular n-gon has n equal interior angles, the number of degrees in each angle is $\frac{180(n-2)}{n}°$, as desired. □

Table 10.2 lists the degree measures for some regular n-gons. Notice how many degrees are in each angle of a regular 1,000-gon. If you tried to draw one of these, where two edges meet at a vertex it would look almost like a straight line, and the 1,000-gon would look almost like a circle.

In this section, we defined polygons, both convex and nonconvex, and determined the number of degrees in the sum of the interior angles of

Number of sides, n	Sum of interior angles in degrees, $180(n-2)$	Number of degrees in each angle, $180(n-2)/n$
3	180	60
4	360	90
5	540	108
6	720	120
7	900	$128\frac{4}{7}$
8	1,080	135
1,000	179,640	$179\frac{16}{25}$

Table 10.2. Degree measures of various regular n-gons.

n-gons. In the next section, we see how to actually construct many of these polygons. We also learn that some of them cannot be constructed! This is important, since these are the "tiles" we'll want to use to study tessellations in Chapter 15.

10.2   Constructions

Constructions of geometric figures date to before the time of Euclid, who first wrote down a complete set of rules for them. His tools were the straightedge and compass. Although today we use calibrated instruments, such as rulers and protractors, to measure lengths and angles, Euclid's rules do not allow these tools. These rules appear in Euclid's *Elements*, written about 300 B.C. Many complicated geometric constructions are possible, but we use only a few here to create the kinds of shapes, lines, and angles that we need in order to work with symmetry.

The easiest construction with a compass alone is the circle. Whenever we draw a circle, we need to know where the center of it is. So, to draw a circle, mark a point that will be the center. Open the compass to a fixed length, put the point of the compass on the center, and draw a circular arc beginning and ending at the same point (Figure 10.16).

One of the first polygons we can construct is an equilateral triangle. Begin by making a circle. Keep the compass open to the size of the

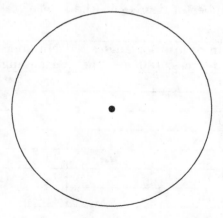

Figure 10.16. A circle drawn by a compass.

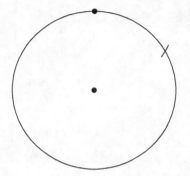 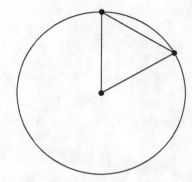

Figure 10.17. Constructing an equilateral triangle.

Figure 10.18. An equilateral triangle.

radius of the circle.[1] Mark any point on the circle, and put the point of the compass there. Use the compass to make an arc intersecting the circle (Figure 10.17).

If we take a straightedge and connect the two points on the circle, and then connect them each to the center, we see that we have an equilateral triangle (Figure 10.18). The measure of each of the interior angles in this triangle is 60°.

Continue this procedure to construct a regular hexagon. Move the point of the compass to the last vertex on the circle, and repeat the procedure. This will make another equilateral triangle (Figure 10.19).

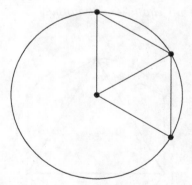

Figure 10.19. Constructing a regular hexagon.

---

[1]Euclid's original rules did not allow keeping the compass to a fixed radius when picking up the compass.

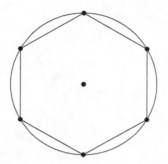

Figure 10.20. A regular hexagon.

We can do this six times in all. The resulting chords will make a regular hexagon, with edges the same lengths as the radius of the circle (Figure 10.20).

Practice Problem 10.4. *Using the measures of the angles, prove that this is a regular hexagon.*

Because the six points are equally spaced around the circle, if we connect every other point, the resulting chords will make a different equilateral triangle (Figure 10.21).

Practice Problem 10.5. *Use the congruence of triangles* AXY, BXZ, *and* CYZ *in Figure 10.21 (see SAS (Side-Angle-Side) in the Appendix) to prove that* XYZ *is an equilateral triangle.*

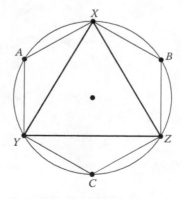

Figure 10.21. Another equilateral triangle.

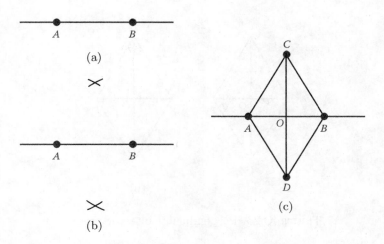

Figure 10.22. Perpendicular bisector construction.

The only other construction we need to learn to make our polygons is the perpendicular bisector. This time, start by drawing a line with the straightedge and marking off a length. Call the endpoints of the line segment A and B (Figure 10.22(a)).

Now open the compass to the width of $\overline{AB}$. Put the point of the compass on A and make an arc above and below the line. Do the same at B (Figure 10.22(b)).

Now use the straightedge to connect the two points where the arcs intersect, calling them C and D (Figure 10.22(c)).

Call O the point of intersection of segment AB and segment CD. How do we know that CD bisects and is perpendicular to AB? Here we use some triangle congruence geometry (see Figure 10.23 and the Appendix).

Triangles ABC and ABD are equilateral and congruent. Therefore, angles CAB, DAB, DBA, and ABC are each 60°, $AC = BC$, and $AD = BD$ (Figure 10.23(a)).

Now we focus our attention on the two triangles on the sides of the figure, triangles CAD and CBD (Figure 10.23(b)). They are congruent because of SAS (Side-Angle-Side; see the Appendix). In particular, this means that angles ACO and BCO at the top of the figure are equal, so triangles ACO and BCO are congruent by ASA (Angle-Side-Angle; see the Appendix). Therefore $AO = BO$, so AB is bisected. Finally, since $\angle AOC = \angle BOC$, they must both be 90°, so we have constructed a perpendicular bisector.

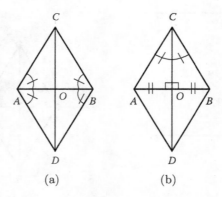

Figure 10.23. Perpendicular bisector proof.

Observe also that this is a way to construct an equilateral triangle without using a circle.

Now, by using our construction of a perpendicular bisector, we can construct a square by starting with a circle. First draw a circle and use the straightedge to draw in a diameter (Figure 10.24(a)).

Then bisect the diameter and connect the points where these two resulting diameters intersect the circle. This makes a square inside the circle.

Practice Problem 10.6. *Use congruences (see the Appendix) to prove Figure 10.24 is a square.*

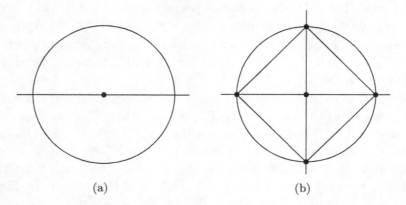

Figure 10.24. Square construction.

Practice Problem 10.7. *Bisect the sides of the square we have constructed to make a regular octagon.*

Once we have made the equilateral triangle and the square, we can use the method of bisecting a line segment to construct any n-gon where n is any power of 2, or 3 times any power of 2.

Practice Problem Solutions and Hints

10.1. A right triangle may be isosceles if the two legs have equal length. A right triangle may be scalene if the two legs have different lengths. A right triangle cannot be equilateral. No scalene triangle is isosceles.

10.2. A square is a rectangle, a rhombus, a parallelogram, a kite, and a trapezoid. A rhombus is a parallelogram, a trapezoid, and a kite. Any parallelogram is a trapezoid.

The illustrated figure is a trapezoid.

10.3. The sum of the degrees in the interior angles of a convex pentagon is 540°.

10.4. Each interior angle of this hexagon measures 120°, and all the side lengths are equal to the radius of the circle. Therefore, this is a regular hexagon.

10.5. Triangles AXY and BXZ are isosceles, so angles AXY and BXZ each measure 30°. Then angle YXZ must be 60°. A similar argument shows angles XYZ and XZY also measure 60°. Congruence shows that sides XY, YZ, and XZ are equal.

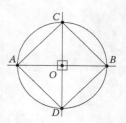

10.6. The midpoint of the diameter is the center of the circle, so $AO = CO = BO = DO$, and triangles AOC, BOC, BOD, and AOD are isosceles. Then because CD is perpendicular to AB, these triangles are right triangles. Therefore, they are all congruent. So $AC = BC = DB = AD$, and all the angles of the inscribed figure are congruent and measure 90°.

10.7. Do it!

Exercises

10.1. The definition of a regular polygon specifies that the figure has both equal angles and equal side lengths. You might wonder whether one of these conditions alone describes the same set of figures. To find out, try to sketch a 3-gon, 4-gon, and 5-gon that have equal sides but different angles. Then try to sketch a 3-gon, 4-gon, and 5-gon that have equal angles but different side lengths. Would either condition alone have implied the other? Does your answer depend on the number of sides in the polygon?

10.2. Suppose that you want to fit together one or more kinds of regular polygons around a single point without gaps or overlaps. You could, for example, put two squares and three equilateral triangles together at a point. Which other sets of regular polygons can fit together around a point?

10.3. Using a calculator or your knowledge of trigonometry, if you have had such a course, complete the following table for the side lengths and perimeters of different regular n-gons inscribed in a unit circle, that is, a circle with radius 1. You may use information from the text as well as your own calculations.

Number of sides, n	Side length of inscribed regular n-gon	Total perimeter of regular n-gon
3		
4		
5		
6		
10		
12		

To what value are the perimeters getting closer as n increases?

10.4. Here is another definition of the word *convex*: A convex set is one in which all segments connecting two points of the set lie entirely in the set. Compare this definition with the definition of a convex polygon given in the text. Are they the same? That is, is any polygon that is convex according to one definition also convex according to the other one? Prove your answer.

10.5. In the process of determining the measure of angles in regular polygons, we used Theorem 10.1, which we know is true only for convex polygons. In order to use this theorem, we need to know that regular polygons are always convex. Explain why this is the case.

10.6. As stated in Problem 10.5, regular polygons are convex. Does having congruent sides imply that a polygon is convex? Find a polygon that has all its sides congruent but is not convex.

10.7. Does having equal angles imply that a polygon is convex? Explain why a polygon that has equal angles must be convex.

10.8. Describe a method for bisecting any given angle using only a straightedge and compass. You need to produce a point whose distance from each ray of the angle is the same. Be sure to explain why your method yields the angle bisector.

10.9. Describe a method for constructing a segment perpendicular to a given line through a given point on the line using only a straightedge and compass. Be sure to explain why your method results in a perpendicular line.

10.10. Construct parallel lines by using the following instructions. Begin with a line l and a point P not on l.
  (a) Construct a perpendicular from P to l and extend it to a line m.
  (b) Construct a second perpendicular line to m through P.

10.11. The answers to parts (a)–(c) below answer the overall question, "What fractions are constructible?"
  (a) Given a segment that is defined to have length 1, can you construct a segment having length $\frac{1}{2}$ with only a straightedge and compass?
  (b) Can you construct a segment with length $\frac{1}{3}$? **Hint:** Mark off a line segment of length 3 on a line. Construct a perpendicular of length 1 at one end of this line segment. Then use similar triangles.
  (c) For which integer values of n can you construct $1/n$?

10.12. The answers to parts (a)–(c) below answer the overall question, "What square roots are constructible?"
  (a) Given a segment that is defined to have length 1, can you construct a segment having length $\sqrt{2}$ with only a straightedge and compass?
  (b) Given a segment having length $\sqrt{n}$, can you construct a segment having length $\sqrt{n+1}$?
  (c) What other kinds of square roots can you construct? For example, can you construct the square root of $\frac{3}{4}$?

10.13. The answers to parts (a)–(c) below answer the overall question, "What angles are constructible?"

(a) Given a 72° angle and a 45° angle, what is the smallest angle with integer degree measure that you can construct from these two? How would you do it?

(b) Given a 60° and a 45° angle, what is the smallest angle with integer degree measure that you can construct from these two? How would you do it?

(c) Using angles that measure $x°$ and $y°$, what is the smallest angle with integer degree measure that you can construct?

10.14. As mentioned in the text, Euclid did not allow picking up the compass and preserving the distance between the point and tip. So, by Euclid's conditions, you could mark out equal distances measured from a particular point, but you could not mark out the same distance from two different points. Does the method of constructing an equilateral triangle given in the text still work under Euclid's conditions? If not, can you modify the construction to work under his conditions? What about the construction of the square?

10.15. Parts (a)–(c) below present definitions of the word *polygon* found in some high school geometry textbooks. Compare each with the definition of a polygon in this chapter. Are there ambiguities in these definitions that do not appear in ours? For example, does any definition allow the pentagram, the square-in-a-square, or the sawtooth figure pictured below to be a polygon?

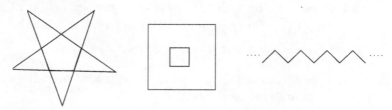

(a) A polygon is a simple closed plane curve formed by joining segments such that no two adjacent segments are collinear.

(b) A polygon is a closed geometric figure in a plane formed by connecting line segments endpoint to endpoint, with each segment intersecting exactly two others.

(c) A polygon is the union of three or more segments in the same plane such that each segment intersects exactly two others, one at each of its endpoints.

10.16. If you extend one edge of a convex polygon from a vertex, you have an *exterior angle* at that vertex. The exterior angle is supplementary to the interior angle at that vertex. Prove that the sum of the exterior angles of the polygon so formed is 360°.

Exterior angle

10.17. We can get the same exterior angle sum for a nonconvex polygon, but we have to define "exterior angle measure" carefully. Here's a way to describe the measure of an exterior angle. Stand on any edge of the polygon with the interior of the polygon on your left and the exterior on your right. Walk toward a vertex. At the vertex, turn to the next edge, keeping the interior of the polygon on your left. Your "turn" is the exterior angle, just as we described in Exercise 10.16.

(a) In the nonconvex quadrilateral shown here, how many turns are clockwise and how many are counterclockwise in the four exterior angles?

(b) If you consider a counterclockwise turn a positive turn and a clockwise turn a negative turn, show that you would turn a total of 360° counterclockwise as you walk along the nonconvex quadrilateral shown here.

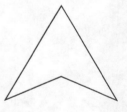

(c) Show that the sum of the degrees in the exterior angles of any n-gon (convex or nonconvex) is 360°.

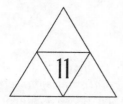

Symmetry Groups

Now we return to the topic of symmetry. Remember that infinitely repeating patterns can have translations and glide reflections, but a bounded plane figure can only have one of two types of symmetry, reflection or rotation.

A *reflection over a line l* moves any point P to the point P' obtained by drawing a perpendicular from P to l and extending it the same distance to the other side of l. If P is on line l, then $P = P'$. Every point on the line of reflection is a fixed point.

A *rotation* about a center point O moves any point P to a point P' the same distance from O as P in the following way. On the circle centered at O with radius OP, start at P and move along the circle counterclockwise to point P' (Figure 11.1). A rotation has the following definitions:

1. A rotation has only one fixed point, the center of rotation.

2. The *angle of rotation*, $\alpha°$, is the angle POP'.

3. The angle of rotation can have any measure from $0°$ to $360°$ counterclockwise. We could as well have defined a rotation from P to P' by moving in a clockwise direction. In this case, the angle of rotation would be $(360 - \alpha)°$. The usual convention in mathematics books is to use rotation in a counterclockwise direction.

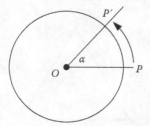

Figure 11.1. Rotation.

A figure has *reflectional symmetry* if it looks the same after you reflect it over a line. A figure has *rotational symmetry* if it looks the same after you rotate it a certain number of degrees around a point.

Let's investigate these two types of symmetry using some letters of the alphabet.

Examples. The letter A has reflectional symmetry through a vertical line. We call this *vertical symmetry*. We can think of this symmetry as a function that flips, or reflects, the letter A over (through) the line of symmetry so that it looks exactly the same.

The letter B has *horizontal* symmetry. We can think of this symmetry as a function that flips, or reflects, the letter B over its line of symmetry so that it looks the same.

The letter Z has *rotational* symmetry. If you imagine a dot right in the center of the letter, and then rotate the letter Z by 180°, it looks the same.

You may have noticed that some letters have lots of symmetry. The letter H has both horizontal and vertical reflectional symmetry and rotational symmetry.

11.1 Symmetric Motions of the Triangle

Here, we want to explore the symmetries of plane figures. We begin by looking at an equilateral triangle. To facilitate the work in this section, use a straightedge and compass to construct an equilateral triangle (Figure 11.2), and then cut it out.

We have seen in the Triangle Game (Chapter 0) that this triangle has both reflectional and rotational symmetry. Now we want to formalize these symmetric motions by labeling them. So we begin by numbering the

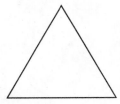

Figure 11.2. An equilateral triangle.

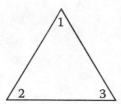

Figure 11.3. Triangle in the initial position.

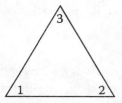

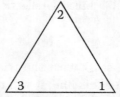

Figure 11.4. Triangle rotated 120° counterclockwise.

Figure 11.5. Triangle rotated 240° counterclockwise.

vertices of the triangle. It is important to number the vertices on the back of the cut-out triangle with the same numbers that are on the front. We begin by placing the number 1 at the top, and continue counterclockwise, with the number 2 at the bottom left-hand vertex and the number 3 at the bottom right.

The triangle labeled this way will be said to be in *initial* position (Figure 11.3). We know that this triangle has rotational symmetry. If we rotate it 120° either clockwise or counterclockwise, the triangle will take up the same space as it did at the beginning, but the numbers will move. For example, if we rotate it 120° counterclockwise, we will see the number 3 at the top, the number 1 at the bottom left, and the number 2 at the bottom right (Figure 11.4).

Put the triangle back in its initial position and this time rotate it 240° counterclockwise. Where are the numbers now? The number 3 has moved to the bottom left vertex, the number 1 is at the bottom right, and the number 2 is at the top (Figure 11.5).

Practice Problem 11.1. *Draw a diagram indicating where the numbers are if you rotate the triangle 120° clockwise or 240° clockwise.*

The triangle also has reflectional symmetry. The easiest line of reflection to see is the vertical line from the top of the triangle through the midpoint of the bottom edge (Figure 11.6).

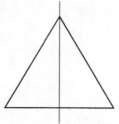

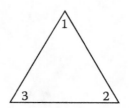

Figure 11.6. Line of reflectional sym-   **Figure 11.7.** Triangle reflected over ver-
metry.   tical line.

If we start with the triangle in its initial position, and pick it up and
flip it over this line, it will occupy the same space it did originally, but the
numbers will move. The number 1 is still at the top, but the two bottom
numbers have traded places. The 2 is at the bottom right and the 3 is at
the bottom left (Figure 11.7).

This triangle has two more lines of reflectional symmetry. One is
through the bottom left vertex and the midpoint of the opposite side
(Figure 11.8).

If we start from the initial position and flip over this line, we see
that the number 2 stays put, but the numbers 1 and 3 trade places (Figure 11.9).

Similarly, if we start from the initial position and use a line through
the bottom right vertex and the midpoint of the opposite side, the 3 stays
put, but the 1 and 2 trade places (Figure 11.10).

What happens to the triangle if we do two of these symmetric motions
in a row? For example, start with the triangle in the initial position. What
if we first rotate it 120° counterclockwise, then flip it over a vertical line

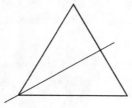

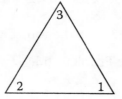

Figure 11.8. Second line of reflectional   **Figure 11.9.** Triangle reflected over sec-
symmetry.   ond line.

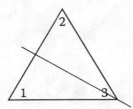

Figure 11.10. Triangle reflected over third line.

through the top vertex? Where will the numbers end up? The first step moves the triangle to the position in Figure 11.11, with the 3 at the top.

If we flip this triangle over a vertical line through the top vertex, the 3 will stay put and the 1 and 2 will change places (Figure 11.12).

A simpler way to reach this arrangement of numbers from the triangle in the initial position is just to reflect it over the line through the bottom left vertex. Compare Figures 11.12 and 11.9.

Now let's describe this situation using the language of multiplication. This is not ordinary multiplication, as occurs in the number systems we have seen. We use the language of multiplication here because we will soon see that this operation has a lot in common with ordinary multiplication. We could say that rotation counterclockwise of 120° times a reflection across the line through the top vertex and bottom midpoint equals a reflection across the line through the left vertex and the midpoint of the opposite side. This is complicated to say in words, but we can make it easier by using symbols for these symmetric motions and for this product. Let's call R_{120} the counterclockwise rotation through 120°; then the counterclockwise rotation through 240° is R_{240}. The flip across the line through the top vertex will be F_{top}. We label the other two flips F_{left} and F_{right} to denote the flips across the lines through the bottom left

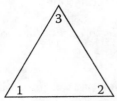

Figure 11.11. Triangle after rotation through 120° counterclockwise.

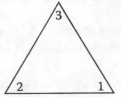

Figure 11.12. Triangle after rotation and flip.

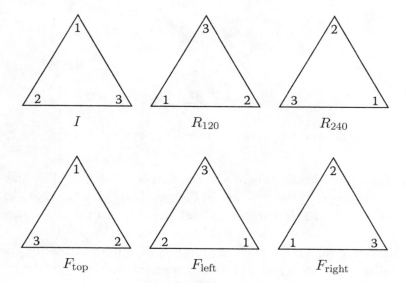

Figure 11.13. The six symmetries of the triangle.

and right vertices, respectively, to the midpoints of the opposite sides. We label the triangle in initial position with an I. The six labeled triangles that result from the six different symmetric motions of the triangle are shown in Figure 11.13.

Next, we write a multiplication table for these motions. We already know that if we rotate by 120° counterclockwise and then flip over the line through the top vertex, we have the triangle that results from the flip over the line through the bottom left vertex. We write this multiplication as $F_{\text{top}} \cdot R_{120} = F_{\text{left}}$. We write the motions in this order because it is the notation that we used for composition of functions in Chapter 1. To write the multiplication table, it doesn't matter which way you choose to write these motions, but you must be consistent every time you write a product. So we read products of motions from right to left.

Practice Problem 11.2. *Use your cutout triangle to check the above statements about combining motions.*

Example. If we want to flip over the bottom left vertex and follow that with a rotation of 240° counterclockwise, we write $R_{240} \cdot F_{\text{left}}$. Use your triangle to check that the result of these two motions will be the same as the symmetry motion F_{right}.

·	I	R_{120}	R_{240}	F_{top}	F_{left}	F_{right}
I	I	R_{120}	R_{240}	F_{top}	F_{left}	F_{right}
R_{120}	R_{120}					
R_{240}	R_{240}			F_{right}		
F_{top}	F_{top}	F_{left}				
F_{left}	F_{left}					R_{120}
F_{right}	F_{right}					

Table 11.1. Symmetric motions of the equilateral triangle: phase 1.

Practice Problem 11.3. *Use your labeled triangle to convince yourself that* $F_{left} \cdot F_{right} = R_{120}$.

Before we proceed, we need to agree that the symbol I means that we should leave the triangle in place. Then, for example, $R_{120} \cdot I$ is the same as R_{120}. Also, $I \cdot R_{120}$ is the same as R_{120}. So for any symmetric motion M of the triangle $M \cdot I = M$ and $I \cdot M = M$.

We can begin to write a table for the multiplication of any of these symmetric motions. To write the table, we first have to decide where to put the product of two motions A and B. If we write $A \cdot B$, we mean that we do B first and A second, and we put the product $A \cdot B$ in A's row and B's column.

The products that we have already calculated are shown as phase 1 in Table 11.1.

We can fill in some other blanks rather easily. For example, what happens if you do the same flip twice in a row? What is $F_{\text{top}} \cdot F_{\text{top}}$? We just have the triangle back in the initial position. So $F_{\text{top}} \cdot F_{\text{top}} = I$. This is true for all three of the flips.

What about the rotations? It is clear that doing R_{120} twice results in R_{240}.

Practice Problem 11.4. *What is the result of doing* R_{240} *twice in a row?*

Now we can fill in more of the table. Compute the product $F_{\text{left}} \cdot R_{240}$. The result is the triangle labeled F_{top}. We can see from Table 11.2 that this is not the same as the product of these two motions in reverse order. So we have found an equation that indicates this multiplication is not commutative.

Practice Problem 11.5. *Use the triangle you have cut out to fill in the remaining entries in Table 11.2.*

·	I	R_{120}	R_{240}	F_{top}	F_{left}	F_{right}
I	I	R_{120}	R_{240}	F_{top}	F_{left}	F_{right}
R_{120}	R_{120}	R_{240}	I			
R_{240}	R_{240}	I	R_{120}		F_{right}	
F_{top}	F_{top}	F_{left}		I		
F_{left}	F_{left}	F_{right}	F_{top}		I	R_{120}
F_{right}	F_{right}					I

Table 11.2. Symmetric motions of the equilateral triangle: phase 2.

Table 11.3 shows the completed table of symmetric motions of the equilateral triangle. We want to make observations about Table 11.3 and the multiplication system that it reveals. We compare them to the Rules of Arithmetic that were in the number theory section.

1. The symmetric motions of the equilateral triangle are closed under the binary operation of multiplication. That is, each pairing of two symmetric motions yields a symmetric motion, and only the six we have described are in Table 11.3.

2. The multiplication is associative. This results from the fact that multiplication of symmetry motions is just composition of functions, which has the associative property.

3. The multiplication is definitely not commutative. Although some pairs of motions commute, such as R_{120} and R_{240}, some do not. For example, $F_{\text{left}} \cdot R_{120} = F_{\text{right}}$, but $R_{120} \cdot F_{\text{left}} = F_{\text{top}}$. So this operation is not commutative.

4. There is an identity. As we have observed, the motion I has the property that for any motion M of the triangle, $I \cdot M = M$ and $M \cdot I = M$.

·	I	R_{120}	R_{240}	F_{top}	F_{left}	F_{right}
I	I	R_{120}	R_{240}	F_{top}	F_{left}	F_{right}
R_{120}	R_{120}	R_{240}	I	F_{right}	F_{top}	F_{left}
R_{240}	R_{240}	I	R_{120}	F_{left}	F_{right}	F_{top}
F_{top}	F_{top}	F_{left}	F_{right}	I	R_{120}	R_{240}
F_{left}	F_{left}	F_{right}	F_{top}	R_{240}	I	R_{120}
F_{right}	F_{right}	F_{top}	F_{left}	R_{120}	R_{240}	I

Table 11.3. Symmetric motions of the equilateral triangle: completed table.

5. The motions all have multiplicative inverses. We can find them on Table 11.3 by looking for the I in each row or column. For example, the inverse of R_{120} is R_{240}, but F_{top} is its own inverse, since $F_{\text{top}} \cdot F_{\text{top}} = I$.

Practice Problem 11.6. *Identify each motion's multiplicative inverse.*

At this point, we can see that the multiplication of symmetric motions of the triangle obeys some of the Rules of Arithmetic. Axioms M1, M3, and M4 all hold. In fact, this multiplication gives a group. This is our first example of a group that is not commutative.

Definition. We call the group made up of the six symmetric motions of the equilateral triangle the *dihedral group* D_6. This group is a noncommutative group.

We have identified the collection of symmetric motions of an equilateral triangle as a group. Is there any connection or relationship between this group and the other groups we discussed in Chapter 6? To begin to answer this question, first notice that there are some parts of the D_6 table (Table 11.3) that occur in blocks. For example, the top three entries in the upper left corner form a three by three square (Table 11.4). This square contains only the three elements I, R_{120}, and R_{240}. If we just focus on this part of the table, we can see that this is a group as well.

Practice Problem 11.7. *Verify that the group axioms hold for the set of motions $\{I, R_{120}, R_{240}\}$ together with the multiplication operation we have been using.*

This set of three motions is a subset of D_6 that forms a group. We call such a set inside a group a *subgroup* (see Exercises 6.32–6.37).

Any one of the flips together with I also forms a subgroup of D_6, as we can see from Table 11.5 using F_{left}. We have met this group before

·	I	R_{120}	R_{240}
I	I	R_{120}	R_{240}
R_{120}	R_{120}	R_{240}	I
R_{240}	R_{240}	I	R_{120}

·	I	F_{left}
I	I	F_{left}
F_{left}	F_{left}	I

Table 11.4. The rotation subgroup of D_6.   **Table 11.5.** F_{left} subgroup of D_6.

·	I	F_{left}
I	I	F_{left}
F_{left}	F_{left}	I

+	0	1
0	0	1
1	1	0

Table 11.6. Comparison of subgroups generated by F_{left} (left) and $\mathbb{Z}_2$ (right).

in another form. The Even-Odd addition table, which we later identified with $\mathbb{Z}_2$, the integers modulo 2 (see Table 6.5), works the same as Table 11.5 for I and F_{left}.

If we look at Table 11.6, the addition table for $\mathbb{Z}_2$ together with Table 11.5, it is easy to see that there is a correspondence between the number 0 and the symmetric motion I, and between the number 1 and the symmetric motion F_{left}.

Practice Problem 11.8. *Identify the numbers 0 and 1 with the appropriate elements in the subgroups made from each of the other flips.*

We see a corresponding structure in Table 11.7, which compares rotations with the group of integers mod 3, $\mathbb{Z}_3$. Here, the correspondence is between the number 0 and I, the number 1 and R_{120}, and the number 2 and R_{240}. In fact, this correspondence is very natural if we think of $R_{120} \cdot R_{120}$ as $(R_{120})^2$. Then $R_{240} = (R_{120})^2$. To simplify the notation further, let's just write R_{120} as R, and R_{240} as R^2. Then the table of the rotations is as shown in Table 11.8.

It is apparent from this notation that this is a cyclic group.

We have said that two groups have the same structure, or equivalently, that their group multiplication tables work in the same way, if we can match up the elements so that the tables are identical. (We may have to rearrange the rows and columns.)

How can we tell when two groups don't have the same structure? We need to show that there is no way to match up the elements. There are

·	I	R_{120}	R_{240}
I	I	R_{120}	R_{240}
R_{120}	R_{120}	R_{240}	I
R_{240}	R_{240}	I	R_{120}

+	0	1	2
0	0	1	2
1	1	2	0
2	2	0	1

Table 11.7. Comparison of subgroups generated by R_{120} (left) and $\mathbb{Z}_3$ (right).

·	I	R	R^2
I	I	R	R^2
R	R	R^2	I
R^2	R^2	I	R

Table 11.8. Table of rotations.

some easier ways of showing that two groups have different structures than checking every possible matching.

First, if two groups have different numbers of elements then they cannot have the same structure. For example, $\mathbb{Z}_3$ and $\mathbb{Z}_2$ cannot have the same structure because they have different numbers of elements. How can we distinguish between groups that do have the same number of elements? For example, both D_6 and $\mathbb{Z}_6$, the group of integers under addition modulo 6, have six elements. Do these groups have the same structure?

Another way to ask the same question is: What properties of a group do not depend upon the names of the elements, but rather depend only on the way that the operation works? One such property is commutativity. Remember that an abelian group is one whose operation is commutative. We check whether a group is abelian by checking whether its group table is symmetric over a diagonal line. This property doesn't change if we rename the elements. Therefore, an abelian group and a nonabelian group always have different structures. Since $\mathbb{Z}_6$ is abelian but D_6 is not, these groups cannot have the same structure. Several other properties also allow us to distinguish between groups with different structures. We study some of them later in this text. See, for example, page 226, where we discuss the structural difference between the cyclic group $\mathbb{Z}_4$ and the Klein-4 group.

Challenge Problem. Cite some other reasons why the two groups D_6 and $\mathbb{Z}_6$ are not structurally the same.

There is a way to write the table of symmetric motions of the triangle by using just a few symbols. We already have seen that if we label R_{120} as R, we can write the rotations as I, R, and R^2. Now let's pick one of the flips, say, F_{top}, and label it F. Then $FR = F_{\text{left}}$ and $FR^2 = F_{\text{right}}$. So we can rewrite the table using just the symbols I, R, and F, and their powers and products. The way to do this is to use the products $RF = F_{\text{right}} = FR^2$ and $R^2F = F_{\text{left}} = FR$ to simplify some of the calculations.

·	I	R	R^2	F	FR	FR^2
I	I	R	R^2	F	FR	FR^2
R	R	R^2	I	FR^2	F	FR
R^2	R^2	I	R	FR	FR^2	F
F	F	FR	FR^2	I	R^2	R
FR	FR	FR^2	F	R	I	R^2
FR^2	FR^2	F	FR	R^2	R	I

Table 11.9. D_6 as products of F and R.

Example. What would be the entry in the table that corresponds to $R^2 \cdot FR^2$? We use associativity to rewrite this as $(R^2 \cdot F) \cdot R^2$. Then, since $R^2 \cdot F = FR$, our equation simplifies to $F \cdot R^3 = F \cdot I = F$.

Table 11.9 shows the table of symmetric motions using only F, R, and I. Here we picked F_{top} as our representative flip to fill in the table. We could just as well have chosen either of the other flips and written Table 11.9 using one of them, together with the rotation R. For any flip we choose to call F, we know that the three products F, FR, and FR^2 will be different. This is because the law of cancelation holds in the group. If we had, say, $FR = FR^2$, canceling the F's would yield $R = R^2$, which is false. Thus, it is possible to write this dihedral group table using only two group elements and their powers.

Practice Problem 11.9. *Write the group table using either F_{right} or F_{left} as F. Show that these are the same group as the group in Table 11.9 after rearranging the rows and columns.*

The equilateral triangle has both reflectional and rotational symmetry. We know other objects that do not have both of these kinds of symmetry. The letter A, for example, has only vertical symmetry. The only symmetric motions for the letter A are I, which leaves the A in its initial position, and a vertical flip, call it V. Table 11.10 shows the symmetric motions of the letter A.

Practice Problem 11.10. *Write the table for the symmetric motions of the letter B, which has only horizontal symmetry.*

The letter Z has only rotational symmetry. If we label this motion R, and call the motion that leaves the Z in its initial position I, then Table 11.11 shows the symmetric motions of the letter Z.

·	I	V
I	I	V
V	V	I

·	I	R
I	I	R
R	R	I

Table 11.10. Symmetric motions of A.   **Table 11.11.** Symmetric motions of Z.

Practice Problem 11.11. *Write the table for the symmetric motions of the letter N, which has rotational symmetry.*

Some objects have rotational symmetry of less than 180°. The equilateral triangle is certainly one of these, and it also has reflectional symmetry. A pinwheel with curved ends and three blades will allow rotational symmetry of 120° and 240°, but no reflectional symmetry (Figure 11.14).

Practice Problem 11.12. *Write the table for the symmetric motions of the pinwheel shown in Figure 11.14. Where have you seen this table before?*

Figure 11.14. Pinwheel.

11.2   Symmetric Motions of the Square

Like the equilateral triangle, the regular 4-gon, or square, allows both rotational symmetry and reflectional symmetry. There are four different rotations, through 90°, 180°, 270°, and 360° (= I). The reflectional symmetries include not only vertical and horizontal symmetry, but also two diagonal symmetries. We want to make a table of the symmetric motions. First, use a compass and straightedge to construct a square and cut it out. Label the four corners of the square with numbers 1, 2, 3, and 4, starting at the top left corner and proceeding counterclockwise (Figure 11.15). Don't forget to label the back of the square.

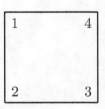

Figure 11.15. Square in the initial position.

We label the rotation through 90° counterclockwise as R. As we noted in the case of the triangle, we can just use powers of R to label the other rotations. Thus, R^2 is the label for the rotation through 180° counterclockwise, and R^3 denotes the rotation through 270° counterclockwise. We label the horizontal and vertical flips with H and V, respectively. The two diagonal flips are labeled D_L and D_R as they use diagonals from the lower left and right corners, respectively. Figure 11.16 shows the eight labeled squares that result from the eight different symmetric motions of the square.

We can fill in some of the table of symmetric motions of the square without much trouble (Table 11.12).

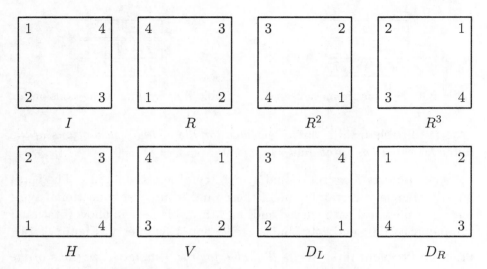

Figure 11.16. The eight symmetries of the square.

Practice Problem 11.13. *Use a cutout and labeled square to fill in the other entries in Table 11.12. Write down some observations about the table as you go.*

Table 11.13 shows the final table of symmetric motions of the square. This is another example of a group. The operation of composition of motions is associative; there is an identity, and every element has an inverse. Again, this is a nonabelian group, the dihedral group D_8.

·	I	R	R^2	R^3	H	V	D_L	D_R
I	I	R	R^2	R^3	H	V	D_L	D_R
R	R	R^2	R^3	I				
R^2	R^2	R^3	I	R				
R^3	R^3	I	R	R^2				
H	H				I			
V	V					I		
D_L	D_L						I	
D_R	D_R							I

Table 11.12. Symmetric motions of the square: preliminary table.

·	I	R	R^2	R^3	H	V	D_L	D_R
I	I	R	R^2	R^3	H	V	D_L	D_R
R	R	R^2	R^3	I	D_L	D_R	V	H
R^2	R^2	R^3	I	R	V	H	D_R	D_L
R^3	R^3	I	R	R^2	D_R	D_L	H	V
H	H	D_R	V	D_L	I	R^2	R^3	R
V	V	D_L	H	D_R	R^2	I	R	R^3
D_L	D_L	H	D_R	V	R	R^3	I	R^2
D_R	D_R	V	D_L	H	R^3	R	R^2	I

Table 11.13. Symmetric motions of the square: final.

As was the case with D_6 for the triangle, this group has some sub-groups. The block of four rotations in the upper left-hand corner of Table 11.13 certainly makes a subgroup. It is structurally the same as $\mathbb{Z}_4$, the group of integers under addition mod 4.

Practice Problem 11.14. *Describe how to make the four rotations correspond to the elements of $\mathbb{Z}_4$ to show that the two groups are structurally the same.*

A subset consisting of any one of the reflections together with the identity motion I also makes a subgroup that is structurally the same as the group $\mathbb{Z}_2$.

There is another way to make a subgroup inside D_8. We can create Table 11.14 by using the symmetric motions R^2, H, V and I.

First, this is in fact a group, since this set is closed under the operation of composition, and the identity and inverses are present. Let's name this group K. This group is abelian. Every element in this subgroup is its own inverse, so no element can generate the group. Have we seen this group before? The only other group of four elements we have seen before is the

·	I	R^2	H	V
I	I	R^2	H	V
R^2	R^2	I	V	H
H	H	V	I	R^2
V	V	H	R^2	I

Table 11.14. Subgroup of D_8.

one that is structurally the same as $\mathbb{Z}_4$, that is, the group of rotations of the square, which is a cyclic group. Because K is not cyclic, it cannot have the same structure as $\mathbb{Z}_4$. Other properties distinguish between these two groups as well. For example, every element of K is its own inverse, whereas $\mathbb{Z}_4$ does not have this property. There is no way to label the elements in Table 11.14 so that this group and the group of rotations have the same table. This is then a new group that we have not seen before. We call this group the *Klein 4-group*.[1] It is interesting to note that the table for this group provides an answer to Exercise 9.7, since $HV = R^2$.

Practice Problem 11.15. *Use $I, R^2, D_L,$ and D_R to form a subgroup of D_8 and decide whether it is structurally the same as $\mathbb{Z}_4$ or as the Klein 4-group.*

Reflections and Rotations

In these groups of symmetric motions, whenever we multiply two rotations together, we end up with a rotation. When we multiply two flips together, we also end up with a rotation. But when we multiply a flip and a rotation, we get a flip. We can see this by looking at the way the tables for D_6 and D_8 split up into blocks of rotations and flips. Let's look at why this happens.

Let's take either the cutout triangle or square that were used to make the corresponding tables. We begin by putting the cutout into the initial position. If we rotate the polygon any number of times, the numbers on the polygon stay in the same order as they were originally. This is why the product of two rotations is another rotation.

If instead, after we put the polygon into the initial position, we then flip it over any line of symmetry, the numbers are in the reverse order from where they started. They have moved from being in counterclockwise order to being in clockwise order. If we do two flips of any kind in a row, one of them will reverse the order of the numbers, and then the second flip will reverse the order a second time, thus returning them to the original order. That is why a flip times a flip will be a rotation.

In the case of a flip times a rotation or a rotation times a flip, one motion will reverse the order of the numbers, and the other will leave the

---

[1]Christian Felix Klein (1849–1925) was a German mathematician who worked in geometry—in particular, in non-Euclidean geometry. He was a professor at the University of Göttingen in Germany and made it a highly successful research center.

order the same. Therefore, the product will have the effect of reversing the order of the numbers, and will be a flip.

A more visual way of seeing why these products have the effects they do is to realize that any rotation leaves the polygon on the front side, but any flip turns it over. Thus, two rotations leave the polygon facing front, and will be a rotation, and two flips will turn the polygon over twice and the polygon will end up facing front again. A flip and a rotation together in either order, however, will have the effect of turning the polygon over just once and will therefore be a flip.

Impossible Motions

In looking at symmetries of the square, we did not allow a motion that would interchange the positions of 1 and 2 while leaving the vertices 3 and 4 fixed, as shown in Figure 11.17.

Even though the "impossible" square does occupy the same space as the initial square, there is a problem with the relationship of the vertices to one another. Since symmetries are a type of congruence, they must preserve distances between vertices. In the original square, vertices 1 and 3 are diagonal, and the distance between them is greater than the distance between adjacent vertices, such as 1 and 2 or 1 and 4 in the original square. Any symmetric motion of the square must preserve these distance relationships among the vertices. Therefore, in the "impossible" square, since the 1 and 3 are now adjacent, this rearrangement of the vertices of the square is not a symmetric motion.

Practice Problem 11.16. *Label the vertices of the square in at least one other way so that the relabeling is not a symmetric motion of the square.*

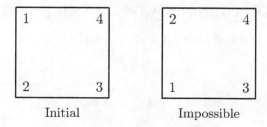

Figure 11.17. An impossible motion.

Economy of Notation Revisited

As we did for the symmetric motions of the equilateral triangle, we can write the table of symmetric motions of the square by using R for a rotation of 90° counterclockwise and powers of R for the other rotations. If we choose any one of the flips as F, then F, FR, FR^2, and FR^3 will be distinct and therefore will represent all the other motions of the square.

Practice Problem 11.17. *Label a specific flip of your square as F and write the table for the symmetric motions of the square using just the symbols I, R, F, and their powers and products.*

11.3   Symmetries of Regular *n*-gons

Figure 11.18. Regular pentagon.

So far, we have worked with the equilateral triangle and the square. Let's look now at a regular pentagon (Figure 11.18) and think about what symmetries it has.

There will be five rotations this time, each through a multiple of 72°. There are also five reflectional symmetries across any line that goes through a vertex and the midpoint of the opposite edge. The group of these motions will be the dihedral group D_{10}.

If, instead, we use a regular hexagon (Figure 11.19), the six rotations will be through multiples of 60°. The reflectional symmetries, like those of the square, will be over a line through two opposite vertices or through the midpoints of two opposite sides. There will be three of each of these reflectional symmetries, and there will be 12 elements in the dihedral group D_{12}.

Figure 11.19. Regular hexagon.

Practice Problem 11.18. *Describe the symmetric motions of a regular n-gon, distinguishing cases when n is even or odd.*

Practice Problem Solutions and Hints

11.1. Here is the triangle rotated 120° clockwise:

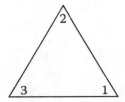

And here is the triangle rotated 240° clockwise:

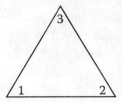

11.2. Do it!

11.3. Do it!

11.4. $R_{240} \cdot R_{240} = R_{120}$.

11.5. See Table 11.3.

11.6.

Motion	Inverse
I	I
R_{120}	R_{240}
R_{240}	R_{120}
F_{top}	F_{top}
F_{left}	F_{left}
F_{right}	F_{right}

11.7. Associativity is inherited from the group D_6. The motion I is the identity, and each element has an inverse, as we can see from Practice Problem 11.6.

11.8. The number 0 always corresponds to the symmetric motion I and the number 1 corresponds to the symmetric motion F, where F is a flip.

11.9. Here is the table using F_{right} as F:

$\cdot$	I	R	R^2	FR	FR^2	F
I	I	R	R^2	FR	FR^2	F
R	R	R^2	I	F	FR	FR^2
R^2	R^2	I	R	FR^2	F	FR
FR	FR	FR^2	F	I	R^2	R
FR^2	FR^2	F	FR	R	I	R^2
F	F	FR	FR^2	R^2	R	I

Manipulate this table by moving the entries of the fourth column into the fifth column, the entries of the fifth column into the sixth column, and the entries of the sixth column into the fourth column. Put the entries of the fourth row into the fifth row, the entries of the fifth row into the sixth row, and the entries of the sixth row into the fourth row. Now this table looks just like Table 11.9.

11.10.

·	I	H
I	I	H
H	H	I

11.11.

·	I	R
I	I	R
R	R	I

11.12. This is the subgroup of rotations of the triangle.

·	I	R	R^2
I	I	R	R^2
R	R	R^2	I
R^2	R^2	I	R

11.13. You might observe that the symmetric motions of a square make a group that is not abelian, and that this group has several subgroups.

11.14. The number 0 corresponds to the motion I, 1 corresponds to R, 2 to R^2, and 3 to R^3.

11.15. Here is the multiplication table for this subgroup:

·	I	R^2	D_L	D_R
I	I	R^2	D_L	D_R
R^2	R^2	I	D_R	D_L
D_L	D_L	D_R	I	R^2
D_R	D_R	D_L	R^2	I

Because each element is its own inverse, this group is structurally the same as the Klein 4-group.

11.16. Here is another impossible motion:

```
1            4



3            2
```

11.17. Let F be a vertical flip (V).

$\cdot$	I	R	R^2	R^3	F	FR	FR^2	FR^3
I	I	R	R^2	R^3	F	FR	FR^2	FR^3
R	R	R^2	R^3	I	FR^3	F	FR	FR^2
R^2	R^2	R^3	I	R	FR^2	FR^3	F	FR
R^3	R^3	I	R	R^2	FR	FR^2	FR^3	F
F	F	FR	FR^2	FR^3	I	R	R^2	R^3
FR	FR	FR^2	FR^3	F	R^3	I	R	R^2
FR^2	FR^2	FR^3	F	FR	R^2	R^3	I	R
FR^3	FR^3	F	FR	FR^2	R	R^2	R^3	I

11.18. In each case there will be n rotations of $\left(\frac{360}{n}\right)^\circ$. There will also be n reflections. If n is odd, the axes of reflection will be through each vertex and the midpoint of the opposite side. If n is even, there will be $\frac{n}{2}$ axes of reflection through opposite vertices and $\frac{n}{2}$ axes of reflection through opposite midpoints.

Exercises

11.1. Joe Shmoe did the exercise of completing the table for symmetries of the equilateral triangle. However, he called his symmetries of the triangle P, Q, R, S, T, and U.

(a) Given Joe's multiplication table below, match up which letter of his corresponds to which letter in our table. **Hint:** Look at the diagonal.

$\cdot$	**P**	**Q**	**R**	**S**	**T**	**U**
P	P	Q	R	S	T	U
Q	Q	P	T	R	U	S
R	R	S	U	T	Q	P
S	S	U	Q	P	R	T
T	T	R	S	U	P	Q
U	U	T	P	Q	S	R

(b) Is there more than one way you could match up the letters?

11.2. Draw an object with the same rotation group as the square, but with no reflections.

11.3. In D_6, which of the six elements $\{I, R, R^2, F, FR, FR^2\}$ is equilateral to each of the following?

(a) R^2FR.

(b) FRF.

(c) $R^{-1}F$.

11.4.

(a) Show that either R or R^3 can generate the group of rotations of the square.

(b) For which integers k, $1 \le k \le n - 1$, does R^k generate the group of rotations of the regular n-gon?

11.5.

(a) For a given n, show that if F is any reflection of the regular n-gon, then $FR = R^{n-1}F$.

(b) In a regular 12-gon, use the fact that $FR = R^{11}F$ to compute the product $R^3F \cdot R^6F$. Express your answer in the form R^k or R^kF for some k, $1 \le k \le 11$.

11.6. In the group D_6, there is only one motion that commutes with every other motion, namely, the identity. In the group D_8, the motion R^2 commutes with every other motion. A concise way to see this is to notice that since R^2 commutes with every rotation, you only need to show it commutes with any reflection.

(a) Find which motions in D_{10} commute with every other motion.

(b) Find which motions in D_{12} commute with every other motion.

(c) If n is even, show that $R^{n/2}$ commutes with every element of D_{2n}.

11.7.

(a) Find all the subgroups in the group of symmetries of the regular hexagon, D_{12}.

(b) What is the order of each of the subgroups?

11.8.

(a) List the symmetries of a nonsquare rectangle and write the group of symmetries.

(b) Is the group in part (a) structurally the same as the group of rotations of the square?

11.9. In the group D_6, will the element RFR^2FR be a reflection or a rotation?

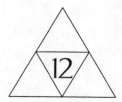

Permutations

We have written the tables for the groups of symmetric motions of regular polygons by representing the motions as they appear geometrically as reflections and rotations. Another way to realize the group tables is by using the idea and notation of permutations. Intuitively speaking, a permutation is a rearrangement of a set of numbers.

Definition. Let $S = \{1, 2, \ldots, n\}$. A *permutation* of S is a one-to-one function from S onto itself.

Examples. The following two functions are permutations.

1. If $S = \{1, 2, 3\}$ and $f(1) = 2$, $f(2) = 3$, and $f(3) = 1$, then f is a permutation.

2. If $S = \{1, 2, 3, 4, 5, 6, 7\}$ and $f(1) = 3$, $f(2) = 5$, $f(3) = 1$, $f(4) = 4$, $f(5) = 6$, $f(6) = 2$, $f(7) = 7$, then f is a permutation.

12.1 Symmetric Motions as Permutations

We want to consider symmetric motions of a regular polygon as permutations of the vertices. We do this by paying attention to the movement of the labeled vertices by keeping track of where the numbers move in relation to their initial position.

Let's use the equilateral triangle and put it in initial position, with the vertices labeled 1, 2, and 3 starting at the top and proceeding counterclockwise, as shown in Figure 12.1.

If we rotate the triangle 120° counterclockwise, we see the numbers on the vertices have moved. The 1 has moved to where the 2 was originally, the 2 is where the 3 was, and the 3 is now where the 1 was (Figure 12.2).

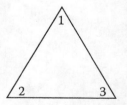

Figure 12.1. Initial position of triangle.

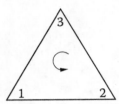

Figure 12.2. Initial triangle rotated 120° counterclockwise.

Figure 12.3. Initial triangle rotated 240° counterclockwise.

We could write this motion by using arrows:

$$
\begin{array}{ccc}
1 & 2 & 3 \\
\downarrow & \downarrow & \downarrow \\
2 & 3 & 1
\end{array}
$$

Example. What would the counterclockwise rotation of 240° look like with this arrow notation? Figure 12.3 is the initial triangle rotated 240° counterclockwise.

Now we have 1 in 3's original position. Here is the labeling of the motion R_{240} with arrows:

$$
\begin{array}{ccc}
1 & 2 & 3 \\
\downarrow & \downarrow & \downarrow \\
3 & 1 & 2
\end{array}
$$

Practice Problem 12.1. *Write all six of the symmetric motions of the triangle using the arrow notation that tells where the numbers go.*

How do these arrows work when we combine two motions of the triangle? We just follow the arrows twice in a row.

Example. The reflection F_{top} has the arrow pattern

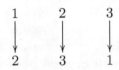

The rotation R_{120} has the arrow pattern

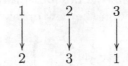

From Table 11.3, we know that the product $R_{120} \cdot F_{\text{top}} = F_{\text{right}}$. The motion F_{right} has the arrow pattern

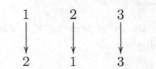

How can we use the arrow notation to multiply the two motions and get the arrow pattern for their product? In computing the product $R_{120} \cdot F_{\text{top}}$, remember that we are following the standard practice for composition of functions; that is, we do F_{top} first and then R_{120}. We simply follow them in succession. For example, the number 2 goes to 3 via F_{top}. Then, we follow 3 in R_{120} to see that it goes to 1. Therefore, the result of doing both motions is to send 2 to 1. In fact, we can see from the arrow pattern of F_{right} that this is what happens to 2 under this symmetric motion

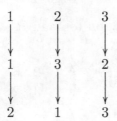

Practice Problem 12.2. *Use the arrow notation to verify the products of the motions* $R_{240} \cdot F_{left}$ *and* $F_{left} \cdot F_{right}$.

These arrows are a bit awkward, so we introduce a slightly different notation that will help to make this easier to write down. The idea of the new notation is to indicate where each number goes in the motion of the triangle without using arrows. We use a two-row list of the numbers 1, 2, and 3 to show the result of a permutation.

238 12 ▷ Permutations

Example. The rotation R_{120} has the effect of moving $1 \to 2$, $2 \to 3$, and $3 \to 1$. The new notation lists the numbers in the first row in the original order, and in the second row in the new order:

$$\begin{pmatrix} 1 & 2 & 3 \\ 2 & 3 & 1 \end{pmatrix}.$$

This notation reads from top to bottom. You can imagine the arrows from the top row of numbers to the bottom. This notation shows that 1 goes to 2, 2 goes to 3, and 3 goes to 1.

Example. In this notation, we represent the 240° counterclockwise rotation of the triangle by the symbol

$$\begin{pmatrix} 1 & 2 & 3 \\ 3 & 1 & 2 \end{pmatrix}.$$

Practice Problem 12.3. *Write the other four motions of the triangle, including the identity motion I, in this notation.*

This notation makes it easier to see how motions of the triangle can be combined. We know from our earlier example that $R_{120} \cdot F_{\text{top}} = F_{\text{right}}$. Let's see how this works using the new symbolic notation:

$$R_{120} = \begin{pmatrix} 1 & 2 & 3 \\ 2 & 3 & 1 \end{pmatrix}, \ F_{\text{top}} = \begin{pmatrix} 1 & 2 & 3 \\ 1 & 3 & 2 \end{pmatrix}, \text{ and } F_{\text{right}} = \begin{pmatrix} 1 & 2 & 3 \\ 2 & 1 & 3 \end{pmatrix}.$$

To multiply, we put F_{top} on top and R_{120} underneath and follow the path of the imaginary arrows:

$$F_{\text{top}} = \begin{pmatrix} 1 & 2 & 3 \\ 1 & 3 & 2 \end{pmatrix}$$

$$R_{120} = \begin{pmatrix} 1 & 2 & 3 \\ 2 & 3 & 1 \end{pmatrix}.$$

Where will the number 2 go in this product? First, by F_{top}, we have 2 goes to 3. Then in R_{120}, we see where 3 goes; it goes to 1. So the result of combining this reflection and rotation will be that 2 goes to 1. Notice that this is the path of the number 2 in the reflection F_{right}. We can follow the number 1 as well. In F_{top}, 1 stays put, then in R_{120}, 1 goes to 2. In the reflection F_{right}, the number 1 goes to 2. You can see where the number 3 goes by following these same steps.

Example. If we compute the product of these same two motions in reverse order, we have

$$R_{120} = \begin{pmatrix} 1 & 2 & 3 \\ 2 & 3 & 1 \end{pmatrix}$$

$$F_{\text{top}} = \begin{pmatrix} 1 & 2 & 3 \\ 1 & 3 & 2 \end{pmatrix}.$$

This shows that 1 goes to 3, 2 goes to 2, and 3 goes to 1. Therefore, this product is the permutation F_{left}, or in our new notation,

$$F_{\text{left}} = \begin{pmatrix} 1 & 2 & 3 \\ 3 & 2 & 1 \end{pmatrix}.$$

Practice Problem 12.4. *Compute the product of any two motions of the triangle using this notation and verify your answer by checking with Table 11.3.*

This notation has the advantage that we can keep track of the symmetric motions of the triangle without having to manipulate an actual model. There is nothing special about the triangle that makes this notation work. We can just as easily use the same idea when we work with the symmetric motions of any regular polygon.

Permutations and the Motions of the Square

We can label the eight motions of the square using this double row notation (Figure 12.4). This time the top row of each motion's symbol will be the numbers 1 through 4, and the bottom row will indicate where each number ends up under that motion.

Again, we can multiply these motions by listing them on top of one another and following a number through both permutations to see where it goes.

Example. To compute $R^2 \cdot H$, we write

$$H = \begin{pmatrix} 1 & 2 & 3 & 4 \\ 2 & 1 & 4 & 3 \end{pmatrix}$$

$$R^2 = \begin{pmatrix} 1 & 2 & 3 & 4 \\ 3 & 4 & 1 & 2 \end{pmatrix},$$

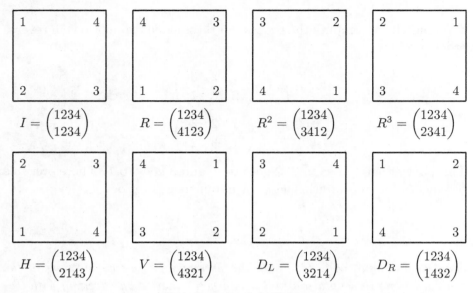

Figure 12.4. Symmetries of the square.

and we get

$$V = \begin{pmatrix} 1 & 2 & 3 & 4 \\ 4 & 3 & 2 & 1 \end{pmatrix}.$$

Practice Problem 12.5. *Using this notation, compute the results of at least two more products of symmetric motions of the square.*

We said at the beginning of this section that a permutation was just a rearrangement of a set of numbers. There are, however, rearrangements of the numbers 1, 2, 3, and 4 that are not among the eight symmetric motions of the square. For example,

$$P = \begin{pmatrix} 1 & 2 & 3 & 4 \\ 2 & 1 & 3 & 4 \end{pmatrix}.$$

This permutation switches 1 and 2, but leaves the numbers 3 and 4 alone. If we look at the numbered square we used to display the symmetric motions, we can see that the permutation P is the motion described in Chapter 11 as the impossible square; it does not represent a symmetric motion of the square. Many more permutations of the numbers 1 through 4 are not represented by symmetric motions of the square.

Practice Problem 12.6. *List at least two other permutations of the numbers 1 through 4 that are not symmetric motions of the square.*

12.2   Counting Permutations and Symmetric Groups

How many permutations, or rearrangements, of the numbers 1, 2, 3, and 4 are there? Every rearrangement will produce a list of the numbers 1, 2, 3, and 4 in some order. For example, one list might be $(3, 2, 1, 4)$, and another could be $(1, 2, 4, 3)$.

Example. We can count the number of possible rearrangements systematically by starting with the number 1. It has four places to go, since it can go to position 1, 2, 3, or 4, for example, $(\_, \_, 1, \_)$. Once 1 is settled, we check where 2 can go. The number 1 has taken one of the four possible places already, so the number 2 has only three places to go, for example, $(2, \_, 1, \_)$. Together, these now make $4 \cdot 3$, or 12, possibilities. Next, the number 3 has only two places to go, and then the number 4 has but one place to go, for example, $(2, \_, 1, 3)$ and $(2, 4, 1, 3)$. Therefore, in all, there are $4 \cdot 3 \cdot 2 \cdot 1$ or 24 possible rearrangements, or permutations, of the numbers 1 through 4.

However, only 8 of the 24 possible rearrangements represent symmetric motions of the physical square. Let's think geometrically about why this is so. In the square, the vertex numbered 1 can go to any of the four vertices in a symmetric motion. Now where can the vertex numbered 2 go? In a symmetric motion, distances must be preserved, so vertex 2 must remain adjacent to vertex 1. Therefore, there are only two, not three, places where vertex 2 can go. After vertex 2 settles into its position, the places of vertices 3 and 4 are determined by their distance from vertex 1. Thus, there are only $4 \cdot 2 = 8$ possible symmetric motions of the square.

Let's go back and think about this idea in relation to the symmetric motions of the triangle. How many possible permutations are there of the numbers 1, 2, and 3? There will be $3 \cdot 2 \cdot 1$, or six of them, and each of these six permutations is represented by the symmetric motions of the triangle. This is the only regular polygon whose symmetric motions correspond to all the permutations of the numbers on its vertices. This is because every two vertices of the triangle have the same distance between them. This is not true of any other regular polygon.

If we look at the set of all permutations of the numbers 1, 2, and 3, we see that they are the same as the elements of the dihedral group D_6. So the set of all permutations of the set of numbers $\{1, 2, 3\}$ forms a group. What about the set of all permutations of the numbers 1, 2, 3, and 4? It turns out that these 24 elements also form a group. Multiplication of permutations is associative because these permutations are functions, and composition of functions is associative. The identity is certainly a permutation, even if it is boring. Every permutation has an inverse that switches the numbers back to their original places. In fact, for any n, we can form a group of the permutations of the set of numbers $\{1, 2, 3, \ldots, n\}$ using the notation and multiplication described above. These groups have a name.

Definition. For any positive integer n, the *symmetric group*, called S_n, is the group of all the permutations of the set $\{1, 2, 3, \ldots, n\}$. There are $n! = n \cdot (n-1) \cdot (n-2) \cdots 3 \cdot 2 \cdot 1$ elements in the group.

12.3   Even More Economy of Notation

Our permutations are a shortcut way of writing the symmetry motions of a figure. There is an even more efficient way to write permutations, however. To understand this, let's return to the symmetry motions of the equilateral triangle. Remember that in the motion R_{120}, vertex 1 moves to 2's position, vertex 2 goes to 3's position, and vertex 3 goes to 1's position. In permutation notation, we write this as

$$\begin{pmatrix} 1 & 2 & 3 \\ 2 & 3 & 1 \end{pmatrix}.$$

In every permutation written this way, we have included in the first line the numbers written in their original order. We would like to make the notation even more efficient and not write the original numbers each time. We have to do this in such a way that, in the new notation, multiplication of permutations still gives the same results as before. To write permutations in just one line instead of two, let's first think about R_{120} and how we might write it in simpler notation. We want the order of the numbers from left to right to indicate where each number goes under the motion R_{120}. So we write $R_{120} = (123)$. The number 1 goes to 2, its right-hand neighbor's spot in the parentheses. The number 2 goes to 3, also its right-hand neighbor's spot in the parentheses. In R_{120}, 3 goes to 1's spot. In this notation, we can think of the last number going to take the place of the first number.

Example. For R_{240}, this notation would give the symbol (132), since 1 goes to 3's place, 3 goes to 2's place, and 2 goes to 1's place.

There is something special about this new notation. What would the symbol (231) mean? If we follow the instructions above, it means that 2 goes to 3's place, 3 goes to 1's place, and 1 goes to 2's place. Notice that this is just the result of doing the motion R_{120}, but starting with 2. The symbol (312) does the same thing, starting with 3.

Practice Problem 12.7. *Write the new notation for R_{240} in two other ways.*

How would we write one of the reflections, say F_{top}, in this new notation? In F_{top}, the number 1 stays put, and the 2 and 3 change places. If we write (23), this will indicate that 2 goes to 3's place and 3 goes to 2' s place. How would we show that we leave the 1 alone? We have two options. We could write (1), indicating that 1 goes to 1's place, or we could just leave it out of this permutation and understand, by its absence, that the permutation F_{top} leaves 1 alone.

Example. For F_{left} we could use either the notation (13)(2) or just (13).

Also, remember that (12) is the same as (21). As a convention, in order to have some notation for the identity permutation, we will write (1) and understand that I leaves every number alone.

Table 12.1 shows the six elements of D_6 or S_3 written in the new notation.

Element	Permutation
R_{120}	(123)
R_{240}	(132)
F_{top}	(23)(1) or just (23)
F_{left}	(13)(2) or just (13)
F_{right}	(12)(3) or just (12)
I	(1)(2)(3) or just (1)

Table 12.1. Group D_6 written in cycle permutation notation.

Now, how do we multiply permutations written this way? We want to go from right to left as we did before. For example, in D_6, if we do F_{top} first followed by R_{120}, we get F_{right}, or $R_{120} \cdot F_{\text{top}} = F_{\text{right}}$. For simplicity, we sometimes omit the dot and write $R_{120} \cdot F_{\text{top}}$ as $R_{120}F_{\text{top}}$. We write the new notation the same way, substituting (123) for R_{120} and (23) for F_{top}, so that $R_{120}F_{\text{top}} = (123)(23)$.

For multiplication in the new notation, the idea is to start with a number in the right-most permutation and first see where it goes in that permutation. Then we follow that new number in the second, or left-most permutation, and see where it goes. That should tell us the end position of the number we started with.

For example, for $R_{120}F_{\text{top}}$, we start with 2 in the notation for F_{top}. In that permutation, 2 goes to 3. Now where does 3 go in the symbol for R_{120}? It goes to 1. So in the product, we will have 2 going to 1. We can begin to write this as a permutation by writing it without the last parentheses as "21" since we haven't yet followed where 1 goes.

Next let's track the number 1. In F_{top}, the number 1 is left alone, so it does not appear in the right-most symbol. Where does 1 go in R_{120}, or the left-most permutation? It goes to 2. So in the product, 1 goes to 2, and in our notation, that means we close the parentheses and get (21), which is the same as (12).

What happens to the number 3 in this product? In F_{top}, or (23), 3 goes to 2. Then in R_{120}, or (123), 2 goes to 3. So 3 goes to 3, or is left alone. Our product is thus (12) or $(12)(3)$, the symbol for F_{right}, which is what we wanted.

Practice Problem 12.8. *Using this permutation notation, show that $F_{\text{right}}F_{\text{left}} = R_{240}$, and that $R_{120}R_{240} = I$.*

There is nothing special about using S_3 to write permutations this way. In general, we can do the same thing in S_n, the group of permutations on n letters.

Definition. Suppose $a_1, a_2, a_3, ...a_k$ are numbers from the set $\{1, 2, ...n\}$. An expression of the form $(a_1, a_2, a_3, ..., a_k)$ is called a *cycle of length k* and indicates the permutation in which a_1 goes to a_2, a_2 goes to a_3, and so on, until a_k goes to a_1. We also call this a *k-cycle*.

A 2-cycle is merely a permutation that interchanges two numbers. We call a 2-cycle a *transposition*. A 1-cycle leaves a number alone.

Element	Permutation
I	$(1)(2)(3)(4)$ or just (1)
R	(1234)
R^2	$(13)(24)$
R^3	(1432)
H	$(12)(34)$
V	$(14)(23)$
D_L	$(13)(2)(4)$ or just (13)
D_R	$(24)(1)(3)$ or just (24)

Table 12.2. Group D_8 written in cycle permutation notation.

Practice Problem 12.9. *Show that any 2-cycle is its own inverse.*

Table 12.2 presents the elements of D_8 using the cycle permutation notation. Notice that all the permutations in this group are either 4-cycles, 2-cycles, products of 2-cycles, or 1-cycles.

Practice Problem 12.10. *Why can't there be any 3-cycles in the group D_8 of symmetry motions of the square?*

In their short form without 1-cycles, we can write the two motions D_L and D_R as *disjoint* permutations, or cycles that have no elements in common.

Practice Problem 12.11. *Show that two disjoint cycles in S_n commute.*

Challenge Problem. Show that we can write every permutation in S_n either as an n-cycle itself, or as a product of disjoint k-cycles, where $k < n$.

Transpositions

For our purposes, we use one other fact about permutations written this way. That is, for $n > 1$, we can write every permutation in S_n as a product of 2-cycles, or transpositions.

Let's try this with R_{120}, one of our permutations from S_3. As a 3-cycle, this is the permutation (123). If we compute the product of the permutations $(13)(12)$, we see that we do get the permutation (123).

Practice Problem 12.12. *Write the 4-cycle $R = (1234)$ from D_8 as the product of transpositions.*

Element	Notation	Product of Transpositions	Even or Odd
I	$(1)(2)(3)(4)$ or just (1)	$(12)(12)$	even
R	(1234)	$(14)(13)(12)$	odd
R^2	$(13)(24)$	$(13)(24)$	even
R^3	(1432)	$(12)(13)(14)$	odd
H	$(12)(34)$	$(12)(34)$	even
V	$(14)(23)$	$(14)(23)$	even
D_L	$(13)(2)(4)$ or just (13)	(13)	odd
D_R	$(24)(1)(3)$ or just (24)	(24)	odd

Table 12.3. Permutations in D_8 as a product of transpositions.

In fact, if $(a_1a_2a_3\ldots a_k)$ is a k-cycle in S_n, we can write it as a product of $n-1$ transpositions:

$$(a_1a_2a_3\ldots a_k) = (a_1a_k)(a_1a_{k-1})\ldots(a_1a_3)(a_1a_2).$$

We can even write the identity as a product of transpositions if we write it in the form $(12)(12)$. Remember that if a number is left out of a permutation, it just means the number is left alone in that permutation.

In this kind of product, the transpositions are not disjoint. However, this way of writing a k-cycle gives us interesting information about the permutations. Table 12.3 shows how to write the permutations in D_8 as a product of transpositions. We also note whether the number of transpositions in the product is even or odd. If the number is even, we say the permutation itself is even; if odd, we call the permutation odd.

Notice that the set of even permutations in this group forms a subgroup of D_8.

Practice Problem 12.13. *Write the permutations of S_3 as products of transpositions and determine which are even and odd. Do the even permutations form a subgroup?*

The French mathematician Cauchy[1] noticed and proved that whenever we write a permutation as a product of transpositions, the number of transpositions in the product will either always be even or always be odd.

---

[1] Augustin-Louis Cauchy (1789–1857) was born in Paris and lived at the time of the French Revolution. He was a professor at the famous École Polytechnique in France, and worked mostly in the area of mathematics known as analysis.

An important observation about the groups S_n of permutations is that all the even permutations form a subgroup. This is a challenging fact to prove, and we include it in the Exercises.

Practice Problem Solutions and Hints

12.1. The motion I is written as

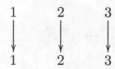

The rotation R_{120} is written as

The rotation R_{240} is written as

The reflection F_{top} is written as

The reflection F_{left} is written as

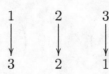

The reflection F_{right} is written as

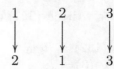

12.2. To check that $R_{240} \cdot F_{\text{left}} = F_{\text{right}}$, write

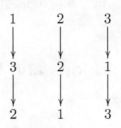

To check that $F_{\text{left}} \cdot F_{\text{right}} = R_{120}$, write

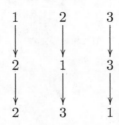

12.3.

$$I = \begin{pmatrix} 1 & 2 & 3 \\ 1 & 2 & 3 \end{pmatrix}; \qquad F_{\text{top}} = \begin{pmatrix} 1 & 2 & 3 \\ 1 & 3 & 2 \end{pmatrix};$$

$$F_{\text{left}} = \begin{pmatrix} 1 & 2 & 3 \\ 3 & 2 & 1 \end{pmatrix}; \qquad F_{\text{right}} = \begin{pmatrix} 1 & 2 & 3 \\ 2 & 1 & 3 \end{pmatrix}.$$

12.4. Do it!

12.5. Do it!

12.6. The motions

$$\begin{pmatrix} 1 & 2 & 3 & 4 \\ 1 & 3 & 2 & 4 \end{pmatrix} \quad \text{and} \quad \begin{pmatrix} 1 & 2 & 3 & 4 \\ 1 & 3 & 4 & 2 \end{pmatrix}$$

are impossible motions.

12.7. $R_{240} = (132)$ may also be written (321) or (213).

12.8. $F_{\text{right}} F_{\text{left}} = (12)(13) = (132) = R_{240}$.
$R_{120} R_{240} = (123)(132) = (1)(2)(3) = I$.

12.9. If we have $(ab)(ab)$, we get $(a)(b)$ = identity.

12.10. A 3-cycle would move only three vertices of the square and leave one fixed. This would be an impossible geometric motion.

12.11. In the product written in either order, a number will appear in just one of the cycles, since the cycles are disjoint. It won't matter which cycle is used first, since the result will be the same in either order.

12.12. $R = (1234) = (14)(13)(12)$.

12.13. $R_{120} = (13)(12)$ is even; $R_{240} = (12)(13)$ is even; $F_{\text{top}} = (23)$ is odd; $F_{\text{left}} = (13)$ is odd; $F_{\text{right}} = (12)$ is odd. $I = (12)(12)$ is even. The even permutations form the subgroup of rotations of the triangle.

Exercises

12.1. Which permutations can be written as disjoint cycles?
   (a) In S_6, write the permutation

$$\begin{pmatrix} 1 & 2 & 3 & 4 & 5 & 6 \\ 1 & 3 & 4 & 5 & 2 & 6 \end{pmatrix}$$

   as a product of disjoint cycles.
   (b) In S_{10}, write the permutation

$$\begin{pmatrix} 1 & 2 & 3 & 4 & 5 & 6 & 7 & 8 & 9 & 10 \\ 2 & 3 & 4 & 5 & 7 & 6 & 5 & 8 & 10 & 9 \end{pmatrix}$$

   as a product of disjoint cycles.
   (c) Show that in S_n you can write any permutation as the product of disjoint cycles.

12.2. Write the 24 permutations in S_4 as disjoint cycles.

12.3. Prove that the order of a k-cycle is k.

12.4. Suppose α is a k-cycle in S_n and β is an m-cycle in S_n. If $(k, m) = 1$, show that the order of the product $\alpha\beta$ is $k \cdot m$.

12.5. If $(a_1a_2a_3)$ is a 3-cycle in S_n, show that the inverse is $(a_1a_3a_2)$.

12.6. Tell whether the following permutations in S_6 are even or odd.
  (a) $(246)(123)$.

  (b) $(12)(345)(16)$.

12.7. Another way to define odd and even permutations is by using cycles. A cycle is even if it has an odd number of elements and is odd if it has an even number of elements. Show this is the same as the definition of even and odd given in the text.

12.8. We can assign a sign to a cycle in the following way:

  ▷ An even cycle has sign $+1$.

  ▷ An odd cycle has sign -1.

The sign of a permutation is the product of the signs of its disjoint cycles.
  (a) What are the signs of (123), (1245), and (56) in S_6?

  (b) What are the signs of $(12)(456)$, $(123)(45)$, and $(12)(34)(56)$ in S_6?

  (c) What are the signs of the permutations in Exercise 12.2?

  (d) Show that a permutation is even if the product of the signs of its disjoint cycles is $+1$ and is odd if the product of the signs of its disjoint cycles is -1.

12.9. Show that the even permutations in S_4 form a subgroup of order 12 in S_4. Is this group structurally the same as D_{12}?

12.10. This exercise is designed to be a group activity project. In S_5, show that the even permutations form a nonabelian subgroup of order 60.

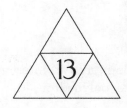

Polyhedra

13.1 Regular Polyhedra

So far, we have been working in two dimensions with regular polygons and discussing their symmetry properties. Now let's move on to three dimensions and consider the objects called polyhedra (plural of polyhedron).

Mathematicians have ideas of what polyhedra look like in three dimensions and have come up with definitions that sometimes include figures that others would not consider to be polyhedra. The following definition suits the purposes of this text.

Definition. A *polyhedron* is the union of a finite set of polygons such that

▷ any pair of polygons meet at only their edges or vertices,

▷ each edge of each polygon meets exactly one other polygon along an edge,

▷ it is possible to travel from the interior of any polygon to the interior of any other along the surface of the polyhedra.

Definition. The following terms are related to polyhedra.

▷ The polygons are called the *faces* of the polyhedron.

▷ The intersection of two faces forms an *edge*.

▷ The intersection of two or more edges forms a *vertex*.

▷ A *diagonal* is a line joining any two vertices of a polyhedron that are not in the same face.

▷ If a line segment joining any two interior points of the polyhedron lies entirely inside the polyhedron, the polyhedron is said to be *convex.*

Challenge Problem. Show that the following definition of a convex polyhedron is equivalent to the one given above: A polyhedron is convex if the intersection of the polyhedron with any plane passing through its interior is a convex polygon.

Definition. A *regular polyhedron* is a convex polyhedron whose faces are congruent regular polygons with the same number of faces meeting at each vertex.

As with polygons, first we will classify the regular polyhedra and then we will study their symmetries. The most familiar regular polyhedron is the cube (Figure 13.1). A cube has faces that are squares, with three meeting at each vertex. There are six faces, twelve edges, and eight vertices.

There are also regular polyhedra whose faces are equilateral triangles. The simplest is one made of just four equilateral triangles as faces, called a *regular tetrahedron* (Figure 13.2).

Three of the triangles meet at each vertex. The regular tetrahedron has four faces, four vertices, and six edges.

How do we create a model of the tetrahedron? One way to begin is to surround a vertex with three equilateral triangles. We start by drawing a diagram with three equilateral triangles sharing a vertex, as shown in Figure 13.3.

If we cut out this figure and fold it along the dotted lines, we get one corner of a regular tetrahedron. To finish the model, we need only to add the last face, as shown in Figure 13.4.

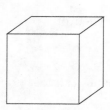

Figure 13.1. Cube.

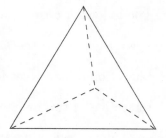

Figure 13.2. Regular tetrahedron.

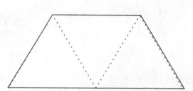

Figure 13.3. Building a tetrahedron model.   **Figure 13.4.** A flat tetrahedron.

For a different regular polyhedron with four equilateral triangles at each vertex, we draw the diagram in Figure 13.5, with four equilateral triangles sharing a vertex.

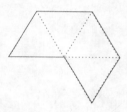

If we cut out this figure and fold it along the dotted lines, we get a square pyramid (Figure 13.6), which is one corner of a *regular octahedron*.

To complete the regular octahedron (Figure 13.7), we make another square pyramid and join the two along their square bottoms.

Figure 13.5. Another flat pyramid.

The regular octahedron has eight faces, six vertices, and twelve edges. Note that in the regular octahedron, four triangles meet at each vertex.

One more regular polyhedron made of equilateral triangles has five at each vertex. This time, we draw a diagram with five equilateral triangles sharing a vertex (Figure 13.8).

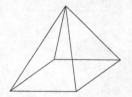

If we cut out this figure and fold it along the dotted lines, we get one corner of a *regular icosahedron*, a polyhedron with 20 faces, 30 edges, and 12 vertices (Figure 13.9).

Figure 13.6. Square pyramid.

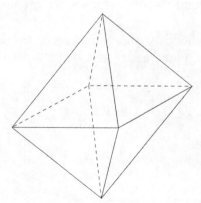

Figure 13.7. Regular octahedron.

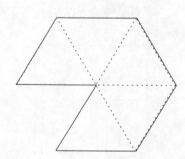

Figure 13.8. A third flat pyramid.

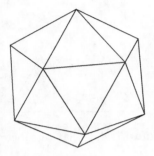

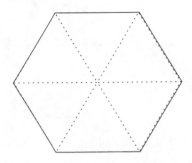

Figure 13.9. Regular icosahedron.   **Figure 13.10.** Unfoldable diagram.

Are there any additional regular polyhedra with equilateral triangles as faces? If we tried to make six triangles meet at a single vertex, we would get the plane (unfoldable) figure of a hexagon (Figure 13.10), because each interior angle has 60°. We can cut out this diagram, but we cannot fold it into a three-dimensional model. Therefore, the tetrahedron, the octahedron, and the icosahedron are the only regular polyhedra whose faces are equilateral triangles.

Practice Problem 13.1. *Figure 13.11 is a picture of two tetrahedra glued together along a face. Explain why this is not a regular polyhedron.*

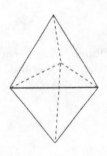

Figure 13.11. Two tetrahedra joined together.

Let's return to the cube, whose faces are squares. If we put three squares together at a single vertex, we get the corner of a cube, also called a regular hexahedron. To make a model, we draw a diagram with three squares sharing a vertex, as in Figure 13.12.

If we cut out this figure and fold it along the dotted lines, we get one corner of a cube (Figure 13.13).

Are there any additional regular polyhedra with squares as faces? If we try to use four squares, we will get the plane figure of a larger square,

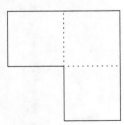

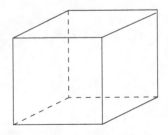

Figure 13.12. Building a cube model.   **Figure 13.13.** Cube.

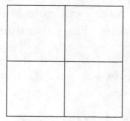

Figure 13.14. Unfoldable diagram with squares.

because each interior angle is 90°, as shown in Figure 13.14. Again, we could cut out this diagram but we couldn't fold it into a three-dimensional figure. Therefore, the cube is the only regular polyhedron whose faces are squares.

Are there any regular polyhedra whose faces are pentagons? If we put three pentagons together at a single vertex, we will get a corner of the regular dodecahedron. To make a model, we draw a diagram with three regular pentagons sharing a vertex, as shown in Figure 13.15. If we cut out this figure and fold it along the dotted lines, we get one corner of the regular dodecahedron, a polyhedron with 12 faces, 30 edges, and 20 vertices (Figure 13.16).

Are there any additional regular polyhedra with pentagons as faces? If we try to make a solid by having four regular pentagons meet at a vertex,

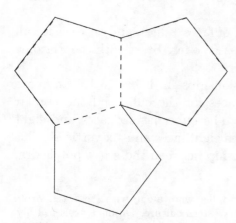

Figure 13.15. Building a model with pentagonal faces.

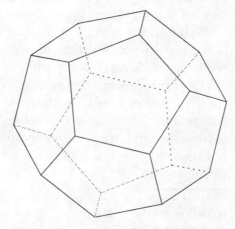

Figure 13.16. Regular dodecahedron.

we will be unsuccessful since the pentagons will overlap and therefore not form a corner of a convex three-dimensional figure.

Practice Problem 13.2. *If you try to join four flat regular pentagons at a vertex, what kind of shape do you get? Can you continue the process of joining four flat regular pentagons at every vertex to form a polyhedron?*

What about other regular polygons as faces for regular polyhedra? If we try to make three-dimensional figures out of hexagons, we find it is impossible. Three hexagons meeting at a vertex create a plane figure since each interior angle is 120°. If we try to use a regular n-gon with seven or more sides, three of them will overlap and thus will not form the corner of a three-dimensional figure. This is because the interior angle of a regular polygon is $180(n-2)/n°$, which is more than 120° for any $n > 6$. Three of these angles will add up to more than 360°. Thus, the only possible regular polyhedra are the five that arise from using equilateral triangles, squares, and regular pentagons.

These five regular polyhedra have been known since the time of the Greeks in the sixth century B.C. They are sometimes referred to as the *Platonic solids*, since Plato[1] was among those who first gave instructions on how to make models of them. Diagrams for constructing the regular polyhedra using cardboard and a little glue or tape are presented at the end of this chapter. All of these use only a straightedge and compass to construct equilateral triangles and squares. In Chapter 15, we discuss the construction of the regular pentagon, which can be used to make a regular dodecahedron.

Table 13.1 is a chart of the number of faces, edges, and vertices for all of the regular polyhedra we have discussed so far. Start with the Platonic solids and look carefully at the numbers.

We can see some interesting relationships in Table 13.1. Compare the numbers of edges, vertices, and faces for the cube and the regular octahedron. Both have twelve edges. The cube has six faces and eight vertices, and the regular octahedron has eight faces and six vertices. This is a consequence of the special relationship between these two polyhedra, called *duality*.

---

[1]Plato (428–348 B.C.) was a Greek philosopher who, along with Aristotle, wrote texts that are the philosophical foundation of Western culture. On his travels to Italy, he discovered the work of Pythagoras and came to appreciate mathematics. Plato wrote about the importance of the idea of proof and laid the foundation for some of the work of Euclid.

Regular Polyhedra	Vertices	Edges	Faces
Tetrahedron	4	6	4
Cube	8	12	6
Octahedron	6	12	8
Dodecahedron	20	30	12
Icosahedron	12	30	20

Table 13.1. Vertices, edges, and faces of regular polyhedra.

Definition. The *dual* of a convex polyhedron is the polyhedron obtained by joining the centers of adjacent faces of the original polyhedron.

The dual of a regular polyhedron is always another regular polyhedron. In this particular case, we can make a correspondence between the vertices of a cube and the faces of a regular octahedron, and between the vertices of a regular octahedron and the faces of a cube. Figure 13.17 clearly shows the idea of duality.

Here, joining centers of adjacent faces of a cube creates a regular octahedron. Similarly, joining centers of adjacent faces of a regular octahedron creates a cube. So we say that the octahedron and cube are dual. In general, if we start with any type of polyhedron, when we take the dual and then the dual of the dual, we get the original type of polyhedron back.

The same kind of duality exists between the regular icosahedron and the regular dodecahedron. Each has 30 edges, and the number of faces and vertices of the icosahedron is the same as the number of vertices and faces of the dodecahedron.

Is there a dual for the regular tetrahedron? In fact, the regular tetrahedron is dual to itself. Figure 13.18 shows the correspondence between the faces and vertices of two regular tetrahedra.

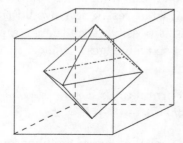

Figure 13.17. An octahedron as the dual of a cube.

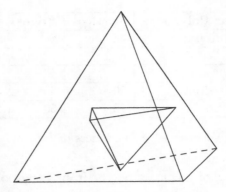

Figure 13.18. A tetrahedron as the dual of itself.

Challenge Problem. Try to construct models of a cube and a regular octahedron such that one fits inside the other and displays the duality relationship. Even more challenging, try to do this with a regular dodecahedron and a regular icosahedron.

13.2   Euler's Formula

Another relationship exists among these numbers of vertices, edges, and faces. The formula, named for Euler, says that in convex polyhedra, the number of vertices plus the number of faces minus the number of edges will always equal 2. We usually write this symbolically as $V - E + F = 2$.

Practice Problem 13.3. *Check that Euler's formula is true for the five regular polyhedra. Use Table 13.1.*

Does this formula work for other kinds of polyhedra? For example, does it hold for the trigonal bipyramid (see Figure 13.11)? If the formula is true for all polyhedra, what restrictions would there then be on V, E, and F? For example, is it possible that there's a polyhedron with eight vertices, eight faces, and sixteen edges? In fact, this would be impossible, and Theorem 13.1 tells why.

Theorem 13.1 (Euler's Formula). *If a convex polyhedron has V vertices, E edges, and F faces, then $V - E + F = 2$.*

Proof: Suppose we begin with a convex polyhedron that has E edges, F faces, and V vertices.

We use a proof technique that is common in mathematics; it is sometimes called the method of "invariants." We would like to show that, in the original polyhedron, $V - E + F = 2$. In order to show this, we describe a few methods for getting a simpler figure without changing the value of $V - E + F$. If we can show that, after doing enough of these simplifications, we get a figure with $V - E + F = 2$, then we know that the original polyhedron also had $V - E + F = 2$.

The first step is to make a two-dimensional "map" of the polyhedron. To do this, imagine that the polyhedron is made of rubber. After cutting out one face, imagine flattening out what's left onto a plane. Stretch the faces or edges to do this, but do not tear any more of the polyhedron. This can be done as long as the polyhedron is convex.

Now draw a map in the plane, indicating the vertices and edges of each of the polygons in the polyhedron. Do the stretching in just the right way, so that the edges are still straight lines. Then the resulting figure is a polygon with extra vertices and edges in its interior, which form the faces of the polyhedron. The plane surrounding this polygon can be thought of as the removed face.

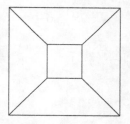

Figure 13.19. A flattened cube.

If we started with a cube, the flattened shape would look like Figure 13.19. How do the number of faces, edges, and vertices of this new figure differ from the original polyhedron? We have removed one face, so the number F is reduced by one. However, the number E and the number V are the same for the figure as for the original polyhedron. So our strategy now is to prove that the formula $V - E + F = 1$ holds for this figure, and then we will know the desired formula also works for our original polyhedron.

Now we remove parts of the figure in a systematic way. At each step, although V, E, and F may change, we show that the value of $V - E + F$ does not change. Our goal is to get a simple figure whose faces, edges, and vertices are easy to count.

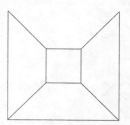

Figure 13.20. Flattened cube after removing an edge.

The first step is to remove any single edge on the outer polygon. If we do this, we also eliminate one of the faces by opening it up to the outside. The number of vertices will not be affected. Figure 13.20 shows what the flattened cube looks like after removing the edge.

We can translate these facts into equations as follows: Let V', E', and F' be the number of vertices, edges, and faces on Figure 13.20. Then $F' = F - 1$, $E' = E - 1$, $V' = V$. Do these new numbers change the value of $V - E + F$? $V' - E' + F' = V - (E - 1) + (F - 1) = V - E + F$. So the new figure has the same value of $V - E + F$ as the original figure.

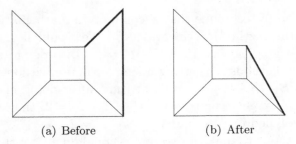

(a) Before   (b) After

Figure 13.21. Flattened cube before and after merging two edges into one.

In this new figure with one face removed, there may be some vertices that are attached to only two edges. Let's see what happens if we remove such a vertex, merging the two edges into one edge. For example, Figure 13.21 shows what happens with the cube.

In this case, we reduce both V and E by 1, without changing F. The result is

$$V' - E' + F' = (V - 1) - (E - 1) + F = V - E + F.$$

So we may remove this kind of vertex anytime we find one without affecting the value of $V - E + F$.

Notice how these two processes work by considering Table 13.2.

Using these two steps, how simple a figure can we get? Well, as long as there is an outer polygon, we may remove an edge. There are two cases when we no longer have an outer polygon:

Case 1. Two polygons separated by an edge (Figure 13.22).

Case 2. A single chain (Figure 13.23).

For case 1, we must work separately on each polygon. By removing an edge from each, we will get a figure that looks like a tree (Figure 13.24).

Move	Change in V	Change in E	Change in F	Change in $V - E + F$
Erasing an outer edge	0	-1	-1	0
Merging two consecutive edges	-1	-1	0	0

Table 13.2. Effects of changes to original figure.

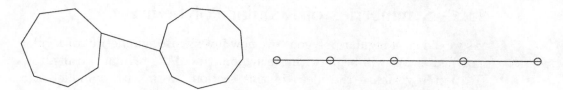

Figure 13.22. Case 1.   **Figure 13.23.** Case 2.

To make this tree into case 2, we remove one vertex and one edge together, which doesn't alter the value of $V' - E' + F'$. We can repeat this process until we are left with a single chain (Figure 13.25).

In case 2, we can merge edges in the chain until we have just a single edge and two vertices (Figure 13.25).

At this final stage, we have $V' - E' + F' = 2 - 1 + 0 = 1$. Since the only step at which we altered the value of $V - E + F$ was the initial removal of one face, we know that the original polyhedron must have satisfied the equation $V - E + F = V' - E' + F' + 1 = 2$.   □

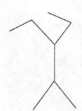

Figure 13.24. A tree.

Figure 13.25. One edge.

Practice Problem 13.4. *Does $V - E + F = 2$ hold for the trigonal bipyramid in Figure 13.11 (the polyhedron formed by two tetrahedrons glued along a face)?*

Practice Problem 13.5. *Our proof that $V - E + F = 2$ holds for convex polyhedra. Does the same equation hold for either of the following non-convex polyhedra?*

(a) A nonconvex polyhedron.

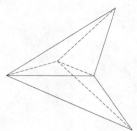

(b) A toroidal polyhedron with a hole in the middle.

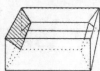

13.3   Symmetries of Regular Polyhedra

As we did with regular polygons, we now investigate symmetries of regular polyhedra. We will begin with the symmetries of the regular tetrahedron. To better visualize the ideas in this section, work with a model of a tetrahedron you construct.

Any symmetric motion of the tetrahedron will move a given vertex of the tetrahedron to one of the four vertices. Once that vertex is fixed, the three remaining vertices form a triangle whose vertices can be arranged in any of six different ways. Thus, there are $4 \cdot 6 = 24$ different symmetric motions of the tetrahedron. We can regard each symmetric motion of the tetrahedron as a permutation of the four vertices. Which permutations will occur?

Rotations of the Tetrahedron

Just as we did with regular polygons, we can rotate the regular tetrahedron in such a way that it will occupy the same space. First, imagine the tetrahedron sitting on a flat surface, and put an axis of rotation from the top vertex through the center of the bottom, or opposite, face. Since the base of this sitting tetrahedron is an equilateral triangle, we can rotate around this axis either 120° or 240° and the tetrahedron will occupy its original space.

If we number the vertices of the tetrahedron starting with 1 at the top, and 2, 3, and 4 at the bottom, the first rotation really is a rotation that fixed vertex number 1 and rotated the other three (see Figure 13.26). In

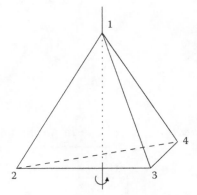

Figure 13.26. Rotating the tetrahedron.

our permutation notation, we can write this as R_1 (indicating that vertex number 1 is left alone by this rotation):

$$R_1 = \begin{pmatrix} 1 & 2 & 3 & 4 \\ 1 & 3 & 4 & 2 \end{pmatrix}.$$

Let's call R_1^2 a rotation of 240° around this axis. In our permutation notation,

$$R_1^2 = \begin{pmatrix} 1 & 2 & 3 & 4 \\ 1 & 4 & 2 & 3 \end{pmatrix}.$$

If we rotate the tetrahedron an additional 120°, we get R_1^3, or the identity I:

$$R_1^3 = I = \begin{pmatrix} 1 & 2 & 3 & 4 \\ 1 & 2 & 3 & 4 \end{pmatrix}.$$

An axis joining any of the four vertices to the center of the opposite face allows these same kinds of rotations. How many different rotations like this are there? In addition to the identity, each of the four vertices allows both a rotation of 120° and a rotation of 240°. So we have the identity plus eight rotational symmetries of the tetrahedron this way.

Practice Problem 13.6. *Write the other six rotational symmetries that fix one vertex using this permutation notation.*

There is another way we can rotate the tetrahedron and have it occupy its original space. Select one of the six edges and imagine an axis through the midpoint of that edge and through the midpoint of the opposite edge, as shown in Figure 13.27.

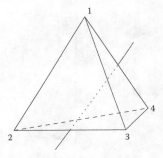

Figure 13.27. Another axis of rotation.

If we rotate the tetrahedron through 180° around this axis, we see that it occupies the same space it did originally. In this rotation, labeled S, vertices 1 and 4 change places, and so do vertices 2 and 3. In permutation notation, this rotation has the form

$$S = \begin{pmatrix} 1 & 2 & 3 & 4 \\ 4 & 3 & 2 & 1 \end{pmatrix}.$$

How many such rotations are there? There are three pairs of opposite edges, and each one allows one of these 180° rotations. This adds three new rotations. Thus, including the identity, we have found twelve rotations of the tetrahedron.

Practice Problem 13.7. *Write these three new rotations using permutation notation.*

Tetrahedron "Flips," or Reflections

When we were working with the equilateral triangle, we drew an axis through the triangle and "flipped," or reflected, the triangle over this axis to get a symmetric motion. Now, since we are using a three-dimensional figure, we will be reflecting it over a plane. For help with visualization, let's again label the vertices of the tetrahedron with the numbers 1, 2, 3, and 4, starting with the top (Figure 13.28).

Imagine a plane that bisects our tetrahedron through the edge joining vertices 1 and 2, and through the midpoint of the edge joining vertices 3 and 4 (Figure 13.29). A reflection through this plane will fix vertices 1 and 2, and will exchange vertices 3 and 4.

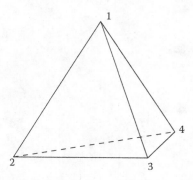

Figure 13.28. Initial position.

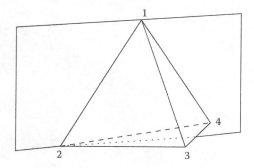

Figure 13.29. A plane of reflection.

If we label this symmetric motion with an F we can write F using permutation notation. In this case:

$$F = \begin{pmatrix} 1 & 2 & 3 & 4 \\ 1 & 2 & 4 & 3 \end{pmatrix}.$$

If you are working with a model and are trying to picture how this reflection or flip actually occurs, you may be having some difficulty. To realize why, think for a minute about the flips we did with the equilateral triangle. The triangle is a two-dimensional object, and when we flipped it, we needed three dimensions to actually perform this flip. Therefore, since the tetrahedron is a three-dimensional object, we would need four dimensions to perform this flip! However, even without using four dimensions to see what happens, we can use permutation notation to write down the result of flipping through a plane that bisects an edge.

Practice Problem 13.8. *Use permutation notation to write down the five additional flips through the other edges and midpoints.*

We now have a total of 18 symmetries of the tetrahedron that come from rotations and reflections.

Economy of Notation

When we were working with the symmetric motions of the triangle and square, we used economy of notation to write the tables using the rotations and just a single flip or reflection and their products to express all the elements of the group. We would like to do a similar thing for the tetrahedron. We have 12 rotations (including the identity). If we select a reflection, say the F illustrated in Figure 13.29, then the products of F with each of the rotations will result in an additional 12 symmetric motions. These account for 24 symmetric motions.

In fact, these represent all of the symmetric motions of the tetrahedron. Remember that if we are permuting four numbers or vertices, then there are precisely 24 possible permutations. Since we have found 24 symmetric motions, we must have them all. Thus, any one vertex of the tetrahedron can first be sent to any of the four vertices. Since the remaining three vertices are adjacent, they can be permuted in six ways, for a total of 24 permutations. The group of these symmetric motions is the group S_4.

Let's investigate some of the products of these reflections and rotations.

Example. If we use the rotation R_1 and flip F, the product $R_1 F$ will be given by

$$F = \begin{pmatrix} 1 & 2 & 3 & 4 \\ 1 & 2 & 4 & 3 \end{pmatrix}$$

$$R_1 = \begin{pmatrix} 1 & 2 & 3 & 4 \\ 1 & 3 & 4 & 2 \end{pmatrix}$$

$$R_1 F = \begin{pmatrix} 1 & 2 & 3 & 4 \\ 1 & 3 & 2 & 4 \end{pmatrix}.$$

This is one of the flips over the plane that bisects the edge connecting vertices 2 and 3.

Practice Problem 13.9. *Compute FR_1, FR_1^2, and $R_1^2 F$ using permutation notation, and identify the resulting motion. You may want to refer to Chapter 12 to recall how this multiplication works.*

Something new happens if we use a different rotation, say, R_4, and this same reflection. Remember that R_4 is the rotation through the axis that contains vertex 4, as shown in Figure 13.30.

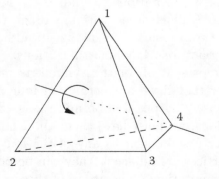

Figure 13.30. The rotation R_4.

This time, we multiply F times the rotation R_4

$$F = \begin{pmatrix} 1 & 2 & 3 & 4 \\ 1 & 2 & 4 & 3 \end{pmatrix},$$

$$R_4 = \begin{pmatrix} 1 & 2 & 3 & 4 \\ 2 & 3 & 1 & 4 \end{pmatrix},$$

and we get a new permutation P:

$$P = \begin{pmatrix} 1 & 2 & 3 & 4 \\ 2 & 3 & 4 & 1 \end{pmatrix}.$$

In this permutation, all four vertices change places in a way we haven't seen before. This is not one of our rotations, nor is it one of the reflections we have described. Here we have a symmetric motion of the tetrahedron that we have not been able to visualize because of our limitations when working in three dimensions.

Challenge Problem. Try to visualize the geometry that leads to the six motions of the tetrahedron that are neither pure rotations nor reflections through a plane.

Rotations of the Cube

Next come the symmetric motions of the cube. Before we begin to label them, let's think about what possibilities there are for permuting the eight vertices. This time, unlike in the case of the tetrahedron, the vertices are not all the same distance apart. The square is different from the triangle because of its diagonals, and the cube differs from the tetrahedron for the same reasons. Any symmetric motion of the cube must preserve the relationship among diagonal vertices and also must keep adjacent vertices adjacent to one another. Thus, to begin with, any vertex can be moved to any of the eight vertices. However, there is a restriction on where the other seven vertices may go. Once a vertex has been moved, its adjacent vertices can be permuted only among themselves. Arranging these vertices determines the locations of the remaining vertices of the cube. Since there are three adjacent vertices, they have six possible permutations. Thus, there will be at most $8 \cdot 6 = 48$ possible symmetric motions of the cube. Let's see how many we can find.

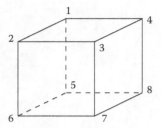

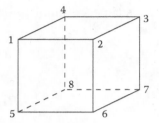

Figure 13.31. Cube in initial position.   **Figure 13.32.** Cube rotated 90°.

We begin again with the rotations because those are easier to visualize. There are several places to put an axis through a cube so that the rotation of the cube around the axis will be a symmetric motion. First, imagine an axis through the centers of opposite faces, say, the top and bottom. Rotating the cube 90°, 180°, or 270° about this axis will all be symmetric motions. There are three pairs of these opposite faces, and each allows these three rotations. Including the identity, these kinds of rotations account for ten symmetric motions of the cube.

Before we look for other rotations, let's think about these rotations as permutations of the eight vertices. Figure 13.31 shows a labeled cube in initial position.

A rotation R of 90° about an axis through the top and bottom face moves the vertices into the position shown in Figure 13.32.

In permutation notation, this is

$$
R = \begin{pmatrix} 1 & 2 & 3 & 4 & 5 & 6 & 7 & 8 \\ 2 & 3 & 4 & 1 & 6 & 7 & 8 & 5 \end{pmatrix}.
$$

Practice Problem 13.10. *Write some of the other rotations of this type in permutation form.*

Because symmetric motions must preserve distance, we observe that any symmetric motion permutes the main diagonals. Let's make explicit how this works using permutation notation. First, we need to realize that the diagonals are the pairs of vertices $A = \{1, 7\}$, $B = \{2, 8\}$, $C = \{3, 5\}$, and $D = \{4, 6\}$. In the permutation R, where does the pair $\{1, 7\}$ move? Permutation R sends vertex 1 to 2 and vertex 7 to 8, so the pair $A = \{1, 7\}$ moves to the pair $B = \{2, 8\}$. Similar things happen to the other diagonals. Thus, R leads to a permutation of the diagonals, $P(R)$,

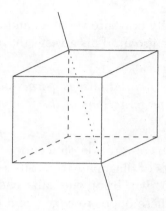

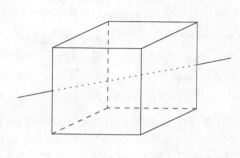

Figure 13.33. Rotation axis through opposite vertices.

Figure 13.34. Rotation axis through opposite edges.

written as

$$P(R) = \begin{pmatrix} A & B & C & D \\ B & C & D & A \end{pmatrix}.$$

Practice Problem 13.11. *Use each of the rotations we have found so far to write a permutation of the diagonals A, B, C, and D.*

The permutation

$$\begin{pmatrix} A & B & C & D \\ B & A & D & C \end{pmatrix}$$

is a permutation of the diagonals but does not explicitly tell you whether vertex 1 in the diagonal $A = \{1, 7\}$ goes to vertex 2 or vertex 8 in the diagonal $B = \{2, 8\}$.

While every permutation of the cube gives a permutation of the diagonals, each permutation of the diagonals comes from two different permutations of the vertices. Exactly one of the two permutations is a rotation.

Let's look for other rotations. Another place to put an axis for rotation is through diagonally opposite vertices (Figure 13.33). A model makes it clear that rotating around this axis by 120° three times will bring the cube back to its starting point.

Note that these rotations are different from the first ten because none of the earlier nonidentity rotations fixes a diagonal.

There are four of these pairs of diagonally opposite vertices, and each contributes two nonidentity rotations to the group. This gives eight more rotations, or 18 so far.

Practice Problem 13.12. *Write these kinds of rotations as permutations of the eight vertices and the corresponding permutations of the four main diagonals.*

There is still another place to put an axis of rotation on a cube. If we use an axis that goes through the midpoints of diagonally opposite edges (Figure 13.34), rotating twice through 180° will bring the cube back to its starting place. There are six pairs of these opposite edges, and each gives one more nonidentity rotation of the cube.

Practice Problem 13.13. *Write these kinds of rotations as permutations of the eight vertices and the corresponding permutations of the four main diagonals.*

So we have found 24 rotations of the cube. How do we know that these are all different symmetries? Two of these rotations might actually have the same effect on the cube. To show that this is not the case, for any particular rotation, we can see which points don't move at all when we do the rotation. There will always be two such "fixed points" for any rotation. These two points are the points where the axis of rotation intersects the cube. We have described three different categories of fixed points of rotations of the cube. One involves midpoints of opposite faces, the second involves pairs of vertices on main diagonals, and the third involves midpoints of diagonally opposite edges. We can accomplish a particular rotation using exactly one category of fixed points.

Once we categorize a rotation by determining its fixed points, we can distinguish it from other rotations in that category by looking at the effect of this rotation on the vertices. Within a category, each rotation is distinct. Therefore, each of the 24 rotations is a different rotation of the cube.

Challenge Problem. Show that the 24 permutations of the main diagonals of the cube form the group S_4.

Cube "Flips," or Reflections

As with the tetrahedron, we cannot physically flip the cube to see how these kinds of symmetric motions work. But if we imagine a plane through

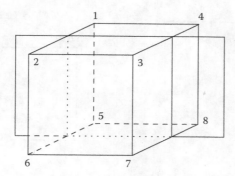

Figure 13.35. A plane of reflection for the cube.

the center of the cube bisecting the cube and parallel to one pair of faces (see Figure 13.35), we can see how the vertices will change places.

Call this reflection F. This reflection as a permutation of the eight vertices is written as

$$F = \begin{pmatrix} 1 & 2 & 3 & 4 & 5 & 6 & 7 & 8 \\ 2 & 1 & 4 & 3 & 6 & 5 & 8 & 7 \end{pmatrix}.$$

As a permutation of the diagonals, we have

$$P(F) = \begin{pmatrix} A & B & C & D \\ B & A & D & C \end{pmatrix}.$$

Practice Problem 13.14. *Use a different plane to get another reflection, and write it as a permutation of the eight vertices and of the four main diagonals.*

Challenge Problem. Which rotation led to the same permutation $P(F)$ of the main diagonals?

Let's count how many symmetric motions of the cube we have described. As was the case with the tetrahedron, when we multiply any reflection—in particular, this F—by each rotation, we get 24 more symmetric motions of the cube. We have 48 symmetric motions in all. These then account for all the symmetric motions of the cube.

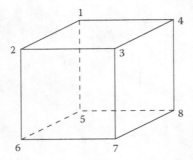

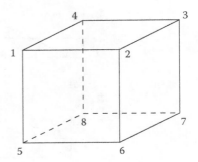

Figure 13.36. Cube in initial position.   **Figure 13.37.** Cube after rotation R.

13.4   Reflections and Rotations

If we take a careful look at the cube, we can identify the difference between arrangements that result from a rotation and arrangements that result from a reflection. To see this, number the vertices of the cube as before (Figure 13.36).

Look at the face with vertices labeled 1, 2, 3, and 4. Looking from above, the numbers are in counterclockwise order. After any rotation, when we look at this same face from above the cube, we see that the numbers 1, 2, 3, and 4 are still arranged in counterclockwise order. For example, look at the cube after the rotation R has been performed (Figure 13.37). We say that this rotation *preserves* the orientation of the vertices.

In contrast, look at the cube after the reflection F has been performed (Figure 13.38). Now if we view the 1-2-3-4 face from above the cube, we find that the numbers are arranged in clockwise order. The reflection F has *reversed* the orientation of the vertices.

There is nothing special about the 1-2-3-4 face. Any rotation will preserve the order of the vertices on every face, and any reflection will reverse the orientation of every face.

Practice Problem 13.15. *Confirm that all rotations preserve the orientation of the vertices and that all reflections reverse orientation of the vertices.*

This helps to explain why in the symmetry group of the cube, a rotation times a rotation will always be a rotation, and why a reflection

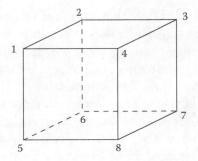

Figure 13.38. Cube after reflection F.

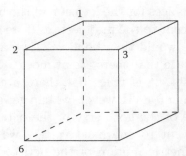

Figure 13.39. Vertex 2 and adjacent vertices.

times a rotation is a reflection. Two reflections combined will restore orientation, and thus will be a rotation.

When we counted the number of possible symmetric motions of the cube, we first counted the eight places any vertex could go. Then we talked about the six possible places the adjacent vertices could go. We focus our attention now on just the upper front corner of the cube (vertex 2) and its adjacent vertices 1, 3, and 6 (Figure 13.39).

If the three adjacent vertices are all permuted in a cyclic fashion, the result will be a rotation, since the orientation of the vertices will be preserved. If, however, only two of these vertices are switched, then the result will be a reflection because the orientation of the vertices will be reversed.

Challenge Problem. Write the 24 symmetric motions of the cube that are rotations, both as permutations of the vertices and as permutations of the diagonals. Show that the resulting permutations of the diagonals form the group S_4.

Symmetries of the Octahedron

The problem of finding the symmetries of the octahedron is really quite simple. Because the cube and the octahedron are dual polyhedra, any symmetry of one will be a symmetry of the other, and vice versa. So, in fact, the same symmetries of the cube will be symmetries of the octahedron.

As we did with the cube, let's try to find an upper limit on the number of symmetric rotations. Since the octahedron has six vertices, there are six choices for the location of the first vertex. Then, there are four vertices adjacent to the first vertex. If we could arrange them in any possible way, there would then be $4! = 24$ ways to arrange these vertices, which would indicate that there are at most $6 \cdot 24 = 144$ symmetries of the octahedron. In fact, it is true that there are no more than 144 symmetries of the octahedron. However, we can get a much better upper limit by considering more of the geometry of the octahedron. If we look at the four vertices adjacent to one fixed vertex, we notice that they form a square. Any symmetric motion of the octahedron must preserve this square. As we saw in our discussion of symmetries of the square, some permutations of the vertices are impossible. Therefore, there are not 24 ways to arrange the four adjacent vertices, but only 8, which is the number of symmetries of the square. Combined with the 6 ways to position the first vertex, this gives 48 possible arrangements of the octahedron—the same number as we found for the cube.

The Dodecahedron and the Icosahedron

The other two regular polyhedra are the dodecahedron and the icosahedron. What are their symmetric motions? These two polyhedra are also duals, so to discover their symmetries, we really need to work with only one. These are much more complicated figures, but seeing some of the motions is easy. Let's work with the dodecahedron.

How many possible symmetries are there? Any symmetric motion of the dodecahedron will move a given vertex to one of the 20 vertices. Then, there are 3 vertices adjacent to the chosen vertex. Permuting these in any possible way results in 6 choices for their arrangement. These choices determine the location of every other vertex of the dodecahedron because distances must be preserved by symmetry. So, there are at most $20 \cdot 6 = 120$ symmetric motions of the dodecahedron.

Now let's see if we can identify all 120 motions. Start by thinking about the dodecahedron sitting on a plane, with pentagons at the top and at the bottom. If we put an axis through the center of the top and bottom pentagons (Figure 13.40), we can see that rotating the dodecahedron five times around this axis through an angle of 72° will bring it back to its original position. There are six pairs of these opposite faces, so, including the identity, there are 25 of these types of rotations.

We could also put an axis of rotation through opposite pairs of vertices (Figure 13.41). Rotating three times through 120° around this axis will

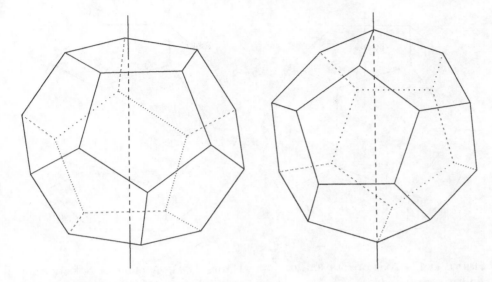

Figure 13.40. Axis of rotation for the dodecahedron (faces).

Figure 13.41. Axis of rotation for the dodecahedron (vertices).

bring the dodecahedron back to its original position. Since there are ten pairs of opposite vertices, these add 20 more rotations.

We could also put an axis through the midpoints of opposite edges (Figure 13.42). We can rotate 180° about these axes. Since there are 15 pairs of opposite edges, these give 15 more rotations. Altogether, we have found 60 rotations so far.

We can also slice the dodecahedron in half with a plane and produce a symmetric motion that is a reflection (Figure 13.43). As with the cube, we must now ask ourselves whether any of these are counted twice. That is, are we counting one symmetry as two different rotations? Just as with the cube, each rotation has exactly two "fixed points" that are not moved by the rotation. The line joining these two points must be the axis of symmetry, and then the effect of the rotation on the vertices tells us precisely what the rotation is.

As with the other three-dimensional figures, multiplying the above reflection times the 60 rotations gives 120 symmetric motions. So we have found them all.

Again, having moved a vertex to another vertex, moving its adjacent vertices can result in either an orientation preserving or reversing motion. Thus, 60 of the motions are rotations, and 60 are not.

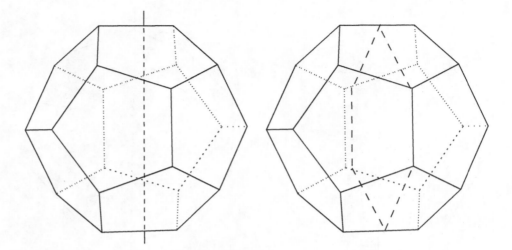

Figure 13.42. Axis of rotation for the dodecahedron (edges).

Figure 13.43. A plane of reflection for the dodecahedron.

13.5   Variations on a Theme: Other Polyhedra

We can make more three-dimensional figures by using more than one type of regular polygon to create a polyhedron. One familiar example is the arrangement of regular pentagons and regular hexagons that fit together to make a soccer ball (Figure 13.44). In this polyhedron, each vertex is surrounded by the same kinds of polygons in the same order. There are two regular hexagons and a regular pentagon meeting at a vertex.

Definition. A *semiregular polyhedron* is a convex polyhedron with faces that comprise more than one kind of regular polygon. Every vertex is surrounded by the same kinds of polygons in the same cyclic order.

We can label these polyhedra by the types of regular polygons that surround a vertex. For example, the soccer ball, which has two regular hexagons and a regular pentagon meeting at each vertex, could be labeled with a 5-6-6. We could just as well call it a 6-6-5 or 6-5-6 pattern. Using this notation, a tetrahedron has pattern 3-3-3, and a cube has 4-4-4.

Practice Problem 13.16. *Label the other Platonic solids by using the notation for arrangement of faces around a vertex.*

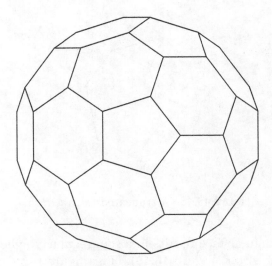

Figure 13.44. A soccer ball.

The semiregular polyhedra are often called Archimedean solids. We know that Archimedes[2] wrote a text about these figures because another Greek author included Archimedes' work in his book. This Greek mathematician, Pappus,[3] was a writer at the Alexandrian library in the fourth century and compiled much of the mathematics known at the time. Even though the original text is lost, Pappus tells us that Archimedes wrote down instructions for making models of these polyhedra.

How might we discover other semiregular polyhedra? We can figure out how to make many of them by beginning with one of the Platonic polyhedra and modifying it in some way. For example, if we begin with a regular tetrahedron and divide each edge in thirds, we can connect the marks to get a hexagon on each face. Removing the corners and replacing

---

[2] Archimedes (287–212 B.C.) was a native of Syracuse and visited the Greek center of learning at Alexandria. He worked in geometry, and wrote a (now lost) treatise that included descriptions of semiregular polyhedra and how to make models of them. He is famous for solving unusual problems, including how to help a ruler decide if a crown was made of pure gold, how to move the earth, and how to calculate the areas and volumes of geometric figures.

[3] Pappus (ca. 290–350) lived in Alexandria and spent his time doing mathematics and compiling results from the earlier mathematicians, such as Euclid and Archimedes. In his work *The Collection*, we see formulas in geometry and ideas about mechanics. He recognized the honeycomb pattern of bees as an efficient way of using space to store honey.

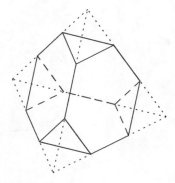

Figure 13.45. A truncated tetrahedron.

each with a triangular face results in a semiregular polyhedron called the truncated tetrahedron (Figure 13.45). This figure has eight faces. Four faces are regular hexagons and four are equilateral triangles; it therefore has the pattern 3-6-6.

Practice Problem 13.17. *Divide the edges of the cube in thirds to make a truncated cube. Tell how many faces there are, what kinds of polygons these faces are, and what the pattern is.*

There are 13 semiregular polyhedra, and we can see how to make each of them by altering one of the regular polyhedra. For example, we can truncate all five Platonic polyhedra by dividing the edges in thirds and get semiregular polyhedra. In fact, the soccer ball is a model of a truncated icosahedron. The truncated version of each of the regular polyhedra has the same axes of rotation as its original polyhedra. We have already seen the truncated versions of the tetrahedron and the cube.

To make the truncated octahedron requires that square pyramids be removed from each vertex (Figure 13.46). The remaining faces become hexagons, and the truncated octahedron has the pattern 4-6-6. This polyhedron is the only Archimedean polyhedron that fills space, leaving no gaps and having no overlaps. The truncated dodecahedron, made by removing tetrahedra from each vertex, has the pattern 3-10-10 (Figure 13.47).

Another method for creating Archimedean polyhedra by removing corners of the Platonic polyhedra uses a slightly different way. For example, we can begin with a cube and mark the midpoint of each edge. Then we connect the midpoints on each face to get a smaller square. If we cut off

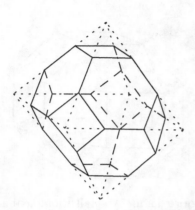

Figure 13.46. A truncated octahedron.

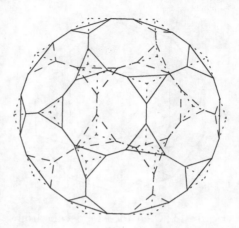

Figure 13.47. A truncated dodecahedron.

the eight corners of the cube, we will create eight faces that are equilateral triangles. The resulting polyhedron, called a cuboctahedron, has 14 faces (Figure 13.48). Six faces are squares and eight are equilateral triangles. It has the pattern 3-4-3-4. Because of duality, the same semiregular polyhedron results from bisecting the midpoints of the edges of an octahedron (Figure 13.49).

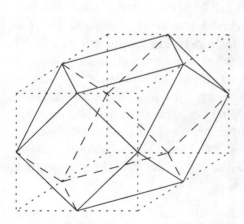

Figure 13.48. A cuboctahedron.

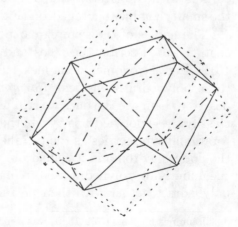

Figure 13.49. A cuboctahedron (in an octahedron).

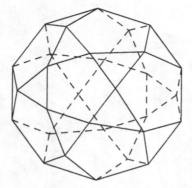

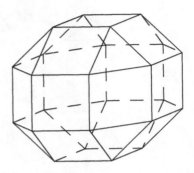

Figure 13.50. An icosidodecahedron.   **Figure 13.51.** A small rhombicuboctahedron.

If we bisect the edges of a dodecahedron and remove the tetrahedra
on the corners, we get a solid called an icosidodecahedron, with the pattern 3-5-3-5 (Figure 13.50). Again, because of duality, we get the same
polyhedron if we bisect the edges of an icosahedron and remove pyramids
with pentagonal bases from the corners.

There is another way to see how to use the Platonic solids to get more
of the semiregular polyhedra. Start with the cube, and imagine that the
edges come apart and the faces expand outward. Into the empty space,
we insert 12 squares corresponding to the 12 edges, and then fill in the
corners with equilateral triangles. This procedure gives us a figure called
the small rhombicuboctahedron (Figure 13.51). It has the pattern 3-4-4-4.

The names of the semiregular polyhedra are long, complicated words
that try to describe the faces and the underlying symmetry type. In
1619, the astronomer Johannes Kepler[4] wrote detailed descriptions of
these polyhedra in his text *Harmonicus Mundi*. The names he used to
describe them are the ones we still use today. One way to understand and
classify these polyhedra is by whether they have the same kinds of axes for
rotational symmetries as the tetrahedron, or the cube and octahedron, or
the dodecahedron and icosahedron. Each will have the symmetry type of
the underlying polyhedra used to make it. Table 13.3 lists the semiregular
polyhedra together with their patterns and symmetry type.

---

[4]Johannes Kepler (1571–1630) was a mathematician as well as an astronomer and a
physicist. His three laws of planetary motion helped establish new ideas in astronomy.
He carefully studied regular polyhedra and made a model of the known universe in
which the planets were in orbits that matched up with the five Platonic solids.

Name	Pattern	Symmetry type
Truncated Tetrahedron	3-6-6	Tetrahedron
Truncated Cube	3-8-8	Cube/Octahedron
Cuboctahedron	3-4-3-4	Cube/Octahedron
Truncated Octahedron	4-6-6	Cube/Octahedron
Small Rhombicuboctahedron	3-4-4-4	Cube/Octahedron
Great Rhombicuboctahedron	4-6-8	Cube/Octahedron
Snub Cube	3-3-3-3-4	Cube/Octahedron
Truncated Dodecahedron	3-10-10	Dodecahedron/Icosahedron
Truncated Icosahedron	5-6-6	Dodecahedron/Icosahedron
Icosidodecahedron	3-5-3-5	Dodecahedron/Icosahedron
Small Rhombicosidodecahedron	3-4-5-4	Dodecahedron/Icosahedron
Great Rhombicosidodecahedron	4-6-10	Dodecahedron/Icosahedron
Snub Dodecahedron	3-3-3-3-5	Dodecahedron/Icosahedron

Table 13.3. Semiregular polyhedra.

Just as the regular polyhedra have duals, so do the semiregular polyhedra. In an Archimedean dual, all the faces are congruent, but they are not regular polygons. Moreover, not all the vertices have the same number of edges, so there is no pattern to describe the vertices. One of these duals is a familiar shape, occurring in crystal form in gems such as the garnet. This shape, the rhombic dodecahedron, is the dual of the cuboctahedron. It has 12 faces, each one a rhombus. This is the only one of the Archimedean duals that can fill space. Another of the duals that may be familiar to game players is the rhombic triacontahedron, the dual of the icosidodecahedron, which makes a 30-sided die.

Prisms and Pyramids

There are two infinite families of semiregular polyhedra, prisms and pyramids.

Definition. A *regular prism* is a polyhedron made of two congruent regular n-gons, $n \geq 3$, in parallel planes, with squares for the remaining faces (Figure 13.52). A regular prism has $n + 2$ faces and a pattern of 4-4-n. The cube is an example of a regular prism.

If we begin with two congruent regular n-gons, $n \geq 3$, in parallel planes and rotate one n-gon by an angle of π/n, we can make an *antiprism* by

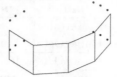

Figure 13.52. A regular prism.   **Figure 13.53.** A regular antiprism.

using equilateral triangles for the remaining faces (Figure 13.53). An antiprism has the pattern 3-3-3-n.

Practice Problem 13.18. *Use a model or a diagram of an octahedron to show it is an antiprism with equilateral triangles as bases.*

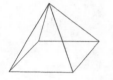

Figure 13.54. A square pyramid.

If we do not insist on a uniform pattern at each vertex, there are even more families of convex polyhedra. A *regular pyramid* is a polyhedron that has a regular polygon for the base and congruent triangles for the remaining faces. Half an octahedron is an example of a pyramid with a square for the base (Figure 13.54). Notice that the vertices at the bases of the pyramid have the pattern 3-3-4, whereas the vertex at the top has the pattern 3-3-3-3.

Other Convex and Nonconvex Polyhedra

Mathematicians have been interested in classifying different kinds of polyhedra since ancient times. The trigonal bipyramid we saw in Section 13.1 (Figure 13.11) is just two tetrahedra glued together along one pair of faces. This polyhedron is an example of a *deltahedron*, a polyhedron whose faces are all equilateral triangles. The tetrahedron, octahedron, and icosahedron are all deltahedra. Altogether there are eight convex deltahedra.

Practice Problem 13.19. *Show that gluing two pentagonal pyramids along their bases makes a deltahedron with ten faces. This is the same deltahedron we get by gluing two "caps" of icosahedra together.*

There are convex deltahedra with 4, 6, 8, 10, 12, 14, 16, and 20 faces. It is instructive to construct models of new ones. One of the nonconvex deltahedra can be made by taking the union of two tetrahedra, one upside-down from the other (Figure 13.55).

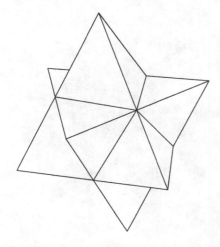

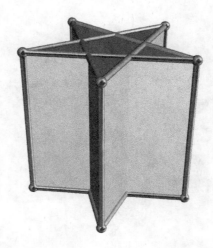

Figure 13.55. Two upside-down tetrahedra.

Figure 13.56. A pentagrammic prism.

We can make more complicated polyhedra by starting with a pair of nonconvex polygons, such as pentagrams, for the bases, and then making a prism with square vertical faces[5] (see Figure 13.56).

We can also attach pyramids to the faces of regular polyhedra to make *stellated polyhedra*. The small stellated dodecahedron (Figure 13.57) arises from attaching pyramids to the faces of a dodecahedron. The great stellated dodecahedron (Figure 13.58) is what we call the polyhedron that results from extending the faces of an icosahedron.

Still another way to describe a nonconvex polyhedron is one whose faces are all regular polygons and has the same number of faces meeting at each vertex, but allowing the faces to intersect each other and allowing the faces to be regular nonconvex polygons, like the pentagram.

Figure 13.59 shows how the small stellated dodecahedron has pentagram "faces." Kepler investigated these *star polyhedra* and found two of them. The other two were not found until the early nineteenth century, and Cauchy proved that there are only four of them.

A *uniform polyhedron* is any polyhedron, convex or not, whose vertices have the same pattern. Perhaps the most amazing of these is one called the Yog-Sothoth, named after a fictional ancient demon. It has 112 faces,

---

[5]Images of a pentagrammic prism, small stellated dodecahedron, and great stellated dodecahedron have been created with Robert Webb's Great Stella software (http://www.software3d.com/Stella.html).

Figure 13.57. A small stellated dodecahedron.

Figure 13.58. A great stellated dodecahedron.

of which 12 are pentagrams, 40 are triangles of one type, and 60 are
triangles of another. It has the same symmetry type as the icosahedron
and the dodecahedron.

Figure 13.59. Another view of the small stellated dodecahedron.

Diagrams (Nets) for Making Polyhedra

Net for a cube:

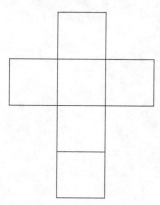

Net for a tetrahedron:

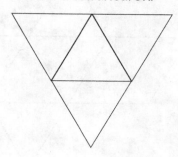

Net for a dodecahedron:

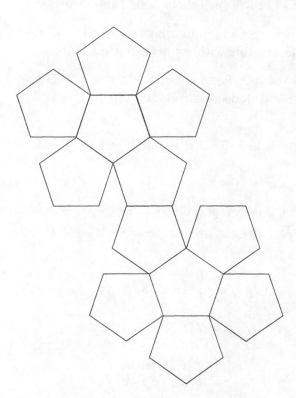

Net for an icosahedron:

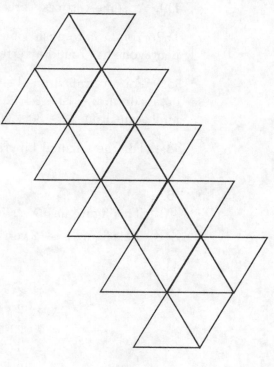

Net for an octahedron:

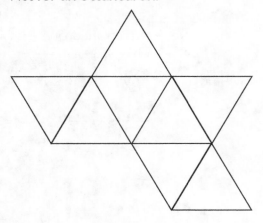

Practice Problem Solutions and Hints

13.1. At some vertices, three faces meet; at others, four faces meet.

13.2. You have to push some pentagons up and others down to make four pentagons fit around a vertex. Try this with some cutout pentagons.

13.3. For a tetrahedron, $4 - 6 + 4 = 2$. For a cube, $8 - 12 + 6 = 2$. For an octahedron, $6 - 12 + 8 = 2$. For a dodecahedron, $20 - 30 + 12 = 2$. For an icosahedron, $12 - 30 + 20 = 2$.

13.4. For the trigonal bipyramid, $V = 5$, $E = 9$ and $F = 6$ so $V - E + F = 2$.

13.5.

(a) This figure has 5 vertices, 9 edges, and 6 faces. So $V - E + F = 2$.

(b) This figure has 12 vertices, 24 edges, and 12 faces. So $V - E + F = 0$.

13.6.

$$R_2 = \begin{pmatrix} 1 & 2 & 3 & 4 \\ 3 & 2 & 4 & 1 \end{pmatrix}, \quad R_2^2 = \begin{pmatrix} 1 & 2 & 3 & 4 \\ 4 & 2 & 1 & 3 \end{pmatrix},$$

$$R_3 = \begin{pmatrix} 1 & 2 & 3 & 4 \\ 2 & 4 & 3 & 1 \end{pmatrix}, \quad R_3^2 = \begin{pmatrix} 1 & 2 & 3 & 4 \\ 4 & 1 & 3 & 2 \end{pmatrix},$$

$$R_4 = \begin{pmatrix} 1 & 2 & 3 & 4 \\ 2 & 3 & 1 & 4 \end{pmatrix}, \quad R_4^2 = \begin{pmatrix} 1 & 2 & 3 & 4 \\ 3 & 1 & 2 & 4 \end{pmatrix}.$$

13.7. The rotation about the axis between the midpoints of the 1-3 and 2-4 edges is

$$\begin{pmatrix} 1 & 2 & 3 & 4 \\ 3 & 4 & 1 & 2 \end{pmatrix}.$$

The rotation about the axis between the midpoints of the 1-2 and 3-4 edges is

$$\begin{pmatrix} 1 & 2 & 3 & 4 \\ 2 & 1 & 4 & 3 \end{pmatrix}.$$

The rotation about the axis between the midpoints of the 1-4 and 2-3 edges is described in the text.

13.8. The flip over the plane through the 1-3 edge and the midpoint of the 2-4 edge is

$$\begin{pmatrix} 1 & 2 & 3 & 4 \\ 1 & 4 & 3 & 2 \end{pmatrix}.$$

The flip over the plane through the 1-4 edge and the midpoint of the 2-3 edge is

$$\begin{pmatrix} 1 & 2 & 3 & 4 \\ 1 & 3 & 2 & 4 \end{pmatrix}.$$

The flip over the plane through the 2-3 edge and the midpoint of the 1-4 edge is

$$\begin{pmatrix} 1 & 2 & 3 & 4 \\ 4 & 2 & 3 & 1 \end{pmatrix}.$$

The flip over the plane through the 2-4 edge and the midpoint of the 1-3 edge is

$$\begin{pmatrix} 1 & 2 & 3 & 4 \\ 3 & 2 & 1 & 4 \end{pmatrix}.$$

The flip over the plane through the 3-4 edge and the midpoint of the 1-2 edge is

$$\begin{pmatrix} 1 & 2 & 3 & 4 \\ 2 & 1 & 3 & 4 \end{pmatrix}.$$

13.9. The motion FR_1 is a flip over the plane through the 1-3 edge and the midpoint of the 2-4 edge:

$$R_1 = \begin{pmatrix} 1 & 2 & 3 & 4 \\ 1 & 3 & 4 & 2 \end{pmatrix}$$

$$F = \begin{pmatrix} 1 & 2 & 3 & 4 \\ 1 & 2 & 4 & 3 \end{pmatrix}$$

$$FR_1 = \begin{pmatrix} 1 & 2 & 3 & 4 \\ 1 & 4 & 3 & 2 \end{pmatrix}.$$

The motion FR_1^2 is a flip over the plane through the 1-4 edge and the midpoint of the 2-3 edge:

$$R_1^2 = \begin{pmatrix} 1 & 2 & 3 & 4 \\ 1 & 4 & 2 & 3 \end{pmatrix}$$

$$F = \begin{pmatrix} 1 & 2 & 3 & 4 \\ 1 & 2 & 4 & 3 \end{pmatrix}$$

$$FR_1^2 = \begin{pmatrix} 1 & 2 & 3 & 4 \\ 1 & 3 & 2 & 4 \end{pmatrix}.$$

The motion $R_1^2 F$ is a flip over the plane through the 1-3 edge and the midpoint of the 2-4 edge:

$$F = \begin{pmatrix} 1 & 2 & 3 & 4 \\ 1 & 2 & 4 & 3 \end{pmatrix}$$

$$R_1^2 = \begin{pmatrix} 1 & 2 & 3 & 4 \\ 1 & 4 & 2 & 3 \end{pmatrix}$$

$$R_1^2 F = \begin{pmatrix} 1 & 2 & 3 & 4 \\ 1 & 4 & 3 & 2 \end{pmatrix}.$$

13.10. A rotation by 90° about an axis through the front and back faces would yield this permutation:

$$R' = \begin{pmatrix} 1 & 2 & 3 & 4 & 5 & 6 & 7 & 8 \\ 5 & 6 & 2 & 1 & 8 & 7 & 3 & 4 \end{pmatrix}.$$

A rotation by 90° about an axis through the two side faces would yield this permutation:

$$R'' = \begin{pmatrix} 1 & 2 & 3 & 4 & 5 & 6 & 7 & 8 \\ 2 & 6 & 7 & 3 & 1 & 5 & 8 & 4 \end{pmatrix}.$$

13.11. Rotations by 90°, 180°, and 270°, respectively, around an axis through the top and bottom faces are written below both as permutations of the vertices and permutations of the diagonals:

$$90° = \begin{pmatrix} 1 & 2 & 3 & 4 & 5 & 6 & 7 & 8 \\ 2 & 3 & 4 & 1 & 6 & 7 & 8 & 5 \end{pmatrix} = \begin{pmatrix} A & B & C & D \\ B & C & D & A \end{pmatrix},$$

$$180° = \begin{pmatrix} 1 & 2 & 3 & 4 & 5 & 6 & 7 & 8 \\ 3 & 4 & 1 & 2 & 7 & 8 & 5 & 6 \end{pmatrix} = \begin{pmatrix} A & B & C & D \\ C & D & A & B \end{pmatrix},$$

$$270° = \begin{pmatrix} 1 & 2 & 3 & 4 & 5 & 6 & 7 & 8 \\ 4 & 1 & 2 & 3 & 8 & 5 & 6 & 7 \end{pmatrix} = \begin{pmatrix} A & B & C & D \\ D & A & B & C \end{pmatrix}.$$

Likewise, rotations by 90°, 180°, and 270°, respectively, around an axis through the front and back faces are written as

$$90° = \begin{pmatrix} 1 & 2 & 3 & 4 & 5 & 6 & 7 & 8 \\ 5 & 6 & 2 & 1 & 8 & 7 & 3 & 4 \end{pmatrix} = \begin{pmatrix} A & B & C & D \\ C & D & B & A \end{pmatrix},$$

$$180° = \begin{pmatrix} 1 & 2 & 3 & 4 & 5 & 6 & 7 & 8 \\ 8 & 7 & 6 & 5 & 4 & 3 & 2 & 1 \end{pmatrix} = \begin{pmatrix} A & B & C & D \\ B & A & D & C \end{pmatrix},$$

$$270° = \begin{pmatrix} 1 & 2 & 3 & 4 & 5 & 6 & 7 & 8 \\ 4 & 3 & 7 & 8 & 1 & 2 & 6 & 5 \end{pmatrix} = \begin{pmatrix} A & B & C & D \\ D & C & A & B \end{pmatrix}.$$

Finally, rotations by 90°, 180°, and 270°, respectively, around an axis through the two side faces are written as

$$90° = \begin{pmatrix} 1 & 2 & 3 & 4 & 5 & 6 & 7 & 8 \\ 2 & 6 & 7 & 3 & 1 & 5 & 8 & 4 \end{pmatrix} = \begin{pmatrix} A & B & C & D \\ B & D & A & C \end{pmatrix},$$

$$180° = \begin{pmatrix} 1 & 2 & 3 & 4 & 5 & 6 & 7 & 8 \\ 6 & 5 & 8 & 7 & 2 & 1 & 4 & 3 \end{pmatrix} = \begin{pmatrix} A & B & C & D \\ D & C & B & A \end{pmatrix},$$

$$270° = \begin{pmatrix} 1 & 2 & 3 & 4 & 5 & 6 & 7 & 8 \\ 5 & 1 & 4 & 8 & 6 & 2 & 3 & 7 \end{pmatrix} = \begin{pmatrix} A & B & C & D \\ C & A & D & B \end{pmatrix}.$$

13.12. The rotations around various diagonals, written as a permutation of the vertices and as a permutations of the diagonals are

$$R_A = \begin{pmatrix} 1 & 2 & 3 & 4 & 5 & 6 & 7 & 8 \\ 1 & 4 & 8 & 5 & 2 & 3 & 7 & 6 \end{pmatrix} = \begin{pmatrix} A & B & C & D \\ A & D & B & C \end{pmatrix},$$

$$R_A^2 = \begin{pmatrix} 1 & 2 & 3 & 4 & 5 & 6 & 7 & 8 \\ 1 & 5 & 6 & 2 & 4 & 8 & 7 & 3 \end{pmatrix} = \begin{pmatrix} A & B & C & D \\ A & C & D & B \end{pmatrix},$$

$$R_B = \begin{pmatrix} 1 & 2 & 3 & 4 & 5 & 6 & 7 & 8 \\ 6 & 2 & 1 & 5 & 7 & 3 & 4 & 8 \end{pmatrix} = \begin{pmatrix} A & B & C & D \\ D & B & A & C \end{pmatrix},$$

$$R_B^2 = \begin{pmatrix} 1 & 2 & 3 & 4 & 5 & 6 & 7 & 8 \\ 3 & 2 & 6 & 7 & 4 & 1 & 5 & 8 \end{pmatrix} = \begin{pmatrix} A & B & C & D \\ C & B & D & A \end{pmatrix},$$

$$R_C = \begin{pmatrix} 1 & 2 & 3 & 4 & 5 & 6 & 7 & 8 \\ 6 & 7 & 3 & 2 & 5 & 8 & 4 & 1 \end{pmatrix} = \begin{pmatrix} A & B & C & D \\ D & A & C & B \end{pmatrix},$$

$$R_C^2 = \begin{pmatrix} 1 & 2 & 3 & 4 & 5 & 6 & 7 & 8 \\ 8 & 4 & 3 & 7 & 5 & 1 & 2 & 6 \end{pmatrix} = \begin{pmatrix} A & B & C & D \\ B & D & C & A \end{pmatrix},$$

$$R_D = \begin{pmatrix} 1 & 2 & 3 & 4 & 5 & 6 & 7 & 8 \\ 8 & 5 & 1 & 4 & 7 & 6 & 2 & 3 \end{pmatrix} = \begin{pmatrix} A & B & C & D \\ B & C & A & D \end{pmatrix},$$

$$R_D^2 = \begin{pmatrix} 1 & 2 & 3 & 4 & 5 & 6 & 7 & 8 \\ 3 & 7 & 8 & 4 & 2 & 6 & 5 & 1 \end{pmatrix} = \begin{pmatrix} A & B & C & D \\ C & A & B & D \end{pmatrix}.$$

13.13. The rotation through the axis connecting the midpoints of the 1-2 and 7-8 edges, written as a permutation of the vertices and a permutation of the axes, is

$$\begin{pmatrix} 1 & 2 & 3 & 4 & 5 & 6 & 7 & 8 \\ 2 & 1 & 5 & 6 & 3 & 4 & 8 & 7 \end{pmatrix} = \begin{pmatrix} A & B & C & D \\ B & A & C & D \end{pmatrix}.$$

The rotation through the axis connecting the midpoints of the 1-4 and 6-7 edges, written as a permutation of the vertices and a permutation of the axes, is

$$\begin{pmatrix} 1 & 2 & 3 & 4 & 5 & 6 & 7 & 8 \\ 4 & 8 & 5 & 1 & 3 & 7 & 6 & 2 \end{pmatrix} = \begin{pmatrix} A & B & C & D \\ D & B & C & A \end{pmatrix}.$$

The rotation through the axis connecting the midpoints of the 1-5 and 3-7 edges, written as a permutation of the vertices and a permutation of the axes, is

$$\begin{pmatrix} 1 & 2 & 3 & 4 & 5 & 6 & 7 & 8 \\ 5 & 8 & 7 & 6 & 1 & 4 & 3 & 2 \end{pmatrix} = \begin{pmatrix} A & B & C & D \\ C & B & A & D \end{pmatrix}.$$

The rotation through the axis connecting the midpoints of the 2-3 and 5-8 edges, written as a permutation of the vertices and a permutation of the axes, is

$$\begin{pmatrix} 1 & 2 & 3 & 4 & 5 & 6 & 7 & 8 \\ 7 & 3 & 2 & 6 & 8 & 4 & 1 & 5 \end{pmatrix} = \begin{pmatrix} A & B & C & D \\ A & C & B & D \end{pmatrix}.$$

The rotation through the axis connecting the midpoints of the 2-6 and 4-8 edges, written as a permutation of the vertices and a permutation of the axes, is

$$\begin{pmatrix} 1 & 2 & 3 & 4 & 5 & 6 & 7 & 8 \\ 7 & 6 & 5 & 8 & 3 & 2 & 1 & 4 \end{pmatrix} = \begin{pmatrix} A & B & C & D \\ A & D & C & B \end{pmatrix}.$$

The rotation through the axis connecting the midpoints of the 3-4 and 5-6 edges, written as a permutation of the vertices and a permutation of the axes, is

$$\begin{pmatrix} 1 & 2 & 3 & 4 & 5 & 6 & 7 & 8 \\ 7 & 8 & 4 & 3 & 6 & 5 & 1 & 2 \end{pmatrix} = \begin{pmatrix} A & B & C & D \\ A & B & D & C \end{pmatrix}.$$

13.14. Use the plane bisecting the 1-4, 2-3, 6-7, and 5-8 edges to get

$$F^* = \begin{pmatrix} 1 & 2 & 3 & 4 & 5 & 6 & 7 & 8 \\ 4 & 3 & 2 & 1 & 8 & 7 & 6 & 5 \end{pmatrix},$$

$$P(F^*) = \begin{pmatrix} A & B & C & D \\ D & C & B & A \end{pmatrix}.$$

13.15. Do it!

13.16. The label for the octahedron is 3-3-3-3; for the icosahedron, it is 3-3-3-3-3; and for the dodecahedron, it is 5-5-5.

13.17. This figure has 14 faces: 6 regular octagons and 8 equilateral triangles. It has the pattern 3-8-8.

13.18. Do it!

13.19. Do it!

Exercises

13.1. Using the nets on pages 285–286, construct a model of each of the regular polyhedra.

13.2. Why is there no regular polyhedron with only two polygons meeting at a vertex?

13.3. The definition of a regular polyhedron specifies that it must be a convex polyhedron. Would removing this requirement actually change the set of regular polyhedra? Find a nonconvex polyhedron that satisfies all the other conditions for a regular polyhedron: its faces must be congruent regular polygons and the same number of polygons must meet at each vertex.

13.4. By joining certain vertices of a dodecahedron, you can form a cube or a tetrahedron. On your own model (or sketch) of a dodecahedron, identify the vertices that form a cube and the vertices that form a tetrahedron.

13.5. The definition of a regular polyhedron has three conditions: (1) the faces are regular polygons, (2) the faces are congruent, and (3) the same number of faces meet at each vertex. Mathematicians, who are always trying to clarify their definitions and prune out unnecessary conditions, would wonder if all three are necessary. One way to answer this question affirmatively would be to find a nonregular polyhedron that satisfies two of the conditions but not all three. Can you find such a polyhedron?

13.6. What is the dual of the polyhedron that looks like two tetrahedra glued together along a face, called a trigonal bipyramid? (See Practice Problem 13.1.) Verify that the dual of the resulting polyhedron is the trigonal bipyramid again.

13.7. Which of the regular and semiregular solids are prisms or pyramids?

13.8. Make a chart of the number of vertices, edges, and faces on the semiregular polyhedra and prisms. Does Euler's formula hold for these polyhedra?

13.9. Most of the polyhedra you've seen so far have been spherical. That is, if you smoothed out the corners, you'd get a sphere. You might imagine a polyhedron that, instead of being spherical, was "doughnut" shaped, such as the a toroidal polyhedron from Practice Problem 13.5, reproduced below.

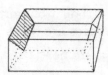

A variation of Euler's formula applies to these types of polyhedra.
(a) Construct, or sketch, models of several of these polyhedra and derive the corresponding formula about faces, edges, and vertices. You might try using different kinds of polygons as faces.

(b) What happens when you consider, instead, "Siamese doughnuts" (a two-holed torus)?

13.10.
(a) Can you find an abelian subgroup of order six in the symmetry group of the cube?

(b) Can you find a nonabelian subgroup of order 12?

13.11. Suppose you try to build a spherical polyhedron out of 1 square, 4 triangles, 1 octagon, and 4 hexagons.
(a) How many faces will the polyhedron have?

(b) How many edges will the polyhedron have?

(c) How many vertices will the polyhedron have?

13.12. In 1990, Professors Smalley and Curl, chemists at Rice University in Houston, Texas, produced a previously undiscovered form of carbon. They received the 1996 Nobel Prize for this work. Before their discovery, graphite and diamond were the only known forms of carbon, each of which was a three-dimensional lattice that extends indefinitely. The form these chemists discovered was different in that it was composed of distinct molecules, each of which was composed of 60 carbon atoms. From other chemical tests, Professor Smalley and his graduate students were able to determine that the molecule was approximately spherical. It was already known that a carbon atom bonds to at least three other carbon atoms and that the angle between these bonds is normally at least 90 degrees. Your job, in this assignment, is to deduce the structure of this C_{60} molecule. Think of the molecule as forming a polyhedron, with each atom being represented by a vertex and each bond being represented by an edge.

(a) How many vertices does the polyhedron have?

(b) Considering that each carbon atom forms three bonds to other carbon atoms, how many edges are there in the polyhedron?

(c) How many faces must there be? (Use the Euler characteristic.)

(d) What's the average number of edges per face? (Remember, each edge is part of two faces.)

(e) What shape(s) are the faces? How many of each type must there be to get the result in part (d)?

(f) At each vertex, how many of each type of face will meet?

(g) What does the polyhedron look like?

13.13. Complete the following chart of Platonic solids.

Name	Vertices	Edges	Faces	Shape of Faces	Dual
tetrahedron					
	8			squares	
		12			cube
		30		triangles	
dodecahedron					

13.14. Let F' be the symmetry of the cube that reflects the entire cube through its center. That is, F' takes each vertex to the other vertex on its main diagonal. Show that if R is any rotation of the cube, then $F'R = RF'$.

13.15. Show that if F' is as in Exercise 13.14, every element of the symmetry group of the cube that is a reflection can be written in the form RF', where R is one of the 24 rotations.

13.16. In estimating the size of the symmetry group of the cube or the dodecahedron, we said that any given vertex could move to any other vertex, and that the three adjacent vertices had to remain adjacent, but could be permuted among themselves. Show that each permutation of these three adjacent vertices actually represents some symmetry of the cube or dodecahedron. That is, write each of these in the form RF' or R for some rotation R and the element F' in Exercise 13.14.

13.17. Show that the rotation group of the dodecahedron is the group of even permutations in S_5.

13.18. If F^* is the reflection through the center of a dodecahedron, show that every element of the group of symmetries is in the form RF^* or R for some rotation.

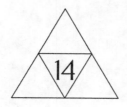

Graph Theory

14.1 Introduction

When we proved Euler's formula for polyhedra in Chapter 13, we used a two-dimensional "map" of the polyhedron that had the same number of vertices and edges as the original polyhedron. We used regions in the plane to represent the faces of the polyhedron. This kind of map, or graph, is an object worthy of study in its own right. Graph theory is useful for solving many kinds of problems that can be more easily represented and understood by these two-dimensional diagrams. It has applications in physics, chemistry, computer science, and many areas of mathematics. Graph theory can serve as a mathematical model for any system that uses a binary operation. We begin this chapter with the basic definitions for graphs as well as some elementary and historical puzzles and problems that lead to deep results.

Definition. A *graph* G is a finite nonempty set V together with a collection E of two-element subsets of V. The elements of V are called *vertices* and the elements of E are called *edges*. We say that two vertices of an edge are *adjacent*. We also say that two edges sharing a common vertex are *adjacent*. Vertices and edges that are joined are called *incident*.

Example. If $V = \{A, B, C, D\}$ and $E = \{\{A, B\}, \{A, C\}, \{C, D\}, \{A, D\}\}$, then $G = (V, E)$ is a graph. Figure 14.1 shows a diagram of the graph G. In this graph, vertices A and D are adjacent, but B and D are not. The edges $\{C, D\}$ and $\{A, C\}$ are adjacent. The edge $\{A, C\}$ and the point A are incident.

It is the custom to represent a graph by means of such a diagram and to refer to the diagram as the graph rather than use the sets and subsets themselves.

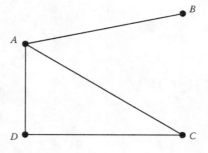

Figure 14.1. The graph G.

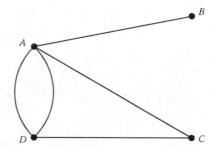

Figure 14.2. The graph G'.

Note that we do not allow a two-element subset such as $\{A, A\}$ in a graph. We do not allow such a line connecting a point to itself, or a *loop*, in a graph. However, multiple edges may exist in a graph. For example, if we use the same set $V = \{A, B, C, D\}$, and if $E' = \{\{A, B\}, \{A, C\}, \{C, D\}, \{A, D\}, \{A, D\}\}$, then the diagram for the graph $G' = (V, E')$ would look like Figure 14.2. The graph G' is called a *multigraph*.

Definition. Two graphs $G = (V, E)$ and $G' = (V, E')$ are called *isomorphic* if there exists a one-to-one correspondence between the vertex sets V and V' that preserves adjacency.

Figure 14.3 shows two graphs G and G' that are isomorphic. In the picture of G, even though the edges AC and BD intersect, the point of their intersection is not a vertex in the graph.

Figure 14.4 shows two nonisomorphic graphs with two vertices. The one on the left consists of two vertices and no edges; the other graph consists of two vertices and the edge joining them.

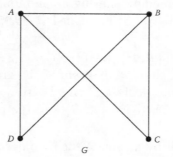
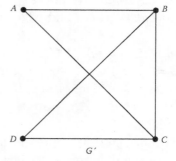

Figure 14.3. Isomorphic graphs.

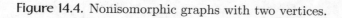

Figure 14.4. Nonisomorphic graphs with two vertices.

Practice Problem 14.1.

(a) *Draw the four nonisomorphic graphs with three vertices.*

(b) *Draw the 11 nonisomorphic graphs with four vertices.*

(c) *Draw two nonisomorphic graphs that each have five vertices and six edges.*

Definition. The *degree* of a vertex v in G is the number of edges adjacent to v. The degree of v is written $\deg v$. A vertex is *odd* or *even* depending on whether its degree is odd or even. A vertex with degree 0 is called *isolated*. A vertex with degree 1 is called an *endpoint*.

Example. In the graph of G in Figure 14.1, vertex A has degree 3 and thus is odd, and vertices C and D have degree 2 and are even. Vertex B is an endpoint.

Theorem 14.1 relates the degrees of the vertices in a graph to the number of edges.

Theorem 14.1. *The sum of the degrees of the vertices in a graph G is twice the number of edges.*

Proof: Since each edge is adjacent to two vertices, it contributes two to the sum of the degrees of all the vertices.   □

Sometimes we want to start with a graph and remove or add vertices or edges to create a new graph. When we proved Euler's formula (Theorem 13.1), we successively removed vertices and edges to reduce the polyhedron to simpler cases.

Definition. A *subgraph* H of a graph G is a graph that has all its vertices and edges in G. Suppose v is a vertex of graph G. The graph $H = G - v$ is the subgraph of G consisting of all the vertices of G except for v, and all the edges of G not incident to v. If v_1 and v_2 are not adjacent in G, and we add the edge $e = \{v_1, v_2\}$ to G, we get a new graph, $G + e$.

Example. Consider the graph G in Figure 14.5. Then $G - v_1$ and $G + v_4 v_5$ are as shown. Note that G is a subgraph of $G + v_4 v_5$.

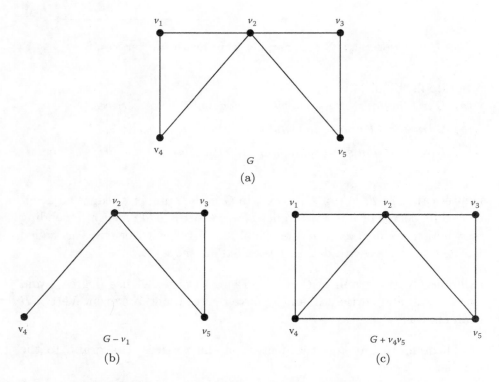

Figure 14.5. Adding and removing vertices or edges: (a) original graph G, (b) removing vertex v_1, and (c) adding edge v_4v_5.

Practice Problem 14.2.

(a) *Starting with graph G in Figure 14.5(a), draw at least two subgraphs of G.*

(b) *If G is the graph in Figure 14.5(a), draw the subgraphs $G - v_2$ and $G + v_4v_3$.*

Stanislaw Ulam[1] made a conjecture about the subgraphs of a graph G. His idea is that if there are two graphs, each with at least three vertices, and each also has the same collection of isomorphic subgraphs, then the original graphs must also be isomorphic. Here is the formal statement:

---

[1]Stanislaw Marcin Ulam (1909–1984) was a Polish mathematician and physicist who participated in the Manhattan project and solved the problem of how to initiate fusion in the hydrogen bomb. He also invented the Monte Carlo method of sampling.

Ulam's Conjecture. *Suppose a graph G has p vertices v_i and a graph H has p vertices w_i, with $p \geq 3$. If, for each i, the subgraphs $G - v_i$ and $H - w_i$ are isomorphic, then the graphs G and H are isomorphic.*

This easily stated conjecture has not been proved.

We can also combine more than one graph. For example, we can form the union of two graphs or the sum of two graphs. The results of each operation are slightly different.

Definition. Suppose G_1 and G_2 are graphs with disjoint vertex sets V_1 and V_2 and edge sets E_1 and E_2, respectively. The *union* $G_1 \cup G_2$ is the graph with vertices $V = V_1 \cup V_2$ and edges $E_1 \cup E_2$. The *sum* $G_1 + G_2$ is the graph with vertices $V = V_1 \cup V_2$ and edges that join every vertex in V_1 with every vertex in V_2.

Example. Start with graphs G_1 and G_2; their union and sum are shown in Figure 14.6.

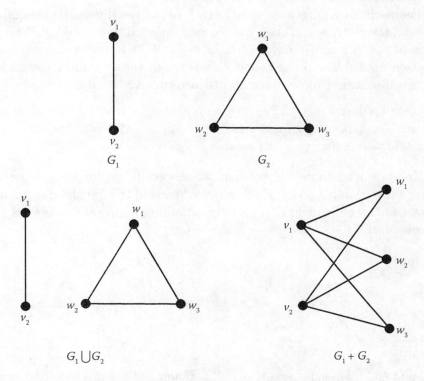

Figure 14.6. The union and sum of two graphs.

Figure 14.7. The city of Königsberg.

14.2 The Königsberg Bridge Problem

Before we move on to more theorems, we present here a historical problem that Euler considered and solved in 1736. The city of Königsberg in Prussia consists of two islands linked to each other and the banks of the Pregel River by seven bridges (see Figure 14.7).

People from Königsberg would stroll along the different parts of the city and the river, and if they walked long enough, they could visit all the parts of the city and cross several bridges. This leads to an interesting question, namely, could someone walk through the city and cross each of the seven bridges exactly once and then return to the starting point?

Practice Problem 14.3. *Try to accomplish this task by tracing a path through the diagram in Figure 14.7 to see if you can tour the city, cross each bridge exactly once, and return to where you started.*

Euler proved that this problem is unsolvable in an early paper on graph theory. In his proof, Euler replaced each of the four land areas by a vertex, and each bridge with an edge adjoining appropriate vertices (see Figure 14.9).

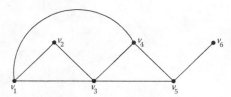

Figure 14.8. An example of trails and circuits.

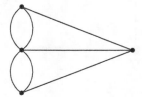

Figure 14.9. A graph of the city of Königsberg.

Solving the Königsberg bridge problem thus becomes a problem of deciding the traversability of a particular graph, or whether you can trace each edge of the graph exactly once, and begin and end at the same vertex. We need to define precisely what is meant by traversing a graph from one vertex to another.

Definition. In a multigraph G, a *trail* (sometimes also called a *path*) is an alternating sequence of vertices and edges of the form

$$v_1, v_1 v_2, v_2, v_2 v_3, v_3, \ldots, v_{n-1}, v_{n-1} v_n, v_n,$$

where all the edges are distinct. Each edge is preceded and succeeded by the vertices adjacent to it. We call the vertex v_1 the *initial vertex* and the vertex v_n the *final vertex*. A trail whose initial and final vertices are the same is called a *circuit*.

To make the notation easier, note that in any trail, the sequence $v_i, v_i v_j, v_j$ always includes the edge indicated by the adjacent vertices. So we use the shorter notation $v_i v_j$ for this sequence.

Example. In the graph G in Figure 14.8, we see that $v_1 v_2 v_3 v_5$ is a trail, and that $v_1 v_2 v_3 v_1$ is a circuit.

Practice Problem 14.4. *Find a circuit in the following graph. Remember that the point where edges $v_1 v_3$ and $v_2 v_4$ intersect is not a vertex of the graph.*

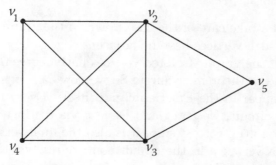

Definition. A graph G is *connected* if every pair of vertices is joined by a trail.

Example. The graph G in Figure 14.10 is not connected since we cannot find a trail between vertices v_1 and v_4.

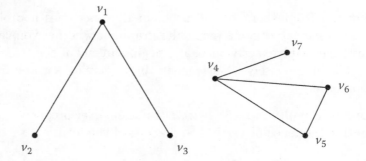

Figure 14.10. A disconnected graph.

Returning to the problem of the Königsberg bridges, we can rephrase it in terms of our definition. Does the graph in Figure 14.9 have a circuit that contains all its edges? In honor of Euler, we call such a circuit an Eulerian circuit.

Definition. In a graph G, a trail that contains all the edges of a graph is called an *Eulerian trail*. A circuit that contains all the edges of G is an *Eulerian circuit*.

Euler answered the Königsberg question with a more general answer about multigraphs and Eulerian circuits. Theorem 14.2 was proved by Euler.

Theorem 14.2. *If a multigraph G has an Eulerian circuit, then every vertex of G has even degree.*

Proof: First, let's take care of the case in which the graph G has isolated vertices. Since such vertices have no adjacent edges, they will not appear in any circuit of the graph. Isolated vertices have degree 0, so they are of even degree, as the theorem requires. So we now can consider only those vertices that appear in the Eulerian circuit itself. Every time a vertex v appears in the circuit, we enter and leave v via distinct edges. This is true even of the initial vertex, since it is also the final vertex. Thus, each appearance of a vertex v in the circuit contributes two to the degree of v. Since every edge of G appears exactly once in the circuit, every vertex must have even degree.   □

If we now look at the graph G in Figure 14.9, we can see that every vertex has odd degree. Thus, according to the theorem, G has no Eulerian circuit.

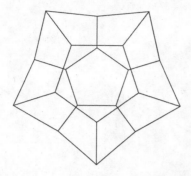

Figure 14.11. Map for *Around the World*.

Practice Problem 14.5. *Decide whether or not the following graphs have Eulerian trails or Eulerian circuits.*

(a)

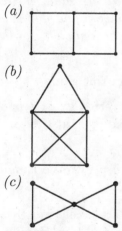

(b)

(c)

    Eulerian paths and circuits include each edge of a graph. We might ask a similar question about the relationship of traversability of a graph and its vertices. Can we find a trail in a graph G so that each vertex is visited exactly once? Can we find a circuit like this that begins and ends at the same vertex? Sir William Rowan Hamilton[2] introduced a game posing this question in 1859. He called his game *Around the World*, and used a map of a dodecahedron (Figure 14.11), with vertices labeled with the names of international cities.

---

[2]Sir William Rowan Hamilton (1805–1865) was an Irish physicist, astronomer, and mathematician. He was named the Royal Astronomer of Ireland at the age of 18. His contributions to mathematics include the discovery of the algebra of the quaternions. He invented his "Icosian Game" in London in 1857.

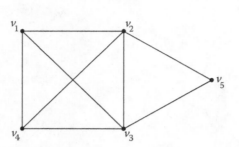

Figure 14.12. An example of a Hamiltonian circuit.

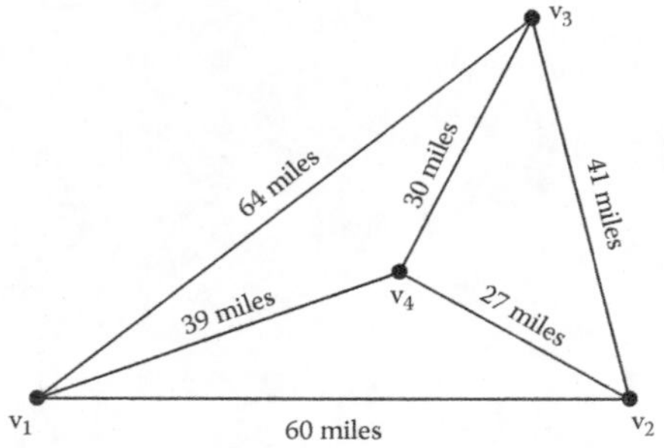

Figure 14.13. A weighted graph.

Practice Problem 14.6. *Find a circuit in the graph G in Figure 14.11 that visits each vertex exactly once and begins and ends at the same vertex. You need not travel across all the edges. Once you have found one such circuit, can you describe how to find others by using the symmetry of the graph?*

A natural definition and some questions arise from this game.

Definition. In a graph G, a trail that includes each vertex exactly once is called a *Hamiltonian trail*. If such a trail begins and ends at the same vertex, it is called *Hamiltonian circuit*.

Example. In the graph G in Figure 14.12, the path $v_1 v_2 v_5 v_3 v_4$ is a Hamiltonian trail, and $v_1 v_2 v_5 v_3 v_4 v_1$ is a Hamiltonian circuit.

Unlike the theorem about Eulerian circuits, there are no elegant theorems that tell us conclusively which graphs do and do not have Hamiltonian circuits. Even so, the problem of Hamiltonian circuits is interesting with many applications. One common application is the "traveling salesman" problem. In this problem, a salesman must visit all of the cities on the route and return to where he started. The graphs representing the salesman's route include distances, and are thus examples of a different kind of graph called a *weighted graph*. The mileage between cities is assigned as a "weight" on each edge, and the problem becomes not only to find a Hamiltonian circuit, but also to find the one with the shortest length.

Example. In the graph G in Figure 14.13, we have indicated weights on the edges. The shortest Hamiltonian circuit is $v_1 v_4 v_3 v_2 v_1$.

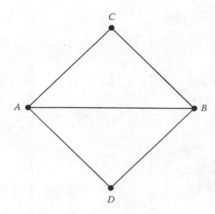

Figure 14.14. A planar representation of Figure 14.3.

14.3   Colorability and Planarity

There can be a point in the diagrams of some graphs where edges cross but the point is not a vertex of the graph (see Figure 14.3, for example). It would be best if we could diagram graphs so that this situation does not occur. For example, we could draw Figure 14.14, the same graph as Figure 14.3, so that no edges cross where there are not vertices. Note that the graphs of Figures 14.3 and 14.14 are isomorphic.

Definition. A graph G is *planar* if it can be diagrammed on a plane so that distinct edges do not cross.

Euler used planar graphs in his study of polyhedra, and in Chapter 13 we have seen how to make diagrams of the Platonic solids as planar graphs. Not all graphs are planar.

The following puzzle is impossible to solve because the graph that represents it is not planar. Imagine three friends live in a city with a strange building code for utilities. The gas, electric, and water lines that run to homes must run along the surface of the ground and must not cross. Each of the friends wants to be connected to each of the three utilities.

Practice Problem 14.7. *Experiment with Figure 14.15 to try to find a way for all three friends to be connected to each of the three utilities with none of the lines crossing.*

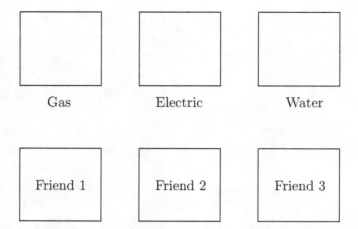

Figure 14.15. The utility problem.

No matter how much you try, you will not be able to satisfy the conditions of this problem. The graph representing the situation is not planar. We call the graph that connects each of the three friends to each of the three utilities $K(3,3)$. Figure 14.16 is a diagram of the graph $K(3,3)$. Notice that there are no triangles in this graph; this fact will be important later. This notation indicates that there are two distinct sets of three vertices, and that each vertex in the first set is connected to each vertex in the second set. This results in nine edges.

The nonisomorphic graphs with two, three, or four vertices are all planar (see Practice Problem 14.1). However, there is a graph with five vertices that is not planar. It is the graph of a pentagram, or a star drawn inside a pentagon (Figure 14.17). This was the symbol associated with the society of the Pythagoreans.

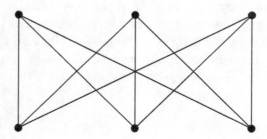

Figure 14.16. The graph $K(3,3)$.

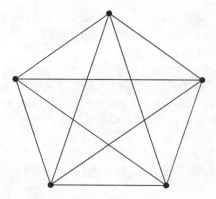

Figure 14.17. The graph K_5.

No matter how hard you try, you will always end up with two distinct edges that cross. We call this graph K_5. This notation indicates that every vertex is connected to every other vertex.

Our next task will be to show that neither $K(3,3)$ nor K_5 is planar. We will use Euler's formula for polyhedra, translated to the plane. Recall that for polyhedra we had $V - E + F = 2$, where V represented the number of vertices, E the edges, and F the faces. If we have a planar graph, we let regions take the place of faces, and include the external region as one of the faces. If we do this, we still have the same formula. For graphs, we use R for regions and write $V - E + R = 2$.

Example. Using the planar graph of Figure 14.14 (reproduced below), we can see that $V = 4, E = 6$, and $R = 4$, so that the equation holds.

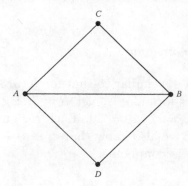

We will now prove that $K(3,3)$ and K_5 are nonplanar. We begin with a counting lemma.

Lemma 14.3. *Let P be a planar graph with E edges and R regions. Suppose every edge of P is in a circuit, and that e_i edges bound each region R_i. Then $e_1 + e_2 + \cdots + e_R = 2E$.*

Proof: Each edge is in two circuits, one of which might be the circuit bounding the exterior of the entire graph. Thus, each edge is counted twice when we sum the e_i's. So the sum of the e_i's is twice the total number of edges in P. So $e_1 + e_2 + \cdots + e_R = 2E$. □

Practice Problem 14.8. *Show that in $K(3,3)$ every edge is in a circuit.*

Now we use this lemma together with Euler's formula to prove the result for $K(3,3)$.

Theorem 14.4. *$K(3,3)$ is nonplanar.*

Proof: We prove this by contradiction. Suppose we have a planar representation of $K(3,3)$. Call this graph P. For $K(3,3)$, and thus for P, we have $E = 9$ and $V = 6$, so Euler's formula gives

$$6 - 9 + R = 2, \text{ or } R = 5.$$

In $K(3,3)$, every edge is in a circuit, and so the same is true for P. So Lemma 14.3 holds and we have $e_1 + e_2 + e_3 + e_4 + e_5 = 2 \cdot 9 = 18$.

Recall that $K(3,3)$ has no triangles. Thus, each region of P has at least four edges in its boundary. By Lemma 14.3, $e_i \geq 4$ for all i. So we also have $e_1 + e_2 + e_3 + e_4 + e_5 \geq 20$. This is a contradiction, so $K(3,3)$ must not be planar. □

Theorem 14.5. *K_5 is nonplanar.*

Proof: The proof for K_5 is similar to that for $K(3,3)$. In K_5, since $V = 5$ and $E = 10$, by Euler's formula, we get $R = 7$. By Lemma 14.3, we also know $e_1 + e_2 + \cdots + e_7 = 2 \cdot 10 = 20$. Each region must be bounded by at least three edges, so we also know that $e_1 + e_2 + \cdots + e_7 \geqslant 21$. This is a contradiction, so K_5 is nonplanar. □

The two graphs K_5 and $K(3,3)$ play a central role in discussing the planarity of any graph. It turns out that if a graph G is nonplanar, it must have a subgraph that "looks like" K_5 or $K(3,3)$. (We omit the precise

Figure 14.18. Map of the continental United States.

statement here.) This deep result was proved by Kazimierz Kuratowski,[3] and the proof will not be given in this text.

One very interesting historical question about planar graphs concerns colorability. It concerns a map of the continental United States (Figure 14.18) in which states are colored with different colors. It is customary to color such a map so that no two states sharing a border are the same color. How many colors would we need to color the US map this way?

One question that arises when we look at the western states is what to do about the "four corners" region where the states of Utah, New Mexico, Colorado, and Arizona come together. Our rule for such states will be that even though Colorado and Arizona share a corner, or vertex, they do not share a border, or edge. The same is true for New Mexico and Utah. So we need only two colors to color these four states.

Practice Problem 14.9. *Look at a map of the continental United States and think about how many colors you would need to color it so that no two states sharing a border have the same color.*

---

[3]Kazimierz Kuratowski (1896–1980) was a Polish mathematician who worked in the areas of set theory and topology. His work in graph theory led to conditions that would assure that a graph was planar. Stanislaw Ulam was one of his doctoral students.

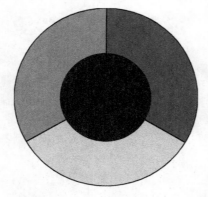 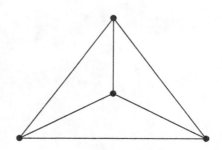

Figure 14.19. A four-colored map.   **Figure 14.20.** The graph for Figure 14.19.

If you work at this problem systematically, you will see that you need only four different colors to color a map of the United States following these rules. Figure 14.19 shows a very simple "map" that also requires four colors. We can turn this map into a graph in much the same way that Euler did for the Königsberg bridge problem. We use a vertex for each region of the graph, and connect two vertices if the two regions in the original map share a common border. Figure 14.20 is the graph for the map in Figure 14.19.

In terms of this graph, the question about colorability is now one of coloring vertices. In the graph, we want to color the vertices so that no adjacent vertices share the same color. It is clear that, in the graph, each vertex is adjacent to three other vertices, so we need four colors to satisfy the conditions for colorability.

It is easy to come up with a graph that needs four colors. A natural question to ask is whether there are any planar graphs that require more than four colors. This question has intrigued mathematicians around the world for centuries, but has only recently been answered.

In 1840, August Möbius[4] posed a problem about a king attempting to divide his kingdom among his five sons. The king wished to divide his kingdom so that each of the sons' regions had a common boundary with the other four. To divide the kingdom in this way would be to create a

---

[4]August Ferdinand Möbius (1790–1868) was a German mathematician and astronomer. He is best known for his discovery of a two-dimensional surface with only one side, called a Möbius strip. He posed an early version of the four color conjecture in 1840.

map that requires five colors. If this were possible, by changing the map of the sons' regions into a graph (as in Figure 14.20), we would get a planar diagram of K_5, and we have already shown this cannot be done. However, this does not solve the four-color conjecture, since there may be some other way to make a map that requires five colors. Thus, while this is a related puzzle, it does not lead to a complete solution to the problem.

An English mapmaker named Francis Guthrie[5] is considered to be responsible for the modern fascination with the four-color problem. In 1852, he made a map of all the counties in England, and noticed that it took only four colors to insure that no two counties sharing a border were the same color. He found maps of other regions where three colors were not enough, but he could not find a map requiring five colors. He wrote to his brother Frederick with the conjecture that four colors were all one needed to color any map this way in the plane, and included an attempt at a proof of this conjecture. Frederick was taking a class at the time with a mathematician named Augustus DeMorgan, who wrote to William Hamilton to share the problem. Unfortunately, Hamilton did not find the problem interesting enough to pursue.

Interest in the problem languished until 1878, when, at a meeting of the London Mathematical Society, Arthur Cayley asked whether there had been any progress toward a solution to the four-color problem. One person attending the meeting, Alfred Kempe, produced a proof in the following year. Kempe was honored for his proof by being elected Fellow of the Royal Society. However, in 1890, another mathematician, Percy John Heawood, published a paper indicating a "defect" in Kempe's proof. Kempe agreed that there was an error, and was unable to correct it.

Through the twentieth century, many mathematicians worked on the problem and created techniques toward a solution that were valuable tools in proving other graph theoretic results. Eventually, by using techniques that reduced the problem to considering a set of particular maps, mathematicians used computer techniques to finish the proof. In 1976, Wolfagand Haken and his student Kenneth Appel announced that, after considering all the remaining maps, the problem had been solved, and the four-color theorem was true.

---

[5]Francis Guthrie (1831–1899) was a student at University College, London, where he studied under DeMorgan when he first noticed that it would require at least four colors to make a map where no two regions sharing a common border had the same color. His profession was as a botanist, and he lived in South Africa where he made important contributions to the classification of native flora.

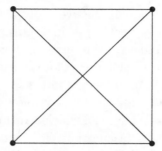

Figure 14.21. The complete graph with four vertices.

14.4   Graphs and Their Complements

The graph K_5 is special in that every vertex in the graph is adjacent to every other vertex. We call such graphs *complete graphs.*

Definition. A graph G is *complete* if every vertex of G is adjacent to every other vertex of G. If G has n vertices, the complete graph is called K_n.

Example. The graph G in Figure 14.21 is the complete graph with four vertices, K_4.

Practice Problem 14.10. *Draw the complete graphs* K_1, K_2, K_3, *and* K_6.

A famous problem of Frank P. Ramsey[6] has an elegant solution using the idea of complete graphs. Suppose a group of six people meet anywhere—a party, an airline flight, or a sporting event, for example. The assertion of Ramsey's problem is that in any group of six people, there are always either three mutual acquaintances (people who all three know one another) or three mutual nonacquaintances (three people who do not know each other). It would be difficult to prove this theorem by considering cases. Instead, we introduce another idea about graphs.

Definition. Let G be a graph. The *complement* of G, written $\overline{G}$, is a graph that has the same vertices as G. Two vertices of $\overline{G}$ are adjacent if and only if they are not adjacent in G; two vertices in G are adjacent if and only if they are not adjacent in $\overline{G}$.

---

[6]Frank P. Ramsey (1903–1930), who died tragically at age 26, made contributions to mathematics, philosophy, and economics. His work in economics focused on probability and decision making.

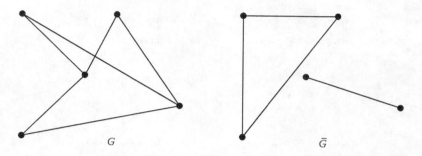

Figure 14.22. A graph and its complement.

Example. Figure 14.22 shows graph G and its complement $\overline{G}$.

Notice that if we superimpose G on $\overline{G}$ (or vice versa), we have the complete graph K_5. This is because in one of the two graphs, every vertex is adjacent to every other vertex. If v is a vertex in graphs G and $\overline{G}$, each with n vertices, then the sum of the degrees of v in G and $\overline{G}$ will be $n - 1$.

Practice Problem 14.11. *Draw the complements of the following graphs.*

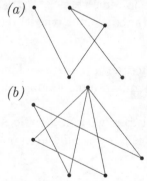

Practice Problem 14.12. *Start with any graph G. If you draw the complement of this graph, you will get a new graph called $\overline{G}$. What is the complement of the new graph $\overline{G}$? Explain your answer.*

Now we use this new definition of complement to prove Ramsey's theorem about six people. We want to formulate the problem of people and acquaintances by using vertices and edges of a graph G. We use the vertices to represent people and the edges to indicate acquaintance. Thus, in the graph of G, two vertices will be adjacent if the people are

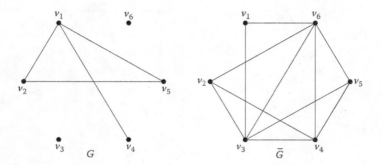

Figure 14.23. An illustration of Ramsey's problem.

acquainted, and will not be adjacent otherwise. What will this mean in the complement $\overline{G}$? In $\overline{G}$, two vertices will be adjacent if the two people are not acquainted, and will not be adjacent otherwise.

Example. In Figure 14.23, we let $v_1, v_2, \ldots, v_6$ represent the six people. In this graph G, v_1 and v_2 are adjacent, as are v_1 and v_4, v_2 and v_5, v_1 and v_5. In the graph $\overline{G}$, we join all vertices that are not joined in G. Thus, for example, v_2 and v_3 are adjacent in $\overline{G}$, but not in G.

Note that in the graph G, we have a triangle joining vertices v_1, v_2, and v_5. This triangle indicates mutual acquaintance among the three people represented by the three vertices.

In the graph $\overline{G}$, we have a triangle joining vertices v_2, v_3, and v_6. This triangle in the complement indicates mutual nonacquaintance among the three people represented by these vertices.

Ramsey's theorem asserts that one of these graphs with six vertices, either G or $\overline{G}$, must have a triangle. Theorem 14.6 is the formal statement of the theorem.

Theorem 14.6 (Ramsey's Theorem). *If G is a graph with six vertices, then either G or $\overline{G}$ must have a triangle.*

Proof: Suppose G is a graph with six vertices. Select any vertex, say, v_1. Since G has six vertices, the sum of the degrees of v_1 in G and $\overline{G}$ is five. So in one of the two graphs, vertex v_1 has degree at least three. Let's assume that v_1 has degree at least three in G. Therefore, it is connected to three vertices in G that we can call g_1, g_2, and g_3 (see Figure 14.24). Now we consider the relationship of these three vertices. If any two of them are adjacent in G, we have a triangle joining the two of them and

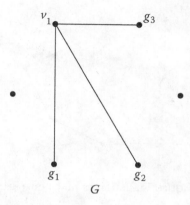

Figure 14.24. Three vertices connected to v_1.

the vertex v_1. If, however, no two of these are adjacent in G, then all three of them will be adjacent in $\overline{G}$. They will then form a triangle in $\overline{G}$, as the theorem requires.   □

Another way of stating Ramsey's theorem that uses only the graph G is to say that a graph G with six vertices must either contain a triangle, K_3, or its complement, $\overline{K}_3$. In the proof of Theorem 14.6, if g_1, g_2, and g_3 are not adjacent in G, they form triangle K_3 in G, but form the complement $\overline{K}_3$ in $\overline{G}$.

Practice Problem 14.13. *Draw a graph G with four vertices so that neither G nor $\overline{G}$ has a triangle. Do the same for a graph H and its complement $\overline{H}$ with five vertices.*

Ramsey's theorem and Practice Problem 14.13 can be combined to show that it requires a graph G with six vertices to ensure that G contains either K_3 or its complement. This leads to a more general definition.

Definition. The *Ramsey number*, $r(m, n)$, is the smallest integer so that every graph G with $r(m, n)$ vertices contains a copy of K_m or $\overline{K}_n$.

According to Ramsey's theorem, using this notation, we have proved that $r(3, 3) = 6$. In general, the determination of Ramsey numbers is one of the unsolved problems of mathematics. In fact, the number $r(5, 5)$ is known only to be between 43 and 49.

Challenge Problem. *Prove that $r(4, 4) = 18$.*

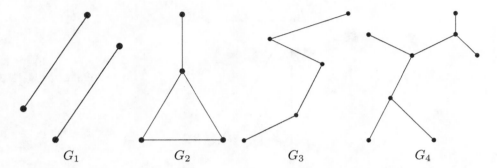

Figure 14.25. Trees and nontrees.

14.5   Trees

One of the most useful kinds of graphs in applications is a tree. Computer programming, chemical element diagrams, and other kinds of network problems can all use trees to represent relationships.

Definition. A connected graph G is a *tree* if G has no circuits.

Example. Graph G_1 in Figure 14.25 is not a tree since it is not connected. Graph G_2 is not a tree since it has a circuit. Graphs G_3 and G_4 are trees.

If G_0 is the graph in Figure 14.26 that consists of a single vertex and no edges, we can see that G_0 is a tree according to our definition. G_0 is a *trivial tree*. There is only one possible connected graph with two vertices, and it is also a tree. There are two possible nonisomorphic connected graphs with three vertices, but one of them, K_3, is not a tree.

Practice Problem 14.14. *Draw the two nonisomorphic trees with four vertices, the three nonisomorphic trees with five vertices, and the six nonisomorphic trees with six vertices.*

Figure 14.26. G_0, two-, and three-point trees.

The situation gets more complicated as we add vertices. There are 11 nonisomorphic trees with seven vertices, and 23 with eight vertices. As we begin to investigate trees, we can make some natural observations. For example, all the nontrivial trees have at least two end vertices (sometimes called leaves), and in any tree, the number of edges is one less than the number of vertices.

We want to establish these and other facts about trees. We do this by proving Theorem 14.7, which gives many equivalent conditions for a graph G to be a tree. Here, *equivalent* means that any one of these conditions defines a tree.

Theorem 14.7. *Let G be a graph with p vertices and q edges. Then the following statements are equivalent:*

(1) G is a tree.

(2) Every two vertices of G are joined by a unique trail.

(3) G is connected and $p = q + 1$.

(4) G has no circuits and $p = q + 1$.

(5) G has no circuits, and if any two nonadjacent vertices of G are joined by an edge e, then the graph $G + e$ has exactly one circuit.

(6) G is connected, is not K_p for $p \geqslant 3$, and if any two nonadjacent vertices are joined by an edge e, then $G + e$ has exactly one circuit.

(7) G is not $K_3 \cup K_1$ or $K_3 \cup K_2$, $p = q + 1$, and if any two nonadjacent vertices are joined by an edge e, then $G + e$ has exactly one circuit.

The proof of this theorem is by a method that is different from other proofs in this text. We begin by assuming the statement in part (1) is true and prove that it implies statement (2). Next, we assume statement (2) is true, and prove that it implies statement (3). We continue in this fashion, and finish by proving that statement (7) implies statement (1). Logically, we will have come full circle and thus have established that all seven conditions are equivalent. We could use any of these seven conditions to define a tree. Here is the proof. Some of the parts in the following proof are quick to do, and others take intricate reasoning. One part involves the use of mathematical induction (see Exercise 2.20).

Proof (of Theorem 14.7): We begin by showing that $(1) \Rightarrow (2)$. Since G is a tree, G is connected. So there exists a trail between every two vertices in G. We have to show that there is a unique trail between any two vertices. Suppose instead that u and v are vertices of G and that two different trails T_1 and T_2 connect them. Start both trails at the vertex u and walk the trails T_1 and T_2. Eventually, we encounter the first vertex w of G such that w is on T_1 and T_2, and its successor on T_1 is different from its successor on T_2. Let w' be the next vertex on T_1 that is also on T_2 (this w' may be the vertex v). The two segments of the trails T_1 and T_2 from w to w' form a circuit. This contradicts the fact that G is a tree.

Proof that (2) $\Rightarrow$ (3). Because every pair of vertices are joined by a trail, G is connected. We prove $p = q + 1$ by induction. This equation is obvious for the connected graphs with one or two vertices. Assume the equation holds for graphs with $p - 1$ vertices and that G is connected and has p vertices. Due to the uniqueness of paths, if we remove any edge e from G, then the result is no longer a connected graph. Call this new graph $G - \{e\}$. It is not connected and therefore has two components. Because each component has fewer than p points, the equation holds for each of the components. Each component has one more vertex than it has edges. Say one component has a vertices and one component has b vertices. Then $p = a + b$. In the components, by the induction hypothesis, the equations give respectively $a - 1$ and $b - 1$ edges. So in G, there are the $a - 1$ edges, the $b - 1$ edges, plus the initial removed edge. This results in $(a - 1) + (b - 1) + 1 = a + b - 1 = p - 1$ edges, so we have the desired result.

Proof that (3) $\Rightarrow$ (4). This time, we have the equation and we just need to show that G has no circuits. We proceed by induction. If we have one vertex, G has no circuits. If $p > 1$, we will show that G is connected and $p = q + 1$ together imply that there exists a vertex of degree 1. Then, by deleting this vertex and the edge adjacent to it, we have a connected graph G' with $p' = p - 1$ vertices and $q' = q - 1$ edges, so $p' = q' + 1$. So, by induction, G' has no circuits. But this means G has no circuits, since all we took away was a vertex of degree 1, which cannot be part of a circuit.

To show that G has a vertex of degree 1, we assume, for the sake of contradiction, that each vertex has degree 2 or greater. (No vertex has degree 0 since G is connected and has more than one vertex.) Then each vertex is adjacent to two or more edges, but each edge is adjacent to only two vertices, meaning there must be at least as many edges as vertices. But this violates $p = q + 1$.

Proof that (4) $\Rightarrow$ (5). Since G has no circuits, each component of G is a tree. Suppose there are k such components. Since each component has one more vertex than it has edges, we must have $p = q + k$. Since we are assuming (4) holds, $k = 1$, and this means G is connected. G is then a tree and there is a unique trail joining any two vertices of G (since (1) $\Rightarrow$ (2)). Now, if we add an edge $e = uv$, that edge together with the unique trail joining u and v, becomes a circuit. The circuit is unique because the trail between u and v was unique.

Proof that (5) $\Rightarrow$ (6). Since K_p contains circuits for $p \geqslant 3$, G cannot be one of these graphs. If G were not connected, an edge e could be added between vertices in two different components, and the resulting $G + e$ would not have any circuits, contrary to (5).

Proof that (6) $\Rightarrow$ (7). For this part of the proof, we need to show only that the equation $p = q + 1$ holds. Our strategy will be to use the work we have already done. We prove G has no circuits, and since we are assuming G is connected, this means G is a tree. But since (1) $\Rightarrow$ (2) $\Rightarrow$ (3), the equation $p = q + 1$ holds.

First, note that G cannot have more than one circuit. If it did, since G is not K_p for $p \geq 3$, we can add in an edge, and the resulting graph would have more than one circuit, contrary to our assumption.

Now, what could the circuit in G look like? Suppose the circuit has four or more vertices on it, as in Figure 14.27. If we add an edge e to G, then $G + e$ has more than one circuit, contrary to our assumption. (Note that e cannot already be there, since G does not have more than one circuit).

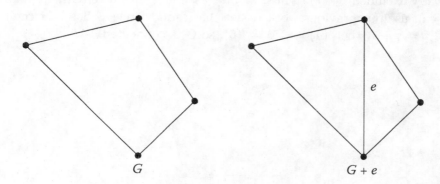

G $\qquad\qquad\qquad\qquad\qquad$ $G + e$

Figure 14.27. A circuit with four vertices.

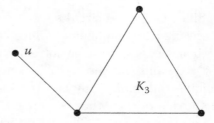

Figure 14.28. The K_3 circuit and another vertex.

Thus, the circuit in question can only have three vertices and is isomorphic to K_3. So K_3 is a subgraph of G. However, K_3 is not all of G, so there is another vertex u of G not on the K_3 circuit, but connected to one of K_3's vertices, as shown in Figure 14.28.

If we add an edge from u to another of the K_3 vertices, there will be two circuits in G, contrary to our assumption. (That edge cannot already be there, since G does not have more than one circuit.)

Proof that $(7) \Rightarrow (1)$. If the graph G has a circuit, then the argument in the previous step shows that the circuit must be K_3 and a component of G. This component has three vertices and three edges. All other components of G must be trees. In order to have $p = q + 1$, there can be only one other component. Suppose this tree contains a trail of length 2. If we add an edge e to this component, then with K_3, we have two cycles (see Figure 14.29).

This is impossible. Thus, the tree must be either K_1 or K_2. This forces one of these two to be the graph G. But each of these graphs is already excluded by (7). Thus, we may assume G has no circuits. Finally, we can use our previous work to get the result. Since G has no circuits, and $p = q + 1$, then $(4) \Rightarrow (5) \Rightarrow (6)$, so G is connected. $\square$

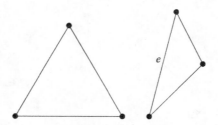

Figure 14.29. Creating a second cycle in G.

With this long proof, we begin to see that the subject of graph theory in general, and of trees in particular, can start with an easily stated set of definitions and lead quickly to more complicated statements and deeper theorems. Graph theory is a relatively simple subject to understand, but we have presented a number of unsolved problems, such as Ulam's conjecture and the determination of Ramsey numbers, that intrigue mathematicians with their complexity. We note the similarity with some aspects of number theory with its unsolved problems that are easy to state but quite difficult to solve.

Practice Problems Solutions and Hints

14.1. Do it!

14.2. Do it!

14.3. This is impossible, but try anyway; it's instructive to see how it fails.

14.4. There are, of course, many circuits; they don't even all have to go through all the edges or vertices. For example, $v_1v_2v_4v_1$ is a circuit that does neither, whereas $v_1v_3v_5v_2v_4v_1$ goes through all the vertices once and misses an edge (which one?). Is it possible to get all the edges and remain a trail (what could go wrong)? And why isn't it possible to go through all the edges and miss a vertex?

14.5. Parts (a) and (b) have Eulerian trails but not Eulerian circuits, since they have vertices with odd degree. Part (c) does have an Eulerian circuit.

14.6. The general pattern is one with which hikers are familiar: switchbacks. You will find the solution easy enough if you try not to get stuck in corners, which is accomplished by turning around a lot. As for the symmetry, a dodecahedron is made of pentagons, so there are a good five rotations through *each* face that will give distinct circuits.

14.7. It's hopeless, but don't let that discourage you. Again, try to figure out what goes wrong, since it will lead you to a more intuitive understanding why it is impossible. Notice that once you've connected two utilities to everyone, you have at least one circle.

14.8. This shouldn't surprise you, since each of the three utilities is connected to each of the three houses; no matter which way you start off from any of them, you can always find a path that loops back and ends up at the beginning.

14.9. Do it!

14.10. Do it!

14.11. Do it!

14.12. You get back G, which shouldn't surprise you. In fact, $\overline{G}$ is all the edges not in G, so in particular, it doesn't contain any of the edges of G, and if there is some edge not in G, it must be in $\overline{G}$. Therefore $\overline{\overline{G}}$, which is all the edges not in $\overline{G}$, can only contain those edges which *are* in G, and of course, it contains all of those.

14.13. Do it yourself, but here is some guidance. First, draw the graphs as a square and a pentagon, respectively, so that things are organized. One way to start, which is very organized, is to draw a random triangle on the graph and then rub out a side; then make the remaining two sides part of G and repeat. Then neither that triangle nor its complement will be in G or $\overline{G}$. If it's possible to construct G to specification, then this will eventually work.

14.14. As you might have noticed, there are no pictures here. When you are doing this problem, you are doomed to frustration unless you have a system: start with the easiest tree (a line with vertices on it) and modify it by moving edges around. Don't move onto a more "complicated" tree until you have convinced yourself that you have already found all the trees of the current "complexity." For extra credit, figure out what "complexity" means here.

Exercises

14.1. Draw a planar graph with nine vertices. Include as many edges as possible.

14.2. How many edges can you draw in a graph with five vertices before it can no longer be planar? With six vertices?

14.3. Is it possible to plan a route on Map 1 that covers each street exactly once? Map 2? On a map that has such a route, can the route begin at any intersection? If not, where should it begin?

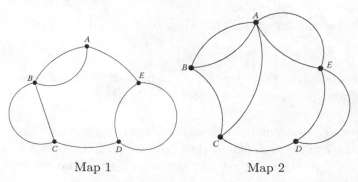

Map 1   Map 2

14.4. Prove the converse of Theorem 4.2: If a multigraph G is connected and every vertex has even degree, G has an Eulerian circuit.

14.5. Prove that if G has an Eulerian trail that is not a circuit, then exactly two vertices of G have odd degree. Prove, conversely, that if G is connected and exactly two vertices of G have odd degree (call them V_B and V_E), then there exists an Eulerian trail beginning at V_B and ending at V_E.

14.6. Find the shortest Hamiltonian *path* in the map below.

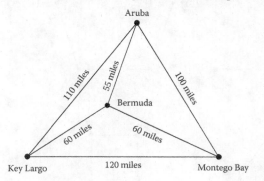

14.7. A certain connected planar graph has five regions and six vertices. How many edges does it have? Sketch such a graph.

14.8. What are the minimum number of colors needed to color the faces of regular polyhedra such that anytime two faces meet on an edge, they have different colors?

14.9. Find a map that requires four colors, in spite of the fact that nowhere in the map are there four countries each sharing a boundary with the other three.

14.10. Find a map on a torus (i.e., the surface of a doughnut) that requires five colors.

14.11. Let n be any positive integer. Is it possible to draw a (not necessarily planar) graph that requires n colors?

14.12. Prove that any map obtained by drawing a finite number of circles in the plane can be two-colored (i.e., colored with two colors such that no two regions meeting at an arc of a circle have the same color).

14.13. Find a "map" that requires five colors, in which just one of the countries is separated into two pieces (i.e., that country requires one color for both its pieces). How does this relate to Exercise 14.10?

14.14. Draw the complements of the following graphs.

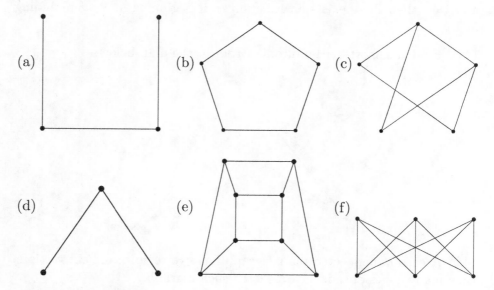

14.15. Draw graphs G and H with six vertices and five edges and six vertices and six edges, respectively. Also draw the complements $\overline{G}$ and $\overline{H}$ of your graphs.

14.16.

(a) Draw the graph

$$G = (\{A, B, C, D, E, F, G\},$$
$$\{AB, AD, AE, AF, BC, BD, BE, CD, CE, CF\}),$$

and also draw its complement $\overline{G}$.

(b) Is graph G connected?

(c) What are the degrees of the vertices?

(d) Find an Eulerian circuit in G.

(e) Prove that G is planar by exhibiting a planar representation of it.

14.17. A graph G is *regular* if every vertex has the same degree. Prove that the complement of a regular graph is also a regular graph.

14.18. Symmetry can be useful in drawing graphs in the right circumstances. Find symmetric drawings of K_4, K_5, and K_6, and from these, figure out how to use the symmetry of K_7 to draw it without doing most of the work.

14.19. The idea in Exercise 14.18 works especially well only when the graphs are drawn in the plane. Produce symmetric drawings of the same graphs in space, where this time the symmetries should be spatial symmetries rather than planar symmetries. Notice that some of them can be drawn so that fewer types of symmetries are necessary than for others. Why is this, and can you find another complete graph that is as symmetric as these?

14.20. Practice drawing trees:

(a) Draw the eleven nonisomorphic trees with seven vertices.

(b) Draw the 23 nonisomorphic trees with eight vertices.

14.21. Suppose that T is a tree and v is an endpoint. Use Theorem 14.7 to prove that $T - v$ is still a tree with $p - 1$ points. How many edges does $T - v$ have?

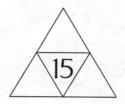

Tessellations

The mathematics of tessellation, or tiling of the plane, is present in art and visual design in our world. The Dutch graphic artist, M. C. Escher,[1] used tessellation in many of his famous prints. Escher first saw these patterns in the tiled and decorated surfaces of the Alhambra, a palace in Grenada, Spain. He sketched these patterns and was inspired to mimic them using animate objects in his repeating designs. If we look carefully at his patterns, we see that he relied on polygons as the basic unit of his designs, and then altered the polygons in such a way that their new edges interlocked.

In the number theory section of this book, we began by focusing on finite numbers and their properties, and then moved on to explore infinity. In this geometry section, we have carefully studied individual objects in two and three dimensions and then analyzed their symmetries. Now we move to the study of infinity in the context of geometry. We investigate two types of patterns. We begin by looking at infinite patterns that cover the plane and then look at infinite patterns that form borders, or friezes, in which a design is repeated in a row. Both of these decorative patterns have beautiful mathematical symmetries that lead to the same kind of group theoretic properties that we saw in the symmetry groups of regular polygons and polyhedra.

---

[1]M. C. Escher (1898–1972) was a Dutch artist who created many interesting tessellations by experimenting with the geometric principles of symmetry. His fascination with tessellations arose after he visited the Alhambra in Spain. This palace, built by the Moors, was decorated with mosaics in interesting patterns. He returned from this trip and began to use animate designs in the fashion of these mosaics. Although not a mathematician by training, he studied mathematics quite a bit, and some of his art represents designs based on non-Euclidean geometry.

Figure 15.1. Kites and darts (no symmetry).

15.1 Tessellating with a Single Shape

We begin with a question: which polygons tessellate the plane? We use our information about polygons to answer this question. In Chapter 9, we talked informally about tessellations. The following is a formal definition of a tessellation by one shape.

Definition. A *tessellation by one shape* is a covering of the plane by congruent copies of a particular plane figure with no overlaps and leaving no gaps. (For a discussion of congruence, see the Appendix.)

We will see many symmetries in the patterns we study, but notice that this definition makes no requirement that the tessellation have symmetry. Figure 15.1 shows a simple example of a tessellation of the plane that does not have symmetry. However, most of the work in this chapter focuses on tessellations that do have symmetry.

Probably the most common tessellation is the kind that uses square tiles (Figure 15.2). The squares fit together neatly at each vertex and leave no gaps at the edges or corners. The pattern repeats in infinite strips one square wide or one square tall, as shown in Figure 15.3.

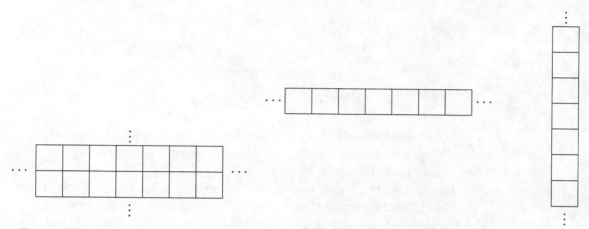

Figure 15.2. Tessellation by squares.

Figure 15.3. Infinite horizontal or vertical strips.

To figure out what polygonal shapes tessellate the plane, we use this idea of infinitely long strips to help decide the answer. The simplest way is to arrange the figure to make an infinitely long strip of fixed width. Then we can make copies of the strip and place them side by side to cover the whole plane.

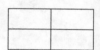

Figure 15.4. Tessellation by rectangles.

Example. Rectangles tessellate the plane. We can fit them together on either edge and make a long strip, then fit the strips together to tessellate the plane.

Figure 15.4 shows one way to do this.

Figure 15.5 shows another way, brick tiling.

Will other quadrilaterals tessellate the plane? Let's try parallelograms next. Since opposite sides are parallel and of equal length, we know that we can fit them together neatly and form one of these strips, as shown in Figure 15.6.

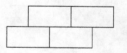

Figure 15.5. Brick tiling.

The fact that all parallelograms tessellate the plane will form the basis for some of the work we do with other polygons and tessellations. Our strategy is to try to fit together shapes to make parallelograms, and then to use strips of them to tessellate the plane.

Figure 15.6. Tessellation by parallelograms.

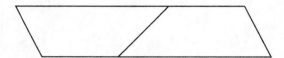

Figure 15.7. Tessellation by trapezoids.

What familiar convex quadrilaterals are left to try? A rhombus is a parallelogram, so it certainly tessellates the plane. However, we still have to look at trapezoids and kites.

Remember that in a trapezoid, two of the sides are parallel. From the Appendix, we know that the adjacent angles at the top and bottom of a trapezoid are supplementary (add up to 180°). We can put two trapezoids together by rotating one of them 180° and matching the sides (Figure 15.7). Two trapezoids positioned like this make a parallelogram! Now we can use strips of these parallelograms to tessellate the plane.

Unfortunately, this same idea does not work for the kite. If we rotate one of the kites 180°, we can fit them together and form a strip (see Figure 15.8), but we may not get parallelograms.

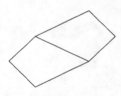

Figure 15.8. Attempt to tessellate kites by using strips.

Practice Problem 15.1. *Under what conditions would two kites form a parallelogram?*

Since making strips won't work, we need another technique to make the kites tessellate. However, before we experiment with ways to do this, let's use the parallelogram-strip method to show how any triangle tessellates the plane. Let's see how this works with an equilateral triangle as an example.

Example. Start with two equilateral triangles, and rotate one of them 180°. Then the two of them form a parallelogram (in fact, it's a rhombus; see Figure 15.9), and we can use strips of them to tessellate the plane (Figure 15.10).

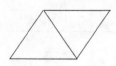

Figure 15.9. Parallelogram of two equilateral triangles.

Right triangles are easy to fit together along the hypotenuse. Two of them form a rectangle, so they also tessellate the plane (Figure 15.11).

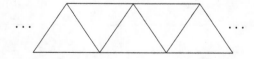

Figure 15.10. Tessellation by equilateral triangles.

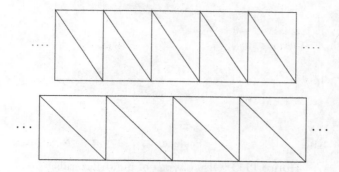

Figure 15.11. Tessellations by right triangles.

Practice Problem 15.2. *Show that two isosceles triangles can be joined along the equal sides to form a parallelogram that tessellates the plane.*

The same procedure also works for scalene triangles. Here, if we put any one of the angles at the top, and then rotate the triangle 180°, we get a parallelogram since opposite sides are parallel and opposite angles are equal (Figure 15.12).

In summary, by using two copies of any triangle to make a parallelogram that tessellates the plane, we have proved Theorem 15.1.

Theorem 15.1. *All triangles tessellate the plane.*

Is the same fact true for quadrilaterals? That is, will any quadrilateral tessellate the plane? Surprisingly, this is true. We have already shown that most of the familiar quadrilaterals do tessellate the plane by forming strips that can be repeated to cover the plane. In the case of convex quadrilaterals, what is left to consider are kites and the generic, unnamed quadrilaterals with four unequal sides and angles and with no parallel sides. We call these *scalene quadrilaterals*. We also have the nonconvex quadrilaterals to consider. These remaining quadrilaterals are shown in Figure 15.13.

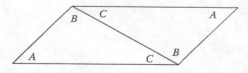

Figure 15.12. Parallelogram of two scalene triangles.

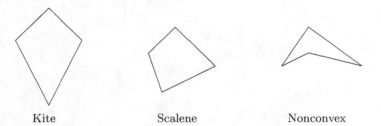

Kite   Scalene   Nonconvex

Figure 15.13. Other types of quadrilaterals.

Without using strips of parallelograms, how can we possibly get these to fit together to tessellate the plane? We can still do it, but it takes more than a few steps.

We just used parallelograms to show that all triangles tessellate the plane. Now we take a detour into a certain family of hexagons to help us investigate tessellation by quadrilaterals.

It pays to look first at how the familiar honeycomb tessellation works. The honeycomb pattern uses regular hexagons. Each of the interior angles in the regular hexagon measures 120°, so three of them fit nicely around a point. We can make strips of them and fit strips together to tessellate the plane. Notice that these strips, unlike the parallelograms, no longer have a fixed width. However, they still fit together in an obvious way, as can be seen in Figures 15.14 and 15.15.

This procedure works for other hexagons as well. Let's start with hexagons in which pairs of opposite sides are parallel and equal (Figure 15.16). We do not require such hexagons to be convex. The same

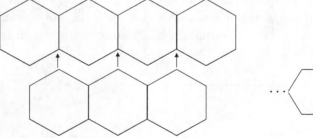

Figure 15.14. Strips of regular hexagons.

Figure 15.15. Tessellation of regular hexagons.

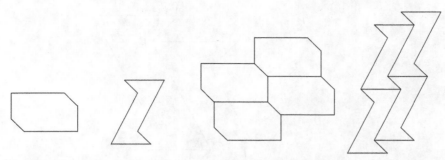

Figure 15.16. Hexagons with opposite sides that are parallel and equal.

Figure 15.17. Strips of hexagons.

procedure for making them tessellate the plane works in both cases. What we do is fit two of these together much like the honeycomb, by matching the equal sides. Like the honeycomb, these form strips that nest together, as shown in Figure 15.17.

How do we know that the strips join together with no gaps or overlaps? Let's show how this works with the convex hexagon. Since the opposite sides are parallel and equal, opposite angles are also equal. Label the three angles A, B, and C (see Figure 15.18). Since the sum of the interior angles in this hexagon is $720°$, the sum $A + B + C = 360°$. Where two strips join together, we have each angle represented, so they fit together exactly around a point.

We can do the same thing for nonconvex hexagons. Again, we make a strip of these hexagons and then fit them together like the honeycomb pattern. If we label the angles A, B, and C, we still have that $A + B + C = 360°$ (see Figure 15.19).

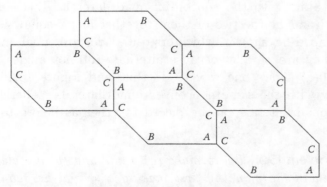

Figure 15.18. Labeled tessellation of hexagons.

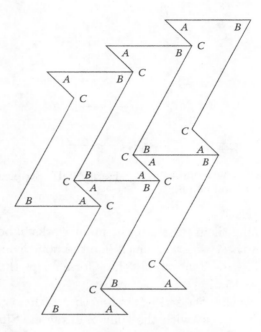

Figure 15.19. Labeled tessellation of nonconvex hexagons.

Thus, if we take any hexagon in which opposite sides are equal and parallel, we can make strips of varying width, and they will fit together to tessellate the plane.

Now we can use these hexagonal tessellations to work with kites and scalene and nonconvex quadrilaterals. The procedure for any one of these quadrilaterals is to fit two of them together to form one of the special hexagons we discussed above, that is, one with opposite sides parallel and equal. Start with the nonconvex quadrilateral. Take another one rotated 180° and fit the two of them together along equal sides. They form a Z-shaped hexagon in which opposite sides are equal and parallel. We know they are parallel because alternate interior angles are equal. Figure 15.20(a) shows one way to do this, and Figure 15.20(b) shows another way with the same nonconvex quadrilaterals. Therefore, these nonconvex quadrilaterals can be pieced together as before to tessellate the plane.

Practice Problem 15.3. *Try to make (a) a kite and (b) a scalene quadri-lateral tessellate the plane by first making a special hexagon and then forming strips that fit together.*

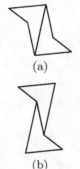

Figure 15.20. Two quadrilaterals form a hexagon, two different ways.

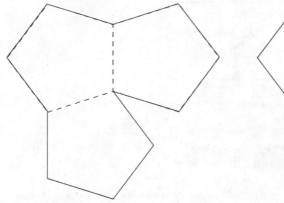
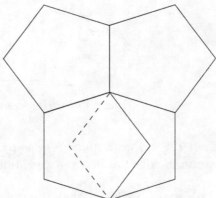

Figure 15.21. Three pentagons around a vertex.

Figure 15.22. Four pentagons around a vertex.

By using strips of parallelograms or strips of hexagons joined together, we have proved Theorem 15.2.

Theorem 15.2. *All quadrilaterals tessellate the plane.*

Now that we have shown that all triangles and all quadrilaterals tessellate the plane, we can ask whether pentagons tessellate the plane. We start with the regular pentagon. We would need to fit them together with no gaps or overlaps. Remember that the measure of each interior angle of a regular pentagon is 108°. Thus, three regular pentagons that meet at a common vertex fill only 324°, leaving a gap between them, as shown in Figure 15.21. Four regular pentagons meeting at a common vertex would take up 432°, thus overlapping, as shown in Figure 15.22.

So there is no way that regular pentagons can tessellate the plane if we insist that adjacent pentagons meet at a vertex and along a common edge. Before we proceed to other pentagons, though, let's consider other regular n-gons. We might ask: which regular n-gons tessellate the plane in this fashion? And Theorem 15.3 gives the answer we might expect.

Theorem 15.3. *Except for $n = 3$, 4, and 6, no regular n-gon tessellates the plane if we insist that adjacent n-gons meet at a vertex and along a common edge.*

Proof: We have already shown that $n = 5$ does not work. If $n \geq 7$, then the measure of each interior angle of the regular n-gon will be greater

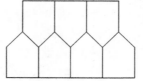

Figure 15.23. House-shaped pentagon tessellation.

than 128°. Two regular n-gons meeting together at a vertex and along a common edge will leave a gap, but three will overlap.   □

If we don't insist that the polygon be regular, there are other n-gons that do tessellate the plane. In fact, there are pentagons that tessellate the plane. One example is the house-shaped pentagon (Figure 15.23).

Practice Problem 15.4. *Try to find a differently shaped pentagon that tessellates the plane.*

So far, we have been forming tessellations where the tiles fit together edge-to-edge, vertex-to-vertex. In the case of pentagons, the assumption that this is how you must fit them together to tessellate the plane led to an interesting situation.

In 1918, a German mathematician named Karl Reinhardt (1895–1941) discovered what he called five classes of convex pentagons that tessellate the plane. He left open the question of whether there were other such classes. In 1968, a physicist at Johns Hopkins University named R. B. Kershner (1968–) published a paper in which he claimed to have found all the different convex pentagons that tessellate the plane. Martin Gardner summarized these results in a popular column he wrote for *Scientific American* in July 1975. This article inspired several amateur mathematicians to try to find all the different kinds of convex pentagons that tile the plane. One man, a computer scientist named Richard James III from California, used a modification of a tiling by octagons and squares to find a tiling of the plane not listed in Kershner's results. James did not assume, as Kershner did, that the pentagons should be edge-to-edge, vertex-to-vertex. Rather, the edge of one pentagon could adjoin the edges of two other pentagons, whose lengths added up to the length of the first side. Marjorie Rice (1923–), a California homemaker, also read Gardner's article and began a search for convex pentagon tilings, developing her own system for classifying them. She also came up with a new tiling not

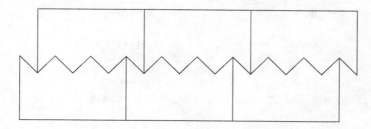

Figure 15.24. Example of n-gons with "teeth."

on Kershner's list, and eventually she found three more. A fourteenth example has since been found. As of the date of publication of this book, the general problem of finding all convex pentagons that tile the plane remains unsolved.

Many other kinds of n-gons can tessellate the plane. In fact, for any $n > 4$, we can make a set of n-gons with "teeth" that fit together to form strips that tessellate the plane (Figure 15.24). These are nonconvex polygons.

For convex polygons, the situation is quite different. Mathematicians have proven a very strong theorem about convex polygons (Theorem 15.4), but the proof is beyond the scope of this text.

Theorem 15.4. *No convex polygon with more than six sides can tessellate the plane.*

See Exercise 15.6 for an indication of the proof of this theorem.

15.2   Tessellations with Multiple Shapes

Polygons can be used in other ways to tile the plane in interesting patterns. We have been doing single-shape tessellations. Another type of tessellation uses more than one shape to make a repeating pattern that tessellates the plane.

We start with two regular octagons lined up side by side. As we said in the last section, a third octagon won't fit together with these two to tessellate the plane. However, the space is just right for a small square to fit in the corner (Figure 15.25).

If we sum the interior degree measures of the two octagons and the square, we get $360°$ to completely surround a point. We can continue with

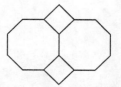

Figure 15.25. A tessellation with octagons and squares.

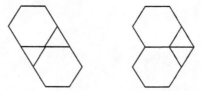

Figure 15.26. The patterns 3-6-3-6 (left) and 3-3-6-6 (right).

a pattern like this to cover the plane with no gaps or overlaps. We have a special name for this kind of tessellation, a semiregular tessellation.

Definition. If two or more regular polygons are fit together edge-to-edge and vertex-to-vertex to tessellate the plane, and if at each vertex the polygons appear in the same order, such a pattern is a *semiregular tessellation*. The pattern will be labeled according to the *n*-gons that fit around any vertex in the tessellation, and that will be the type of the tessellation.

Example. In the case of the octagon and squares, that tessellation is semiregular of type 8-8-4 since two octagons and a square surround a point. We could just as easily call this type 8-4-8 or type 4-8-8. This is the same kind of notation we used with the semiregular polyhedra.

Practice Problem 15.5. *Use a regular hexagon and four smaller equilateral triangles to make a 6-3-3-3-3 semiregular tessellation.*

To make a semiregular tessellation, we must have the same number of each type of polygon surrounding a point. It is possible to make a semiregular tessellation by using two hexagons and two triangles in the pattern 3-6-3-6, but not in the pattern 3-3-6-6, as we see in Figure 15.26.

In each case, we can fit the four polygons together to surround a point. However, in the first case, it is possible to repeat the pattern so that every vertex in the plane will have the 3-6-3-6 pattern. In the second case, if we try to extend the pattern, we will quickly find that a vertex of type 3-6-3-6 would be necessary to cover the plane (see Figure 15.27).

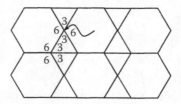

Figure 15.27. The 3-6-3-6 tessellation.

A triangle, two squares, and a hexagon can be used to surround a point in two different ways, 3-4-4-6 or 3-4-6-4, but only one of these makes a semiregular tessellation.

Practice Problem 15.6. *Show which of the* 3-4-4-6 *or* 3-4-6-4 *designs makes a semiregular tessellation.*

15.3   Variations on a Theme: Polyominoes

Another variation on tessellating with regular polygons is to use them to create nonconvex tiles to tessellate the plane. One way to use squares to tessellate the plane is by using a fixed number of squares to form *polyominoes*. These are based on the idea of a *domino*, which is just two squares joined together, edge-to-edge and vertex-to-vertex. Dominoes can be matched up to tessellate the plane (Figure 15.28).

Figure 15.28. A tessellation of dominoes.

Definition. If we join n squares together, edge-to-edge and vertex-to-vertex, we call the shape an *n-omino*.

A domino is thus a 2-omino. An n-omino can be thought of as being n squares cut out from a chessboard so that a rook can move from any square in the n-omino to any other in a finite number of moves.

What are the 3-ominoes, or triominoes? These are all the shapes we can make by matching up three squares edge-to-edge, vertex-to-vertex. There are two basic shapes for these. One is three squares in a row, and the other is an L-shaped form (Figure 15.29).

Figure 15.29.   The triominoes.

A T-shaped form (Figure 15.30) is not a triomino because the squares are not matched edge-to-edge and vertex-to-vertex.

Is the backward L-shaped polyomino different from the L-shaped triomino (Figure 15.31)? In fact, we can rotate one through 90° about the "inner" corner vertex to obtain the other. We call the L-shaped and backward L-shaped polyominoes the same because we can rotate one to get the other.

Figure 15.30. Not a triomino.

In general, we call two polyominoes the same if we can rotate or reflect one to get another. Thus, the straight and L-shape are the only two different triominoes.

Do the triominoes tessellate the plane? We can approach this question in two ways. Each one alone will tessellate the plane by forming parallel strips that can be copied to tile the plane, as shown in Figure 15.32.

Figure 15.31. Equivalent triominoes.

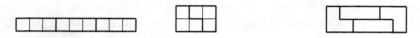

Figure 15.32. Tessellating triomi-   **Figure 15.33.** Tessellating with both
noes.   triominoes.

But we can also find a pattern that uses both shapes together to make a rectangle and use it to tessellate the plane, as shown in Figure 15.33.

Next, what are the 4-ominoes? We call these the tetrominoes, and they may be familiar shapes to students who play the game Tetris. There are five different tetrominoes (Figure 15.34). Remember that, unlike in the game Tetris, the L-shaped and backward L-shaped forms are considered to be the same here (Figure 15.35).

Practice Problem 15.7. *Verify that Figure 15.34 shows the only five tetrominoes.*

Practice Problem 15.8. *Which of the tetrominoes can be used alone to tessellate the plane? Can you put two or more tetrominoes together to tessellate the plane?*

Some interesting problems that are historical in nature have to do with filling checkerboards with polyominoes. One of the easiest to answer is the following one.

Question. Start with an 8 × 8 checkerboard and 31 dominoes. Remove two squares from opposite corners of the checkerboard, as shown in Figure 15.36. Is there an arrangement of the 31 dominoes that covers this new checkerboard?

At first, many students think this should be possible. But after a few tries, we may have covered all but two diagonal squares on the board. Why won't this work? If we consider this problem and think about the coloring of the alternate squares in the checkerboard (see Figure 15.37),

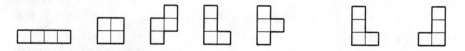

Figure 15.34. The 4-ominoes.   **Figure 15.35.** Equivalent 4-
ominoes.

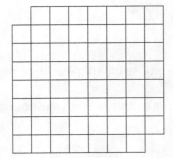

Figure 15.36. Modified checkerboard.

Figure 15.37. Modified checkerboard (colored).

the problem becomes easier to answer. When we remove squares from opposite corners of the checkerboard, we are removing two squares of the same color. Each domino covers one square of each color, no matter where we place it on the board. Thus, the 31 dominoes cannot cover the remaining squares on the board.

Another question about checkerboards is found in H. E. Dudeney's puzzle book, *The Canterbury Puzzles*, from 1907. This problem involves pentominoes, which are made of five squares. There are 12 pentominoes.

Practice Problem 15.9. *Find the 12 pentominoes. Remember that if two are congruent, they are the same, so don't count the same ones twice.*

The context for the puzzle is a story about a chess game in which the loser angrily broke the board into pieces. The 13 pieces were the 12 pentominoes and the square tetromino.

Practice Problem 15.10. *Put the 13 pieces together to form the 8×8 chess board.*

In addition to tessellation questions, polyominoes can be used to work on other kinds of problems about filling space. For example, in the 12 pentominoes, there are a total of 60 squares. Can we use each of the pentominoes exactly once and fill a rectangle that is 1×60 squares? This is obviously not possible, since all but one of the pentominoes is at least two squares wide. What about a 2×30 rectangle or a 4×15 rectangle?

Practice Problem 15.11. *The five tetrominoes fill a total of 20 square units, but they cannot be arranged to cover a 4×5 rectangle. Why not?* **Hint:** *Consider a checkerboard tiling on the rectangle.*

Practice Problem 15.12. *Try to make the 12 pentominoes cover a* 6×10 *rectangle.*

Other regions of 60-unit square shapes are used to pose pentomino puzzles. They are included in the exercises.

We have seen that there are five tetrominoes, and twelve pentominoes. How many hexominoes are there?

Practice Problem 15.13. *Try to find all the hexominoes.*

Figure 15.38. Is this a septomino?

We can now ask similar questions about tiling the plane or filling a rectangular region with the various hexominoes. The question of how many n-ominoes there are for a given n is not easily answered. In addition to the problem of figuring out which n-ominoes are congruent, beginning with $n = 7$, a new difficulty arises that makes the question more complicated. Figure 15.38 shows a shape that is made of seven unit squares. Should we include it as a septomino?

Opinion is divided on whether to use this kind of shape as an n-omino, since it has a hole in the center. For $n = 7$, there are either 107 or 108 septominoes, depending on how we count. So how many n-ominoes are there for a given n? There is no known formula that gives the answer to this question.

15.4   Frieze Patterns

We next want to look at a different kind of infinitely repeating design. These are patterns that form borders, or friezes, in which a design is repeated in a row. Suppose a designer wants to create a repeating pattern for a wallpaper border, or a ribbon for a dress, or an architectural detail. What kinds of motifs might make interesting repeating designs? Which motifs might work if the design needs to turn a corner?

Like tessellations, these decorative patterns have beautiful mathematical symmetry. Understanding the underlying mathematical properties inherent in these patterns due to their symmetry makes this kind of task easier to understand. Because they repeat in only one direction, we will be able to identify and work with their symmetries more easily than with those in two dimensions. Once we see how the symmetry groups arise for these patterns, we will be able to move on to the two-dimensional tessellations and explore their symmetries.

To explore this topic, we need to establish some rules for this kind of repeating design so that the symmetry questions make sense. Some of the ideas about symmetry that we saw with the groups for triangles, squares, and other polygons reappear here, but there are new symmetries to consider as well.

To begin, we select a design unit and make copies that repeat in a long, continuous band. We want to repeat copies of the design unit so that the repeats look exactly like the original object. For example, if the design unit we want to repeat is the word MATH, we could have a strip that looks like this:

$$\text{...MATH MATH MATH MATH MATH MATH...}$$

We imagine that this line of words extends infinitely in both directions; that's the reason for the three dots at the beginning and end of the printed pattern. In addition, we want to be sure that there is the same amount of space between each repeated design unit.

Let's investigate what kinds of symmetry such a design has. With the single triangle or square, we imagined picking it up and rotating or reflecting it until it occupied the same space it did originally. We can do something similar here, although it may be more difficult to visualize because this pattern is infinitely long. Imagine that we have an infinitely long strip of tracing paper so that we can trace over the design. How can we move the traced copy of the design so that it fits exactly over the original design?

One way would be to pick up the copied design, and move it one word to the right or left. Each individual unit—here, the word MATH—will be in a new place but will be occupying the same space as an original design unit. We call this kind of movement of the design units a *translation*. This pattern has *translational symmetry*. There are many ways to translate this design. We could move any design units to the right or left, as long as the pattern sits exactly on top of the original. If we call T the movement of the pattern one design unit to the right, then T^{-1} moves the pattern one unit to the left.

This movement indicated by T is the minimum possible translation for the pattern. If you tried to move the pattern $\frac{1}{2}T$, for example, the words wouldn't line up. This is similar to the situation where a rotation of 120° works for an equilateral triangle, but a rotation of 60° doesn't.

Practice Problem 15.14. *Describe geometrically what T^2 and T^{-3} would do to the pattern.*

Figure 15.39. Footprints.

Every infinitely repeating design made with a single design unit will have translational symmetry by its very nature. There is another kind of symmetry such a repeating pattern can have, and it will be a familiar one. Imagine a line of footprints in the sand (Figure 15.39), extending infinitely forward and backward.

This pattern has translational symmetry, as long as you consider the design unit to be a complete set of two footprints, one left and one right. These footprints have another kind of symmetry we call a *glide reflection*. This symmetry arises because we have right and left feet here. Think about a single footprint as the repeated object; unlike our first example, if we simply translate a left footprint one unit to the right, we will have a left footprint where, logically, a right footprint should be. That violates the rule about having our design unit occupying the same space as the original; thus, this kind of translation of a single footprint is not a symmetric motion of the pattern.

To avoid this difficulty, we would have to change a left footprint into a right footprint. Fortunately, there's a natural way to do this, namely, by reflecting the footprint over a horizontal line that runs through the middle of our pattern. This works beautifully

This symmetry is called *glide reflection symmetry*. The name helps you see that this is a two-step symmetry. First we take the single footprint and translate it one unit (the glide), and then we reflect the footprint over a horizontal line that bisects our pattern (the reflection).

Practice Problem 15.15. *Is a glide reflection the same as a "reflection-glide"? That is, if you translate and reflect, is that the same as reflecting first and then translating?*

Label by t the motion that translates the pattern of only one foot, not the whole design unit, to the right. In this case, remember that t is not a symmetric motion of the pattern, but $t^2 = T$ is. We label the reflection over the horizontal line H. Notice that the product Ht is a symmetric motion, and the same as the product tH. We call this product G, for glide reflection.

Challenge Problem. *Write G^{-1} in terms of the motions H and t.*

This discussion brings up some mathematical questions that would be interesting to explore. For example, do these motions make a group? If we use a different repeating pattern would we get different symmetric motions? Can we make a pattern that uses only a specified number of these motions?

Both the word MATH and the set of footprints that we used to illustrate translation and glide reflection symmetry, respectively, were themselves asymmetric design units. The group of symmetries of the word MATH, for example, or of the pair of bare footprints would just be the identity, if we were considering them as we did the objects in Chapter 11. What happens if we take an object with some internal symmetry and repeat it?

Consider the pattern made by repeating the word COB in an infinitely long strip.

$$\ldots\text{COB COB COB COB COB COB COB}\ldots$$

Here the design unit is the word COB. The word COB itself has horizontal symmetry. Now, imagine that we can trace the pattern onto our infinitely long piece of tracing paper, and try to move the tracing so that it sits on top of the original. We can see that a translation T is still a symmetry of the pattern, as it was in other examples. In this case, though, there is another way to move the copied pattern so that it sits on top of the original. Imagine a horizontal line through the middle of the pattern, and reflect the pattern over this line:

$$\overline{\ldots\text{COB COB COB COB COB COB COB}\ldots}$$

We can see that the pattern still occupies the same space it did originally, but each individual unit of the design is now upside-down. This is a symmetry of the entire pattern, which comes from a symmetry of the design unit COB. We can call this symmetric motion of the pattern H, for horizontal reflection.

If we perform H again, the pattern will not only occupy the same space it did originally, but each individual design unit will be exactly in its initial position. So, in the language of groups, H is its own inverse.

Next, consider the pattern made by repeating the word WOW in an infinitely long strip:

$$\ldots\text{WOW WOW WOW WOW WOW WOW}\ldots$$

Here, the design unit is WOW, which has its own vertical symmetry. This pattern also allows vertical reflection as well as a translation T. We can reflect the pattern over a vertical line through the middle of any O in order to get the pattern to lie on top of the original:

$$\ldots\text{WOW WOW W}\vert\text{OW WOW WOW WOW}\ldots$$

We will label this kind of vertical reflection V.

Practice Problem 15.16. *There is another place to put a vertical axis so that when you flip the WOW pattern over that axis, the pattern lies on top of the original. Where should that vertical line be?*

Practice Problem 15.17. *Convince yourself that there is no way to rotate the copied patterns of MATH, COB, or WOW and have them occupy the same space as the original pattern.*

So, we have found frieze patterns that have translational and reflectional symmetries, but none has had rotational symmetry. Are there any frieze patterns that have rotational symmetry?

The word SIS has rotational symmetry; if you place a dot in the center of the letter I and rotate the word by 180° it will occupy the same space it did originally:

$$\ldots\text{SIS SIS SIS SIS SIS SIS}\ldots$$

So if we make an infinite strip of SISs, we can do something similar. The individual design unit in this pattern is the word SIS. We can trace the pattern, select a single design unit, and put a dot in the middle of that unit. Rotating the pattern 180° either clockwise or counterclockwise around that point will put the pattern exactly on top of the original pattern. This rotation, called R, results in the tops of the original Ss being on the bottom, and vice versa. If we repeat the rotation, the pattern is exactly back to where it started in the original position.

Practice Problem 15.18. *Around what other points does this frieze pattern have rotational symmetry?*

This pattern has translation as well as rotational symmetry, but allows neither a horizontal nor a vertical reflection.

Symmetry Groups

So far, we have discussed four symmetric motions that can occur in friezes: translation, glide reflection, reflection (about a horizontal or vertical axis), and rotation by 180°. These motions of friezes can be combined in ways to form groups, much like the symmetry groups of regular polygons and regular polyhedra. The main difference here is that the groups will have an infinite number of elements in them due to the translations the patterns allow. Suppose we label our translation T for an infinitely repeating pattern. Then T^2, T^3, T^4, and so on will all be elements of the group, since each of them puts the pattern back into the original space it occupied. The same is true for the glide reflection G and its powers G^2, G^3, G^4, and so on.

As we saw in Chapter 6, the group $\mathbb{Z}$ of integers under addition is an example of an infinite cyclic group. A generator of this group is the number 1. We can think of the group generated by the translation T as being like the integers $\mathbb{Z}$.

It is interesting and useful to consider the compositions, or products, of the four symmetric motions. Using a strip of paper, or any other method, experiment with doing two symmetric motions in order. It should not come as a surprise that the composition VH of a vertical reflection and a horizontal reflection is a rotation R, just as it is in the symmetry group of a square. However, some unexpected results can come up by examining other possible combinations of symmetries.

Practice Problem 15.19. *Try to describe as specifically as you can the symmetry that results when performing the following compositions: VT; TV; VV', where V and V' are reflections about different vertical axes; RR', where R and R' are rotations by 180° about different centers; GH; and GR.*

Convince yourself that Theorem 15.5 is true.

Theorem 15.5. *Let T be a translation, and V and V' be vertical reflections about different vertical axes. Then:*

1. *VT is a vertical reflection, and $VT = T^{-1}V$.*

2. *VV' is a translation by twice the distance between the axes of reflection for V and V'.*

Other results of compositions of elements are discussed in the solution to Practice Problem 15.19 and later in this section.

Recall that we wrote all elements of the dihedral group D_{2n} as a combination of F and R, where F has order 2 and R has order n. For frieze patterns, the order of V is 2, and the order of T is infinite. We say the infinite dihedral group, D_∞, is a group made up of the elements T and V, along with their products and inverses. Notice that the relation $VT = T^{-1}V$ is reminiscent of the relation $FR = R^{-1}F$ observed in the study of the dihedral group.

The Four Symmetric Motions

Now, let's address whether there are any other symmetric motions of these infinitely repeating patterns. For finite figures in the plane, such as regular polygons, the proof that there are only two allowable symmetric motions, namely, the reflections and rotations that we saw when we worked with the regular polygons, is attributed to Leonardo da Vinci.[2]

How can we prove that only the four symmetries mentioned before (translation, glide reflection, reflection about a horizontal or vertical axis, and rotation by 180°) occur in friezes? We should begin by imagining an infinite, blank band with no pattern drawn on it (see Figure 15.40). Since any symmetry of a pattern will be a symmetry of this band, it suffices to find all the symmetries of the band.

Figure 15.40. The infinite band.

The first things to consider are the top and bottom lines of the band. If we call the height of the band 1, then the top and bottom are a pair of lines that are parallel and a distance 1 apart. It is easy to see that to preserve this relationship under a symmetry, we have only two choices: either we send the top line to the top line and the bottom to the bottom, or we switch them, sending the top line to the bottom and the bottom line to the top.

Let's investigate the first of these two options. This means that any point on the top line is sent to another point on the top line, and the

---

[2]Leonardo da Vinci (1452–1519) was one of the greatest inventor-scientists in history. Born in 1452, da Vinci was sent to Florence in his teens to apprentice as a painter under Andrea del Verrocchio. Later, da Vinci became the court artist for the Duke of Milan. He was also a civil engineer, architect, and military planner. He is perhaps best remembered as the painter of the *Mona Lisa*.

Figure 15.41. Point A on the top line sent to point A' on the top line.

Figure 15.42. The image of point B.

same for the bottom line. Let's look at a point A on the top line, and suppose it is sent under a symmetry to another point on the top line A', as in Figure 15.41.

Suppose B is some point on the band on the same vertical line as A, which is sent to a point B' under this symmetry (Figure 15.42). What are the choices for B'? We know that a symmetry preserves angles, and since $\overline{AB}$ is perpendicular to the top line, the same must be true of $\overline{A'B'}$. Furthermore, since the distance from A to B is the same as the distance from A' to B', it is easy to see that there is only one choice for B'.

In this way, we see that the symmetry depends only on where we send points on the top line. In other words, each symmetry of the band (sending the top line to itself) is completely determined by a symmetry of the top line. So the natural question to ask is: What are the symmetries of a straight line? The answer is quite simple and is the content of Theorem 15.6.

Theorem 15.6. *The only symmetries of a straight line are translation and reflection.*

Proof: Suppose we have a symmetry of the straight line. First we will analyze the case where the symmetry fixes a point A (that is, A is not moved by the symmetry). If we pick another point B on the line, which is sent to B', we know that the length of $\overline{AB}$ equals the length of $\overline{AB'}$. Then we only have two choices for B': B itself, or the point on the other side of A an equal distance away from A. See Figure 15.43.

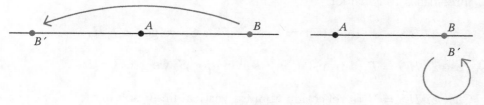

Figure 15.43. Choices for B'.

If we keep B where it is, then we have to keep every other point on the line where it is, also. If, say, there were a third point C such that C and its image C' were on opposite sides of A, then $\overline{CB}$ and $\overline{C'B}$ would have different lengths, and this is not allowed. By the same token, if we move B to the other side of A, we have to do the same thing to every other point. So our symmetry is either reflection about A or the identity (which is a translation by 0 units).

The second case is when the symmetry does not fix any points. Suppose S is a symmetry of the line sending the point A to A'. Now let T denote a translation of the line that sends A' to A. Then the composition TS fixes A. So we have either $TS = I$ (the identity) or $TS = V$, where V is reflection about A. In each case we can "solve" for S by composing with the symmetry T^{-1}, which is translation in the opposite direction of T. In the first case we get $S = T^{-1}$, so S is a translation. In the second case, we have $S = T^{-1}V$, and S is a reflection.   □

Practice Problem 15.20. *Show that $T^{-1}V$ is a reflection. What point on the top line is the fixed point of this reflection?*

Returning to our earlier discussion, we said that a symmetry of the band sending the top line to itself was determined by a symmetry of the top line. By Theorem 15.6, we know that this symmetry is either a translation or a reflection of the line. This corresponds to either a translation of the band or a reflection of the band about a vertical line.

To finish our discussion of symmetries of the band, we are concerned with symmetries that send the top line to the bottom line and the bottom line to the top line. We use a technique very similar to what we used in the proof of Theorem 15.6. Suppose that S is a symmetry that switches the top and bottom lines. Recall that H denotes a reflection about the horizontal line that bisects the band. Then the composition HS fixes the top line and the bottom line. So, by our previous work, we have only three cases to consider:

Case 1. $HS = I$ (the identity): solving for S, we get $S = H$.

Case 2. $HS = T$ (a translation): solving for S, we get $S = HT$.

Case 3. $HS = V$ (a reflection about a vertical line): solving for S, we get $S = HV$.

Practice Problem 15.21. *In each of cases 1, 2, and 3, S is one of the four symmetries: translation, reflection, rotation, or glide reflection. Which kind is it in each case?*

This finishes our proof of Theorem 15.7.

Theorem 15.7. *There are exactly four types of symmetries that can appear in frieze patterns: reflection (over a vertical or horizontal line), rotation (by 180°), translation, and glide reflection.*

The curious reader may wonder what the possibilities are for symmetries of the whole plane. The answer is essentially the same.

Theorem 15.8. *There are exactly four types of symmetries of the plane: reflection, rotation, translation, and glide reflection.*

See the Exercises for a discussion of the proof of Theorem 15.8.

Classification of Friezes

If we have two frieze patterns, we say they have the same *symmetry type* if all symmetries of one are symmetries of the other. For instance, if one has horizontal reflectional symmetry, so does the other, and so on.

If we use a motif to create a frieze pattern, we know that the pattern will have translational symmetry. There are four other possible isometries that our pattern might allow: horizontal reflection, vertical reflection, rotation of 180°, or glide reflection. If we simply count the different ways that these four isometries can be combined with a translation, how many combinations will there be?

Practice Problem 15.22. *Explain why there are 16 possible combinations of isometries that, together with translations, are available for repeating one-dimensional patterns.*

Table 15.1 lists the 16 potential symmetry types for frieze patterns. It is remarkable that there are at most only 16 potential symmetry types of frieze patterns. In fact, the result (Theorem 15.9) is even more surprising.

Theorem 15.9. *There are exactly seven one-dimensional possible symmetry types. Their isometries are listed in Table 15.2.*

	Translation T	Vertical Reflection V	Horizontal Reflection H	Rotation of 180° R	Glide Reflection G
Type 1	Yes	No	No	No	No
Type 2	Yes	Yes	No	No	No
Type 3	Yes	No	Yes	No	No
Type 4	Yes	No	No	Yes	No
Type 5	Yes	No	No	No	Yes
Type 6	Yes	Yes	Yes	No	No
Type 7	Yes	Yes	No	Yes	No
Type 8	Yes	Yes	No	No	Yes
Type 9	Yes	No	Yes	Yes	No
Type 10	Yes	No	Yes	No	Yes
Type 11	Yes	No	No	Yes	Yes
Type 12	Yes	Yes	Yes	Yes	No
Type 13	Yes	Yes	Yes	No	Yes
Type 14	Yes	Yes	No	Yes	Yes
Type 15	Yes	No	Yes	Yes	Yes
Type 16	Yes	Yes	Yes	Yes	Yes

Table 15.1. Potential symmetry types of frieze patterns.

Proof: We need to show two things: that the nine missing types from Table 15.1 of possible one-dimensional patterns cannot occur, and that the seven listed in Table 15.2 actually do occur. Here we consider each of the nine and discover why each of them is not a possible isometry. In each

	Translation T	Vertical Reflection V	Horizontal Reflection H	Rotation of 180° R	Glide Reflection G
Type 1	Yes	No	No	No	No
Type 2	Yes	Yes	No	No	No
Type 3	Yes	No	Yes	No	No
Type 4	Yes	No	No	Yes	No
Type 5	Yes	No	No	No	Yes
Type 12	Yes	Yes	Yes	Yes	No
Type 14	Yes	Yes	No	Yes	Yes

Table 15.2. Symmetry types of frieze patterns.

case, we use the product of two included motions to force the inclusion of a third, and inspect the table to eliminate cases that don't follow these rules.

First, we know that the product HV is the same as a 180° rotation, R. So in the table, if a type includes the symmetries H and V, it must also include R. By inspecting the original table, we can see that types 6 and 13 are not possible.

By using products of these same symmetry motions, we can eliminate two more types from Table 15.1. Suppose that a pattern allows both a horizontal reflection and a rotation of 180°. We know that $HV = R$. For the product HR, substituting HV for R yields

$$HR = H(HV) = V.$$

So by a similar argument, if a type includes the symmetries H and R, it must also include V. By inspecting Table 15.1, we can see that types 9 and 15 are not possible.

Next, let's look at the case where a type includes both G and H. The product GH must also occur, but this is just the motion we called t, a translation of the pattern just half the distance of the minimum translation T. This must not occur, since T is indeed the minimum translation the pattern allows. By inspecting Table 15.1, we can see that types 10 and 16 are not possible.

The rest of this argument proceeds in exactly the same way. For a discussion of eliminating types 7, 8, and 11, consult Exercise 15.19.

We now move to the second part of this theorem: illustrating the seven types that are possible.

The Seven Symmetries

We need an asymmetric motif to create each of these seven distinct symmetry types. There are lots of choices; we could use the word MATH as we did before, or we could use a single footprint. However, we like the lightning bolt (Figure 15.44) for its exciting qualities, so we use the lightning bolt motif to create each of the seven symmetry patterns in one dimension.

Figure 15.44. Lightning bolt.

Type 1 (includes T): This pattern (Figure 15.45) allows only translational symmetry. So we just need to take our asymmetric lightning bolt and repeat it along the infinite band.

Figure 15.45. A type 1 frieze.

Type 2 (includes T and V): This pattern allows translational symmetry and vertical reflectional symmetry. A row of valentine hearts has this kind of symmetry (Figure 15.46).

Figure 15.46. A type 2 frieze.

How can we create a pattern allowing these two symmetries, using our lightning bolt motif? Figure 15.47 shows a set of two lightning bolts reflected over a vertical axis.

To make a pattern for type 2 symmetry, we need to take this pair of reflected motifs and repeat them across the infinite band. Figure 15.48 shows a sample of such a pattern.

Practice Problem 15.23. *This pattern in Figure 15.48 has two types of vertical reflection axes. Identify them.*

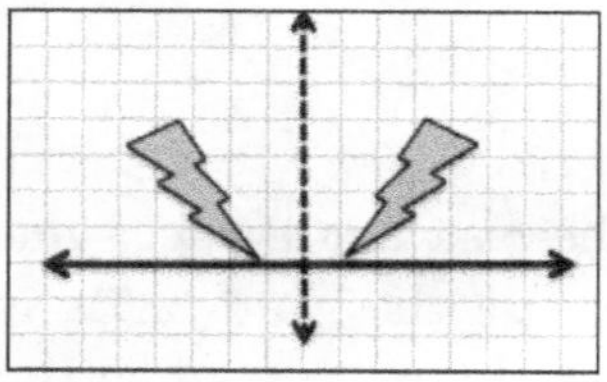

Figure 15.47. Vertically reflected lightning bolts.

Figure 15.48. A type 2 frieze using lightning bolts.

Type 3 (includes T and H): This pattern allows translational symmetry and horizontal reflectional symmetry. A row of crescent moons has this kind of symmetry (Figure 15.49).

Figure 15.49. A type 3 frieze.

How can we create a pattern allowing these two symmetries, using our lightning bolt motif? Figure 15.50 shows a set of two lightning bolts reflected over a horizontal axis.

To make a pattern for type 3 symmetry, we need to take this pair of reflected motifs and repeat them across the infinite band. Figure 15.51 shows a sample of such a pattern.

Figure 15.50. Horizontally reflected lightning bolt.

Figure 15.51. A type 3 frieze using lightning bolts.

Type 4 (includes T and R): This pattern allows translational symmetry and 180° rotational symmetry. A row of repeated letter Ss has this kind of symmetry (Figure 15.52).

The lightning bolt motif doesn't have any rotational symmetry itself, so we need to introduce a rotation to make this pattern visible. Figure 15.53 shows the result of rotating our motif 180° about that origin where the two axes meet.

To make a pattern for type 4 symmetry, we need to take this pair of reflected motifs and repeat them across the infinite band. Figure 15.54 shows a sample of such a pattern.

Practice Problem 15.24. *There are two types of places to put the center of rotation for this pattern. Identify them.*

S S S S S S S

Figure 15.52. A type 4 frieze.

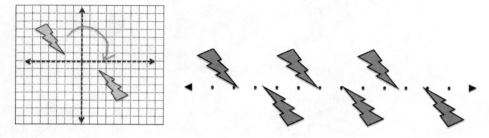

Figure 15.53. Rotated   **Figure 15.54.** A type 4 frieze using lightning bolts.
lightning bolt.

Type 5 (includes T and G): This pattern allows translational symmetry and glide reflection symmetry. This pattern is familiar from bare footprints in the sand. There are no simple horizontal reflections, but glide reflection symmetry is present. We need to treat our lightning bolt motif as if it were a bare foot. Figure 15.55 shows a pair of lightning bolts, with one translated and then reflected over a horizontal line.

To make a pattern for type 5 symmetry, we need to take this pair of reflected motifs and repeat them across the infinite band. Figure 15.56 shows a sample of such a pattern.

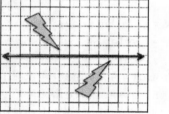

Figure 15.55. Glide re-   **Figure 15.56.** A type 5 frieze using lightning bolts.
flected lightning bolt.

Type 12 (includes T, V, H, and R): This pattern allows translational symmetry, and both horizontal and vertical reflectional symmetry, and therefore rotational symmetry. Figure 15.57 shows an example of the lightning bolts arranged to have all these symmetries.

To make a pattern for Type 12 symmetry, we need to take this set of reflected motifs and repeat them across the infinite band. Figure 15.58 shows a sample of such a pattern.

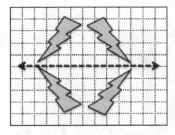

Figure 15.57. Lightning bolts reflected horizontally and vertically.

Figure 15.58. A type 12 frieze using lightning bolts.

Type 14 (includes T, V, R, and G): This pattern allows translational symmetry, as well as vertical reflectional symmetry, rotational symmetry, and glide reflection symmetry. This may seem like a complicated pattern to make, but it is actually very familiar. This pattern looks like the common zigzag line across our translation line. We combine the motifs to be sure there are both a glide reflection and a rotation in our pattern (Figure 15.59).

To make a pattern for type 14 symmetry, we need to take this pair of reflected motifs and repeat them across the infinite band. Figure 15.60 shows a sample of such a pattern.

In summary, we now have created patterns of the seven symmetry types listed in Table 15.2, and we have proven that the remaining nine of the sixteen possible patterns cannot exist. Thus, we have shown that in all the samples of frieze groups, these seven are the only possible patterns.□

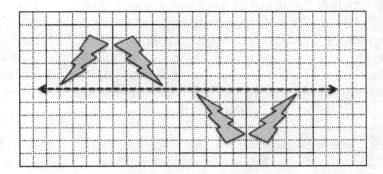

Figure 15.59. Lightning bolts with V, R, and G symmetries.

Figure 15.60. Lightning bolts with T, V, R, and G symmetries.

15.5   Infinite Patterns in Two and Three Dimensions

Each of the seven different types of frieze patterns allows a translation as one of its symmetric motions. Tessellations of the plane often also allow symmetric motions. Patterns that cover the plane often allow translations in two nonparallel directions. We can study tessellations of the plane in the same way as frieze patterns and try to discover how different patterns give rise to different combinations of symmetric motions. Again, the symmetric motions allowed by each type of pattern determines what group is associated with each pattern.

Example. Consider the honeycomb pattern (Figure 15.61). If we copy the design onto tracing paper and then translate this pattern horizontally, vertically, or "diagonally," we can move the pattern so that it fits exactly over the original design.

Figure 15.61. Honeycomb pattern.

Practice Problem 15.25. *Does the honeycomb pattern allow any reflections, rotations, or glide reflections?*

Tessellations, like frieze patterns, can allow all four of the types of symmetries we have discussed. One way we can label the patterns is to identify which of the four symmetries is allowed by a particular pattern. We also know that various polygons are used for the basic design unit in tessellations. So in addition to specifying the symmetries of a pattern, we can also characterize these patterns by what type of polygon is used as the basis for the design. Designs can use triangles, squares, parallelograms, or hexagons as their basic units. The honeycomb design, for example, has a hexagon as its repeating unit. Earlier, we saw an example of a design that uses parallelograms for the repeating design.

Thus, we can use two kinds of notation to identify a pattern that allows translation in two nonparallel directions. One will indicate whether it allows specific symmetries, and the other will indicate what the basic design unit is. Historically, the labels that we use to identify plane

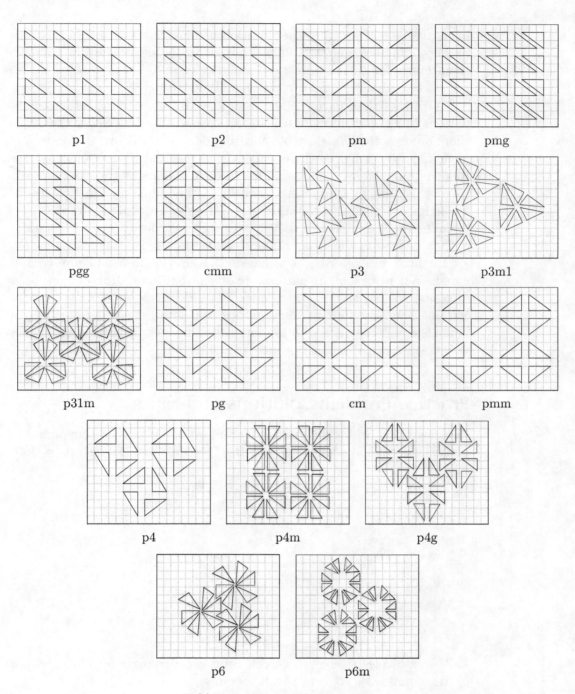

Table 15.3. The wallpaper patterns.

patterns have come from the notation of nineteenth century crystallographers. These scientists studied crystals in the natural world and gave them labels for their symmetry type. For example, they labeled a glide reflection with a "g" and a rotation of 120° with a "3."

With the frieze patterns, we were able to find seven different combinations of symmetric motions. In the case of two-dimensional patterns, there are exactly 17 classes of repeating designs. It is interesting to look at designs by Escher to classify them as one of the 17 different patterns. Readers interested in doing this might want to look at the excellent article on this topic by Doris Schattschneider, "The Plane Symmetry Groups: Their Recognition and Notation" (*AMMonthly* 85 (1978), 439–450). We include here Table 15.3 of the 17 patterns, together with their crystallographic labels, to show the different kinds of designs.

In three dimensions, the situation becomes quite a bit more complicated. When we look at patterns in space that allow three independent directions of translation, there arise 230 different groups of symmetries. Various scientists look at these groups, including crystallographers, and biologists and chemists who are interested in the symmetries and the internal structure of molecules such as DNA.

Practice Problem Solutions and Hints

15.1. Two kites will form a parallelogram if the top and bottom angles (A and C) are equal, because we need $B + C = 180°$ and $A + 2B + C = 360°$.

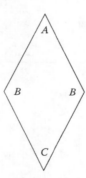

15.2. In isosceles triangles, we can rotate one triangle by 180° and pair up the equal sides to form parallelograms.

15.3. (a) 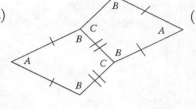 (b)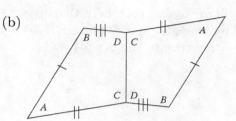

15.4.

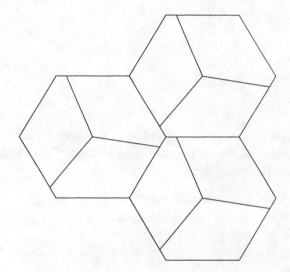

15.5.

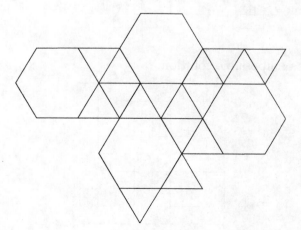

15.6. You can see here that if you start with a hexagon and surround it by triangles and squares in a 3-4-4-6 arrangement, a portion of the tiling is forced into the 3-4-6-4 pattern.

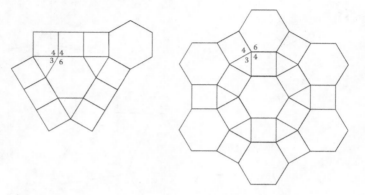

15.7. Do it!

15.8. Each tetromino alone can tessellate the plane (try it!) The point of the computer game Tetris is to use them to form a rectangle. Once this is done, you can also tessellate the plane.

15.9. Here are the 12 different pentominoes:

15.10. Here is one possible solution:

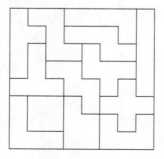

15.11. Every shape except one—the T—will cover two squares of each color. The T covers three squares of one color. This means that the five tetrominoes can cover shapes only if their checkerboard tiling contains 11 squares of one color and 9 squares of the other. A 4×5 rectangle has 10 squares of each color.

15.12. Although there are $2,339$ ways to do this, finding one isn't easy. Here's one way:

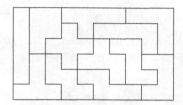

15.13.

You are very good if you found all 35 of them.

15.14. The movement T^2 moves the pattern two units to the right, and T^{-3} moves the pattern three units to the left.

15.15. Try it! (The answer is "yes.")

15.16. The vertical line should be between two WOW words, midway between two W's.

15.17. Do it!

15.18. A point midway between any two individual SIS words will also produce a symmetry of the pattern.

15.19. Some of these compositions are discussed in Theorem 15.5. For the others: RR' is a translation by twice the distance between the centers of rotation, GH is a translation, and GR is a vertical reflection.

15.20. One way to see that $T^{-1}V$ is a reflection is by Theorem 15.5. Alternatively, notice that $T^{-1}V$ fixes the midpoint of A and A' (why?). Then, since $T^{-1}V$ is not the identity, it must be a reflection about the midpoint of A and A'.

15.21. In case a), S is a horizontal reflection. In case b), S is a glide reflection. In case c), S is a rotation by $180°$.

15.22. For each isometry we have two choices: either the pattern has it, or it doesn't. Two choices for each of four isometries yields $2^4 = 16$ combinations.

15.23. A vertical line that meets the lightning bolts at the tips, and a vertical line between two pairs of lightning bolts.

15.24. The midpoint of the tips of two adjacent lightning bolts (arrows a and b) forms a center of rotation for a design unit made of two bolts. The midpoint between two design units (arrow c) forms a center of rotation of two design units (four bolts).

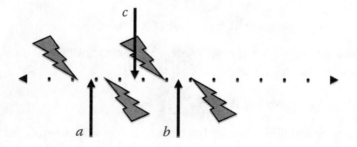

15.25. The honeycomb pattern allows the same kinds of reflections and rotations as the individual hexagonal units. There are also glide reflections.

Exercises

15.1. Two methods were described for tessellating the plane with quadrilaterals: joining two to form a parallelogram or joining two to form a hexagon. Under what conditions is each possible? Are there quadrilaterals that can be joined in either way, depending on how you choose to arrange them?

15.2. Check that two copies of a trapezoid fit together to form a parallelogram by rotating one 180°. This will require that you show that the resulting figure is a quadrilateral with opposite angles equal.

15.3. Show that a convex pentagon with two parallel sides must tessellate the plane.

15.4. The following diagrams show eight convex pentagons (five determined by Reinhardt and three additional pentagons determined by Kershner) that tessellate the plane.

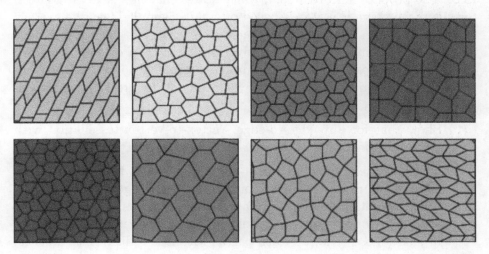

(a) Make copies of each of these and show that each pentagon can tessellate the plane.

(b) Try to determine a convex pentagon that tessellates the plane and is not in any of the eight classes shown above.

15.5. The following diagrams show three convex hexagons (determined by Reinhardt) that tessellate the plane.

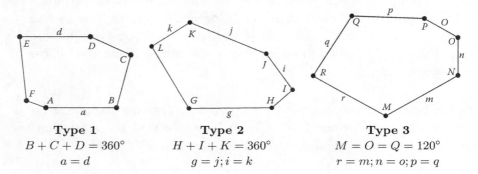

Type 1	Type 2	Type 3
$B + C + D = 360°$	$H + I + K = 360°$	$M = O = Q = 120°$
$a = d$	$g = j; i = k$	$r = m; n = o; p = q$

(a) Make copies of each of these hexagons and show that each can tessellate the plane.

(b) Show that there are no additional classes of convex hexagons that tessellate the plane.

15.6. The proof of Theorem 15.4, which states that no convex n-gon can tessellate the plane if $n > 7$, is based on a fairly simple idea: the average of the angles in an n-gon, $n > 7$, is more than 120°, but the average angle in a tessellation made of convex polygons is less than or equal to 120°. The following steps will guide you through a formal proof of this theorem.

(a) Show that an n-gon is not convex if any of its interior angles measures more than 180°.

(b) Show that for any edge-to-edge, vertex-to-vertex tessellation with convex n-gons, at least three n-gons meet at every vertex.

(c) The *diameter d* of a convex n-gon is defined as the length of the longest line connecting two points within the n-gon. Suppose this n-gon tessellates the plane. Take a disk of radius $32d$ and place it somewhere on top of the tessellation. Say that the number of interior n-gons (those lying at least partly within the disk) is k and the number of boundary n-gons (n-gons that intersect the circumference of the disk) is l. Find an upper bound for $\frac{l}{k}$ by calculating areas. Hint: All boundary n-gons lie in the region bounded by circles of radius $31d$ and $33d$, centered at the same point as the original disk.

(d) Assume the tessellation is edge-to-edge, vertex-to-vertex. Calculate the sum of all the angles in all interior polygons in two different ways, and derive a contradiction. **Hint**: First, add up the angles in every n-gon. Second, count the angles around every vertex. Add 120° for every angle that is not on a boundary polygon, and 180° for every angle that is on a boundary polygon. This gives an upper bound for the sum of the angles.

(e) If you remove the requirement that the tessellation be edge-to-edge, vertex-to-vertex, you get a tessellation with *improper junctions*: points that are vertices of some n-gons, but nonvertex edge points of other n-gons (a *junction* is any point in the tessellation that is a vertex of some n-gon). The number of improper junctions on interior polygons is m; the number of improper junctions of boundary n-gons is m^*. Modify your calculation from part (d) so that you count 360° around every junction. **Hint**: For nonboundary n-gons with an improper junction, you need to add only 120° to account for the improper junction. Why?

15.7.

(a) What arrangements of more than one kind of regular polygon can you fit around a vertex without gaps or overlaps? You should be able to list at least 12 arrangements.

(b) Which of these arrangements can be used to make a semi-regular tessellation of the plane? You should be able to list eight.

15.8. A tiling of a region with dominoes is said to be *simple* unless a subset of at least two, but not all, of the tiles forms a rectangle in the tiling, or the tiling contains only one or two tiles. Find a simple tiling of the plane with dominoes.

15.9. Show that no rectangle has a simple tiling with dominoes.

15.10. Can you tile a 5×5 square with eight L-triominoes and one monomino?

15.11. Can you tile a 10×10 square with straight tetrominoes?

15.12. Can you tile a 10×10 square with L-triominoes?

15.13. Can you tile a 10×10 square with T-tetrominoes?

15.14. Use the 12 pentominoes to form, simultaneously, a 3×5 rectangle and a 5×9 rectangle.

15.15. Use the 12 pentominoes to form the pattern shown here.

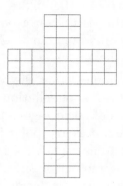

For more information on problems of this type, see Golomb's *Polyominoes*.

15.16. In the infinite dihedral group D_∞, what is the order of the element TH? What about the element HT?

15.17. This exercise proves the "Three Reflections Theorem," which states that any isometry of the plane is just the composition of three reflections.

(a) Suppose ABC and $A'B'C'$ are any two congruent triangles in the plane. Show that you can, by using three reflections, move ABC onto $A'B'C'$. **Hint:** Make the first reflection the perpendicular bisector of AA'. Then this reflection moves A onto A'.

(b) Let ABC be a triangle in the plane that is sent to $A'B'C'$ under an isometry. Explain why, for any other point D in the plane, you can figure out exactly where D' is. **Hint:** The distances DA, DB, DC must equal the distances $D'A'$, $D'B'$, $D'C'$. Why? In other words, show that any two isometries sending ABC to $A'B'C'$ must be the same.

(c) Use parts (a) and (b) to finish the proof of the theorem.

15.18. This exercise proves the result stated in Theorem 15.8.

(a) Show that a reflection about two parallel lines is a translation by twice the distance between the lines.

(b) Show that a reflection about two intersecting lines is a rotation by twice the angle between the lines.

(c) By Exercise 15.17, any isometry of the plane is a composition of three reflections. Divide this exercise into four cases: either there are three parallel lines, three lines that all intersect in a point, two parallel lines cut by a transversal, or three lines that all intersect each other at different points. With the help of parts (a) and (b), analyze each of these cases and prove Theorem 15.8.

15.19. This exercise finishes the proof that certain symmetry types are impossible.

(a) Show that the composition of a vertical reflection with a rotation is a horizontal reflection, and eliminate type 7.

(b) Write a glide reflection as $G = Ht$, and use Theorem 15.5 to show that the composition of a vertical reflection and a glide reflection is a rotation. This eliminates type 8.

(c) Show that the composition of a glide reflection and a rotation is a vertical reflection. This eliminates type 11.

15.20. Sketch a frieze whose symmetry group is generated by a glide reflection.

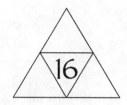

Connections

Number theory and geometry are each, by themselves, rich sources of mathematical ideas and problems. We have seen how some ideas are common to both of these areas. Group theory, in terms of congruences and symmetries, has been a unifying factor in our presentation. Whenever two or three different areas of mathematics meet in a single topic, they reinforce and enrich one another. Throughout the history of mathematics, the interplay among different areas has provided many interesting and deep results. We end the book by presenting two areas where the ideas of geometry and number theory blend together in interesting ways.

16.1 The Golden Ratio and Fibonacci Numbers

The Greek sculptor Polycleitus[1] used the *golden ratio* or *golden mean* when he made careful measurements of human proportion and used it as the basis for his designs. The golden ratio has been a guide to proportion in art since that time.

The Golden Ratio and Geometry

How does the golden ratio arise in geometry? Draw a divided line segment with the longer length labeled a and the shorter length labeled b, as shown in Figure 16.1.

If the ratio of the segment's entire length $(a + b)$ to the longer length (a) is the same as the ratio of the longer length (a) to the shorter length (b), then the ratio is *golden*.

---

[1]Polycleitus (ca. 500 B.C.) was a Greek sculptor and author. In his book *The Canon*, he described the ideal human proportions and how to use these in sculpting a human figure. None of his works survived antiquity.

Figure 16.1. Divided line segment.

The proportion for the golden ratio is $\frac{a+b}{a} = \frac{a}{b}$. Rearranging this equation, we have $\frac{a}{a} + \frac{b}{a} = \frac{a}{b}$, or

$$1 + \frac{b}{a} = \frac{a}{b}.$$

If we rename $\frac{a}{b} = \tau$, this equation becomes $1 + \frac{1}{\tau} = \tau$ or $\tau + 1 = \tau^2$. This leads to the quadratic equation

$$\tau^2 - \tau - 1 = 0.$$

The quadratic formula yields $\tau = \frac{1 \pm \sqrt{5}}{2}$ as roots. The positive root is the golden ratio. Its value is approximately 1.618.

Constructing the Golden Ratio

Now that we know the numerical value of τ, let's look at a method for constructing this length using straightedge and compass. To do this, we must construct two segments, one of length $\frac{1}{2}$ and one of length $\frac{\sqrt{5}}{2}$. If we add these two lengths together, we will have a segment of length $\tau = \frac{1 + \sqrt{5}}{2}$.

Since we know how to construct a perpendicular bisector given a unit segment, we can construct a segment of length $\frac{1}{2}$. What about constructing a segment of irrational length? How can we construct a segment of length $\frac{\sqrt{5}}{2}$?

Let's think about a segment with a more familiar irrational length, namely, $\sqrt{2}$. If we begin with a unit length and use the method of construction of bisectors, we can construct a square with side length 1. By the Pythagorean theorem (see the Appendix), the diagonal of this square has length $\sqrt{1^2 + 1^2} = \sqrt{2}$.

More generally, we know that if a right triangle has legs of length a and b, then its hypotenuse will have length $\sqrt{a^2 + b^2}$. So if we have $a = 1$ and $b = \frac{1}{2}$, then the length of the hypotenuse c will be $\sqrt{a^2 + b^2} = \frac{\sqrt{5}}{2}$. So to construct the golden ratio, we need to construct a right triangle with legs of length 1 and $\frac{1}{2}$ so that the hypotenuse will have length $\frac{\sqrt{5}}{2}$. We can do that inside a unit square, as follows:

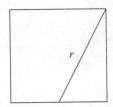

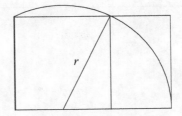

Figure 16.2. Unit square with length r.   **Figure 16.3.** Extending the sides of the unit square.

Construct a square with unit side length, bisect the bottom edge of the square and draw the diagonal of the half-square. Call the length of this diagonal r (Figure 16.2). By our argument above, r has length $\frac{\sqrt{5}}{2}$.

Now we can easily construct a rectangle of width 1 and length τ. First, we use the straightedge to extend the bottom and top edges of the square. Then we place the point of the compass at the midpoint of the bottom edge of the original square. We draw an arc through the top vertices of this square, so that the arc intersects the line containing the bottom edge of the square. From the point where that arc intersects the extended bottom edge of the square, drop a perpendicular from the extended top edge (Figure 16.3).

We now have a rectangle of width 1 and length $\frac{1}{2} + r = \frac{1}{2} + \frac{\sqrt{5}}{2} = \tau$ (Figure 16.4). Any rectangle whose sides are in the ratio τ is called a *golden* rectangle. If we remove the original square from this larger rectangle, the remaining smaller rectangle is also golden.

Practice Problem 16.1. *Show that the small rectangle remaining if you remove the original square is golden; that is, that the ratio of the length to the width of this small rectangle is also the golden ratio.*

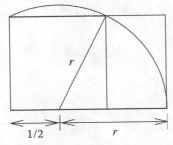

Figure 16.4. A golden rectangle.

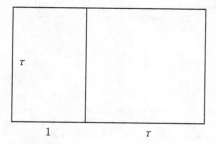
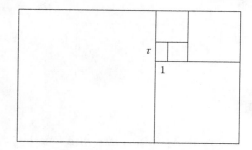

Figure 16.5. A larger golden rectangle.   **Figure 16.6.** Increasingly large golden rectangles.

We could also construct a new golden rectangle from the one we just made by adding a square of side length τ to the longer side, as shown in Figure 16.5.

The resulting rectangle is golden. We can continue this process indefinitely, each time obtaining a larger golden rectangle (see Figure 16.6).

Fibonacci Numbers

In Chapter 5, we introduced the Fibonacci numbers, which we defined as follows.

Definition. The *Fibonacci sequence* $f_1, f_2, f_3, \ldots, f_n, \ldots$ is defined recursively by starting with $f_1 = 1$, $f_2 = 1$, and $f_n = f_{n-1} + f_{n-2}$ for $n \geq 3$.

There is a connection between this sequence and the golden ratio. To see this connection, we form a new sequence of numbers based on the Fibonacci sequence.

If we form the ratios r_n of consecutive Fibonacci numbers f_n and $f_n + 1$, so $r_n = \frac{f_{n+1}}{f_n}$, we get the sequence

$$r_1 = \frac{1}{1} = 1,$$

$$r_2 = \frac{2}{1} = 2,$$

$$r_3 = \frac{3}{2} = 1.5,$$

$$r_4 = \frac{5}{3} = 1.66\ldots,$$

$$r_5 = \frac{8}{5} = 1.6,$$

$$r_6 = \frac{13}{8} = 1.625,$$

$$r_7 = \frac{21}{13} = 1.615\ldots,$$

$$r_8 = \frac{34}{21} = 1.619\ldots,$$

$$r_9 = \frac{55}{34} = 1.617\ldots,$$

$$r_{10} = \frac{89}{55} = 1.618\ldots,$$

$$r_{11} = \frac{144}{89} = 1.6179\ldots,$$

$$r_{12} = \frac{233}{144} = 1.6180\ldots,$$

$$r_{13} = \frac{377}{233} = 1.61802,\ldots$$

Notice that successive terms of the sequence alternately get bigger and smaller.

Challenge Problem. Show that any odd term in this sequence is smaller than any even term in this sequence, that the odd terms of this sequence increase ($r_1 < r_3 < r_5 < \ldots$), and that the even terms of this sequence decrease ($r_2 > r_4 > r_6 > \ldots$).

Also, the ratios of the Fibonacci numbers get closer and closer together. There is precisely one number that is larger than all the odd numbered terms and smaller than all the even numbered terms. This is the golden ratio.

We can see this by starting with the definition for r_{n+1} and then rewriting using the Fibonacci relations to get

$$r_{n+1} = \frac{f_{n+2}}{f_{n+1}} = \frac{f_{n+1} + f_n}{f_{n+1}} = 1 + \frac{f_n}{f_{n+1}}.$$

Written in terms of ratios, this equation becomes

$$r_{n+1} = 1 + \frac{1}{r_n}.$$

The difference between successive even and odd terms gets smaller and smaller. It turns out that all the ratios approach a single number L. Then we can rewrite the equation as $L = 1 + \frac{1}{L}$. If we multiply both sides of this equation by L and rearrange terms, we get the familiar quadratic equation $L^2 - L - 1 = 0$. Again, the quadratic formula gives $\frac{1 \pm \sqrt{5}}{2}$ as roots. Since all our ratios are positive, we choose the positive root, and set

$$L = \frac{1 + \sqrt{5}}{2}.$$

This is the golden ratio. It has been known for centuries that the Fibonacci sequence has this property.

16.2   Constructible Numbers and Polygons

In the geometry section of this book, we used a straightedge and compass to construct geometric objects, including several regular polygons. In Section 16.1, we used the same tools to construct the golden ratio, a number that arises from a geometric relationship. The method of using a straightedge and compass to construct numbers leads to some interesting questions about which real numbers can be constructed in this way. We conclude the text by discussing the related topics of constructible numbers and constructible polygons, and stating a theorem that connects the two ideas.

What do we mean by a constructible number? We say a positive number α is *constructible* if we can construct a line segment of length α in a finite number of steps, starting with a given segment of unit length and using only a straightedge and compass. Remember that we already know how to construct perpendicular bisectors of line segments, and we also know how to construct a line parallel to a given segment through a given point (see Exercise 10.10). Suppose we know how to construct two line segments of lengths α and β. What other lengths can we construct using these two given lengths?

First, we can construct the new length $\alpha + \beta$. We start by extending the line segment of length α with the straightedge. Then, using the

Figure 16.7. Segment OA.

compass with the point at the end of the segment of length α, we mark off an extension of length β. This construction gives us a segment of length $\alpha + \beta$.

Practice Problem 16.2. *Given segments of length α and β, with $\alpha > \beta$, show how to construct a segment of length $\alpha - \beta$.*

Next, we want to construct a segment with length equal to the product $\alpha\beta$ of our two numbers α and β. (If you first thought to construct a rectangle with side lengths α and β, the *area* of this rectangle would be $\alpha\beta$, but what we want is a segment with length $\alpha\beta$.) Using similar triangles (see the Appendix), we can construct segments with any given ratio. We plan a construction here that uses the equation $\frac{1}{\alpha} = \frac{\beta}{\alpha\beta}$ to get a segment with length $\alpha\beta$.

Start by marking off a segment OA of length α, and extending the segment to the right (Figure 16.7).

Use the straightedge to draw any line through O that does not contain A. Mark points U and B on this line so that OU has length 1 and OB has length β (see Figure 16.8).

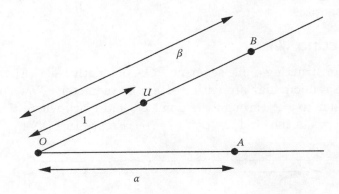

Figure 16.8. Adding specific points U and B.

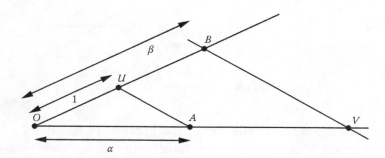

Figure 16.9. Constructing the point V.

Use the straightedge to draw in the segment UA. Now construct the line through B parallel to UA that intersects the extended segment OA at the point V (see Figure 16.9).

Now we have that triangles OAU and OVB are similar triangles. If we say $OV = \gamma$, this gives the ratio $\frac{1}{\alpha} = \frac{\beta}{\gamma}$, or $\gamma = \alpha\beta$, which implies that segment OV has length $\alpha\beta$.

Practice Problem 16.3. *Given segments of length α, β and 1, construct a segment of length $\frac{\alpha}{\beta}$, provided that $\beta \neq 0$.*

What does this process tell us about which numbers we can construct? Certainly, given a segment of unit length, we can construct any positive integer. Because we can also construct quotients of constructible numbers, we can also construct any positive rational number. What about constructing irrational numbers? We have already constructed two, namely, $\sqrt{2}$ and the golden ratio. Now let's find a method for constructing $\sqrt{\alpha}$ if α is constructible.

Constructing $\sqrt{\alpha}$

Begin by constructing a line segment POA of length $1 + \alpha$ (Figure 16.10).

Make a semicircular arc with PA as the diameter. Now construct a perpendicular to PA through the point O, and mark where this perpendicular intersects the circle as point Q (see Figure 16.11).

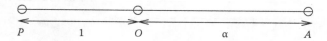

Figure 16.10. Segment POA.

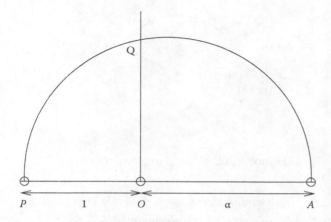

Figure 16.11. Constructing Q.

Since any triangle inscribed in a semicircle is a right triangle, we can see that triangles OPQ and OQA are similar. From these similar triangles, we get the ratio $OQ/OA = OP/OQ$. This means $OQ^2 = \alpha$, or $OQ = \sqrt{\alpha}$. Since we could carry out this procedure for any constructible number α, we see that square roots of constructible numbers are constructible.

Constructible Polygons

Now that we have constructed more numbers, we want to see the connection to the construction of regular polygons. We have already shown in Chapter 10 how to construct equilateral triangles, squares, and regular hexagons. Using methods of bisection, we have also shown how to use these basic constructions to construct regular n-gons for any n that is a power of two or is three times a power of two. Now let's construct the regular pentagon. We show several different methods of construction.

Our first approach will be to construct a regular decagon inscribed in a circle, similar to the way we constructed a regular hexagon. We first need to know the side length of a regular decagon inscribed in a unit circle. Then we connect every other vertex of the decagon to get a regular pentagon from the regular decagon. The proof that this construction results in a regular decagon involves solving quadratic equations and working with proportions.

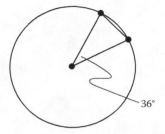

Figure 16.12. Central angle of a decagon.

Theorem 16.1. *If a regular decagon is inscribed in a circle of radius 1, then the length of each side of the decagon is $\frac{-1+\sqrt{5}}{2}$.*

Proof: First, if we have a regular decagon inscribed in a unit circle and connect the center of the circle to the edges, each central angle will be $36°$ (see Figure 16.12).

We want to focus our attention on one of these interior triangles, say OAB, where the angle O is $36°$ (Figure 16.13). Since this triangle is inside a circle of radius 1, sides OA and OB have lengths 1. Therefore, this is an isosceles triangle, and the measure of angle A and angle B is $72°$. Our goal is to calculate the length of AB.

Consider a bisector of angle B, and call D the point where the bisector meets OA (see Figure 16.14). Notice that $\angle ABD = \angle DBO = \angle DOB = 36°$, and $\angle ADB = \angle OAB = 72°$, so triangles ABD and BDO are both isosceles. So, if we call x the length of AB, then $BD = x$ as well. Then we also have $DO = x$. Since AO has length 1, this means the length of $AD = 1 - x$.

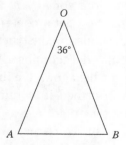

Figure 16.13. Triangle OAB.

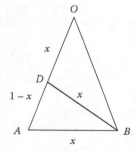

Figure 16.14. Lengths of segments in triangle OAB.

To summarize, we have the following lengths:

$$AO = OB = 1; AB = BD = OD = x; AD = 1 - x.$$

We now exploit the similarity of the two 36°-72°-72° triangles, namely, ABD and AOB. We have the following proportion: $\frac{AO}{AB} = \frac{AB}{AD}$, or $\frac{1}{x} = \frac{x}{1-x}$.

This simplifies to the quadratic equation $x^2 + x - 1 = 0$. We can use the quadratic formula to get the two solutions $x = \frac{-1+\sqrt{5}}{2}$ and $x = \frac{-1-\sqrt{5}}{2}$. Since one of these is a negative number, we discard it, and see that the length of the side AB of the decagon inscribed in a unit circle must be $\frac{-1+\sqrt{5}}{2}$.   ☐

So we see that to construct a regular decagon we only need to construct a segment of length $\frac{-1+\sqrt{5}}{2}$. Notice that the length of the side AB looks similar to the golden ratio, but it is not the golden ratio, which is $\tau = \frac{1+\sqrt{5}}{2}$.

Practice Problem 16.4. *Construct a segment of length $\frac{-1+\sqrt{5}}{2}$.*

Then, to make the decagon, first construct a unit circle and the length $\frac{-1+\sqrt{5}}{2}$. Open the compass to the length $\frac{-1+\sqrt{5}}{2}$, start anywhere on the circle, and mark off ten of these lengths (see Figure 16.15). We know there will be ten of these lengths since each central angle is 36°. Then to make the regular pentagon, just inscribe a regular decagon in a circle and connect every other vertex.

Practice Problem 16.5. *Using congruent triangles, prove that connecting every other vertex of a regular decagon inscribed in a unit circle gives a regular pentagon.*

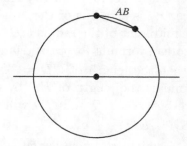

Figure 16.15. Constructing the decagon.

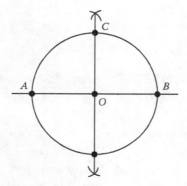 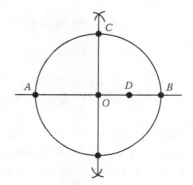

Figure 16.16. Gauss's pentagon con-   **Figure 16.17.** Gauss's pentagon construction, step 1.   struction, step 2.

Gauss's Construction of a Regular Pentagon

Gauss's construction of a regular pentagon does not require solving any equations or knowing any lengths. However, the proof that what he constructed is in fact a regular pentagon depends on the same facts we used in our construction of a regular decagon.

We begin with a diameter AB of a unit circle and bisect it. Label the bisector's intersection with the top of the circle as C (Figure 16.16).

Next, bisect the radius OB. Call the midpoint of this radius D (Figure 16.17).

Now open the compass to the length of CD and swing an arc from D across the diameter AB. Call the point where this arc crosses the radius E (Figure 16.18).

Observe that OE is the side length of a regular decagon inscribed in a circle of radius 1. The algebra to justify this claim follows. The length of OC is 1, since OC is a radius of the unit circle. The length of $OD = \frac{1}{2}$, since D is the midpoint of the radius OB. As we did previously, we can use the Pythagorean formula for right triangles in triangle OCD (Figure 16.19) to see that the length of $CD = \frac{\sqrt{5}}{2}$. We know that the length of CD is the same as the length of ED by construction. On the diameter, the difference between ED and OD will be the length of OE. Thus, $OE = ED - OD = ED - \frac{1}{2}$, or $\frac{\sqrt{5}}{2} - \frac{1}{2}$. This difference, or the length of OE, is $\frac{-1+\sqrt{5}}{2}$, which we know is the side length of a regular decagon inscribed in a unit circle.

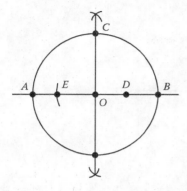

Figure 16.18. Gauss's pentagon construction, step 3.

Figure 16.19. Gauss's pentagon construction, step 4.

To continue, open the compass to the unit radius of the circle and put the point of the compass at E. Make an arc from E and label the point where this arc crosses the circle as F (Figure 16.20).

Now we focus attention on the triangle EFO (Figure 16.21). Side OF is of length 1 because it is the radius of a unit circle. Side EF is of length 1 by construction. Therefore, triangle EFO is an isosceles triangle. From the previous steps, we know that the length of OE is $\frac{-1+\sqrt{5}}{2}$, which is the length of the base of an isosceles 72°-72°-36° triangle whose legs have length 1. Therefore, angle EOF must be 72°. In a regular pentagon, the central angle is also 72°, so if we draw the segment AF, it will be the length that we need for the side of a regular pentagon. Using the

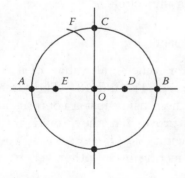

Figure 16.20. Gauss's pentagon construction, step 5.

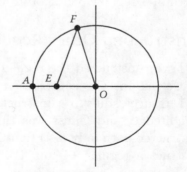

Figure 16.21. Gauss's pentagon construction, final step.

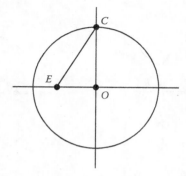

Figure 16.22. Ptolemy's pentagon construction.

compass, mark off lengths equal to the arc AF all around the circle, then join these points by using the straightedge to draw the pentagon.

Practice Problem 16.6. *Once you have constructed the regular pentagon, use the method of bisection to construct a regular decagon.*

A variation of this same basic construction allows us to get the side length of a pentagon without using the decagon construction. This method was written down by Ptolemy in Egypt in A.D. 150. In the same construction that Gauss used, instead of using the point F, Ptolemy claimed that the length EC is the correct length for the side of a regular pentagon (see Figure 16.22).

Challenge Problem. Prove algebraically that the length EC is the side length of a regular pentagon inscribed in a unit circle.

Constructing Other Regular n-gons

We have constructed the regular n-gons for some small values of n, but we are missing the construction for $n = 7, 9, 11$, and 13. The Greeks tried unsuccessfully to construct each of these using just a compass and straightedge, and Gauss was the first to prove that they could *not* be constructed with only a straightedge and compass. Of course, if we could construct a regular 14-gon, then by connecting every other vertex, we could construct a regular 7-gon.

The Greeks knew how to construct the regular 15-gon by combining the constructions of a regular pentagon and an equilateral triangle. To understand how this works, we need to know that one side of a regular 15-gon inscribed in the unit circle marks off an arc that measures $\frac{2\pi}{15} = 24°$. Recall that a side of a regular pentagon inscribed in a unit circle marks off an arc of length $\frac{2\pi}{5}$, and the side of an equilateral triangle gives an arc of length $\frac{2\pi}{3}$. Two arcs of length $\frac{2\pi}{5}$ minus an arc of length $\frac{2\pi}{3}$ give an arc of length $\frac{4\pi}{5} - \frac{2\pi}{3} = \frac{12\pi - 10\pi}{15} = \frac{2\pi}{15}$. This is the arc length we need to construct a regular 15-gon.

This kind of construction works to make a regular n-gon and a regular m-gon combine to make a regular mn-gon precisely when m and n are relatively prime. We prove this now. First, we need the following additional fact about numbers that are relatively prime, which can be derived from the proof of Theorem 4.3.

Fact. Suppose two positive integers a and b are relatively prime. Then there are integers s and t so that $sa + tb = 1$.

Suppose we know how to construct a regular m-gon and a regular n-gon using a straightedge and compass. The steps to construct the regular mn-gon if we assume that $(m, n) = 1$ is as follows. By constructing both a regular m-gon and a regular n-gon, we have been able to construct arcs on the unit circle with measures of $\frac{2\pi}{m}$ and $\frac{2\pi}{n}$. If s and t are integers that make the equation $sm + tn = 1$ true, then the arc whose measure is $\frac{2\pi(sm+tn)}{mn} = \frac{2\pi}{mn}$ is precisely the arc required to construct a regular mn-gon. We can construct this arc by first constructing arcs with measures of $\frac{2\pi s}{n}$ and $\frac{2\pi t}{m}$ by simply marking off copies of the original arcs on the circle.

For example, suppose we want to construct a regular 15-gon. Since $(3, 5) = 1$, there are integers s and t that make the equation $3s + 5t = 1$ true. Let's use $s = 2$ and $t = -1$, for which the equation is true. Then we need to make arcs of length $\frac{2\pi s}{n} = \frac{2\pi \cdot 2}{5} = \frac{4\pi}{5}$ and $\frac{2\pi t}{m} = \frac{2\pi \cdot (-1)}{3} = \frac{-2\pi}{3}$. Because we can construct a regular pentagon, we can construct an arc of length $\frac{2\pi}{5}$. Since $s = 2$, we mark off two of these arcs going counterclockwise around the circle. Because we can construct an equilateral triangle, we can also construct an arc of length $\frac{\pi}{3}$. Since $t = -1$, we begin at the $\frac{4\pi}{5}$ spot and mark off a $\frac{2\pi}{3}$ arc *clockwise* around the circle. Our ending point will be at an arc whose measure is $\frac{4\pi}{5} - \frac{\pi}{3} = \frac{2\pi}{15}$ from the starting point. This is the arc we need to construct a regular 15-gon.

By using our regular n-gon constructions together with bisections, we can make a list of the sets of regular n-gons that the Greeks knew how to construct:

$$n = 4, 8, 16, 32, \ldots,$$
$$n = 3, 6, 12, 24, \ldots,$$
$$n = 5, 10, 20, 40, \ldots,$$
$$n = 15, 30, 60, \ldots.$$

Small values of n not on this list are $n = 7$, 9, 11, 13, 14, and 17. Despite many attempts, the Greeks were unable to come up with a construction of any of these regular n-gons. In 1796, at the age of 19, Gauss came up with a construction of the regular 17-gon. This discovery convinced him to study mathematics for his life's work. Moreover, his work with this problem led him to Theorem 16.2, stating precisely which regular n-gons could be constructed with only a straightedge and compass.

Theorem 16.2. *The values of n for which a regular n-gon can be constructed with straightedge and compass are those $n \geq 3$ that are a power of 2 or can be written as a product of any power of 2 and one or more primes of the form $2^{2^m} + 1$, for $m = 0, 1, 2, \ldots$.*

We have seen the kind of primes in this theorem before. These are the Fermat primes. Recall that only five such numbers are known to be primes, for $m = 0$, 1, 2, 3, and 4. The next Fermat prime after 5 is 17, and Gauss had a proof of the constructibility of the regular 17-gon. Since 7, 9, 11, and 13 are not among the numbers described in Theorem 16.2, none of these regular n-gons can be constructed.

Until Gauss's work with the Fermat numbers, there was no way to tell which regular n-gons were constructible without coming up with the actual construction. The next Fermat primes are 257 and 65,537. A method for constructing the regular 257-gon has been known since 1832, and a method for the regular 65,537-gon construction is also known.

Practice Problem Solutions and Hints

16.1. The length of the remaining rectangle is 1. The width is

$$r - \frac{1}{2} = \frac{\sqrt{5}}{2} - \frac{1}{2}.$$

The ratio of length to width is

$$\frac{1}{\frac{\sqrt{5}}{2} - \frac{1}{2}} = \frac{1}{\frac{\sqrt{5}-1}{2}} = \frac{2}{\sqrt{5}-1}.$$

Using conjugates and multiplying, we get

$$\frac{2}{\sqrt{5}-1} \cdot \frac{\sqrt{5}+1}{\sqrt{5}+1} = \frac{2(\sqrt{5}+1)}{4} = \frac{\sqrt{5}+1}{2} = \tau.$$

16.2. Mark off length α. With the point at the initial point of the segment of length α, mark off length β on the segment of length α. Then the remaining segment length is $\alpha - \beta$.

16.3. Using Figure 16.8, OU has length 1, OB has length β, OA has length α, and OV has length γ. By similar triangles we have the ratio $\frac{\beta}{\alpha} = \frac{1}{\gamma}$, so AV has length $\frac{\alpha}{\beta}$.

16.4. Do it!

16.5. Triangles ABC and CDE are congruent isosceles triangles by SAS, so lengths AC and CE are equal. Because ABC and CDE are isosceles and the angle at B is 144°, angles ACB and ECD are each 18°. So angle ACE is 144° − 36° = 108°, which is the measure of the interior angle of a regular pentagon.

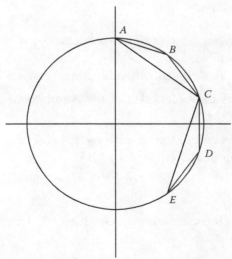

16.6. Do it!

Exercises

16.1. Show that the ratio a/b in the five-pointed star formed by extending the sides of a regular pentagon is the golden ratio.

16.2. If τ is the golden ratio, show that $\tau^3 = 2\tau + 1$. Show that $\tau^4 = 3\tau + 2$.

16.3. Show that $\tau^n = f_n\tau + f_{n-1}$, where f_n is the nth Fibonacci number.

16.4. Describe a method for bisecting any angle using only a straightedge and compass.

16.5. Show that $\frac{1+\sqrt{5}}{2}$ and $\frac{-1+\sqrt{5}}{2}$ are reciprocals. These are the golden ratio τ and the edge length of the decagon we inscribed in the unit circle.

16.6. Suppose that rectangle R, with length l and width w is golden. If a $w \times w$ square is cut off the end of rectangle R, show that the remaining rectangle is also golden.

16.7. Use the following instructions to construct a regular 17-gon with straightedge and compass.

(a) In a circle O, draw perpendicular diameters AB and CD.

(b) Construct the midpoint, M, of OC, and then construct the midpoint, E, of OM.

(c) Draw a circle centered at E with radius EA, and let F be the point of intersection of the circle with CD.

(d) Bisect the arc AF to get point N, then bisect the arc FN to get point G.

(e) Let H be the intersection of EG with OA. Draw I on OB so that angle IEH is 45 degrees.

(f) Draw the circle with diameter AI and let J be its intersection with the radius OC.

(g) Draw the circle with center H and radius HJ. This intersects AB at two points, K and L.

(h) Draw lines perpendicular to AB through K and L, these intersect the circle at points X and Y, respectively, and angle XOY measures $720°/17$, twice the desired central angle. So simply bisect angle XOY to finish the construction.

16.8. Prove that the above construction yields a regular 17-gon.

16.9. Consider the sequence derived from the Fibonacci sequence by taking Fibonacci numbers mod 2. This yields the sequence $1 \equiv 1 \pmod 2$, $1 \equiv 1 \pmod 2$, $2 \equiv 0 \pmod 2$, $3 \equiv 1 \pmod 2$, $5 \equiv 1 \pmod 2$, $8 \equiv 0 \pmod 2$, $13 \equiv 1 \pmod 2$, so we get the sequence $1, 1, 0, 1, 1, 0, \ldots$.

(a) Explain why, in this new sequence, each next term can be obtained by adding the two preceding terms mod 2.

(b) Argue that the sequence is periodic with period 3.

(c) Suppose we did this for mod 3 instead of mod 2. (The first few terms are $1, 1, 2, 0, 2, \ldots$.) Show that this sequence is also periodic. What is its period?

(d) Prove that this sequence is periodic mod n for all n. Try to find a formula for the length of the period.

For even more interesting facts about the golden ratio and nature, see the book *Discovering Geometry* by Michael Serra. For an alternative view, you might also enjoy the article "Misconceptions about the Golden Ratio" by George Markowsky.[2]

---

[2]George Markowsky. "Misconceptions about the Golden Ratio." *College Mathematics Journal* 23 (January 1992), 2–19.

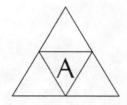

Appendix:
Euclidean Geometry Review

This Appendix has two parts. In Part 1, we state Euclid's basic postulates, prove the three fundamental theorems about congruent triangles, and state two theorems about similar triangles. In Part 2, using the three basic theorems on congruence, we present a pictorial view of the elementary theorems about angles in geometry.

Part 1

Euclid's *Elements* begins with definitions, common notions, and postulates, from which he deduces all of his geometry. The definitions he used are fairly familiar, and will be included if they are necessary. The common notions are descriptions of the arithmetic that can be done on figures (so, for example, the first common notion is transitivity: if $a = b$ and $b = c$, then $a = c$.) The postulates are:

Postulate 1. Two points determine a line.

Postulate 2. Straight lines extend infinitely.

Postulate 3. A circle is defined by its center and radius.

Postulate 4. All right angles are equal.

Postulate 5. If line A falls on two lines B and C, B and C intersect on the side of A where the sum of the interior angles made by A and B and A and C is less than $180°$.

We begin with some simple theorems about triangles.

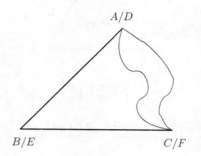

Figure A.1. Congruence by SAS.

Definition. Two triangles are *congruent* if all of their corresponding parts (angles and sides) are equal.

Immediately we can see the SAS theorem.

Theorem A.1 (SAS). *If two triangles have two pairs of congruent sides, and the angles between the sides are congruent, the triangles are congruent.*

Proof: Say there are two triangles with $\angle B \cong \angle E$, $\overline{AB} \cong \overline{DE}$, and $\overline{BC} \cong \overline{EF}$. Now imagine that one triangle is placed on top of the other, so that $\angle E$ coincides with $\angle B$, $\overline{DE}$ is on $\overline{AB}$, and $\overline{EF}$ is on $\overline{BC}$ (Figure A.1). Then Postulate 1 says that there is only one line between A/D and C/F.□

Definition. A triangle is isosceles if two of its sides are congruent.

Theorem A.1 lets us conclude Theorem A.2.

Theorem A.2. *In an isosceles triangle, the opposite angles are congruent.*

Proof: In isosceles triangle BAC, with $AB = AC$, bisect $\angle A$ and say it intersects $\overline{BC}$ at D (Figure A.2). Now apply SAS to $\triangle ADB$ and $\triangle ADC$ to see that $\angle B \cong \angle C$.   □

Then we can use Theorem A.2 to prove Theorem A.3.

Theorem A.3 (SSS). *If two triangles have three corresponding congruent sides, the triangles are congruent.*

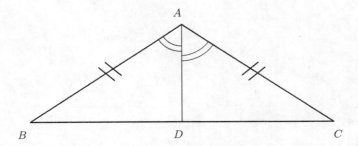

Figure A.2. Isosceles triangle.

Proof: This is a proof by contradiction. Suppose the two triangles CAB and DBA aren't congruent, but they share a base. (So $\overline{CA} \cong \overline{DA}$, $\overline{CB} \cong \overline{DB}$, and $\overline{AB} \cong \overline{AB}$; see Figure A.3.) Now draw a line between C and D. Observe that $\angle BCD$ is smaller than $\angle ACD$, and $\angle ADC$ is smaller than $\angle BDC$. However, $\triangle CDB$ and $\triangle CDA$ are both isosceles, so $\angle BDC \cong \angle BCD$ and $\angle ACD \cong \angle ADC$. This is not possible; therefore, the triangles must be congruent. $\square$

Now we can use Theorem A.1 again to prove Theorem A.4.

Theorem A.4 (ASA). *If two triangles have two congruent angles, and the sides between the angles are congruent, the triangles are congruent.*

Proof: Extend side $\overline{CA}$ of triangle ABC so that $\overline{BC''} \cong \overline{B'C'}$ (Figure A.4). Then by SAS, $\triangle ABC'' \cong \triangle A'B'C'$. Thus, $\angle ABC'' \cong \angle A'B'C' \cong$

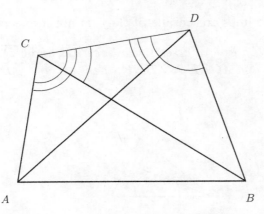

Figure A.3. Two triangles, $\triangle CAB$ and $\triangle DBA$, with a common base.

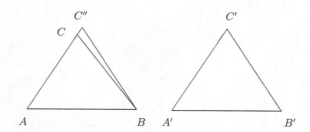

Figure A.4. Two triangles ABC and $A'B'C'$ with $\angle A \cong \angle A'$, $\angle B \cong \angle B'$, and $AB \cong A'B'$.

$\angle ABC$. But then $\overline{BC''} \cong \overline{B'C'}$, so $\overline{AC} \cong \overline{A'C'}$, and by SAS, $\triangle ABC \cong \triangle A'B'C'$. $\qquad\square$

Definition. An exterior angle of a triangle is an angle formed by one side of the triangle and the extension of another side.

Theorem A.5. *The exterior angle at a vertex of a triangle is greater than either opposite interior angle.*

Proof: We want to show that $\angle BCF$ is greater than $\angle B$ in Figure A.5. Let D be the point that bisects $\overline{BC}$. Now draw line $\overline{AD}$ and extend it to point E so that $\overline{AD} \cong \overline{DE}$. Draw line segment EC. $\triangle ADB \cong \triangle EDC$ by SAS and the fact that vertical angles are equal. So $\angle B \cong \angle BCE$, which is obviously less than $\angle BCF$. $\qquad\square$

Next, we look at parallel lines.

Definition. Two lines are parallel if they do not intersect.

Theorem A.6. *If a line intersects two lines $\overline{AB}$ and $\overline{CD}$ at points E and F, respectively, and $\angle CFE \cong \angle BEF$, then $\overline{AB}$ and $\overline{CD}$ are parallel.*

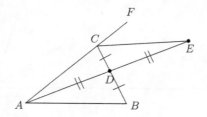

Figure A.5. Triangle ABC with exterior angle BCF.

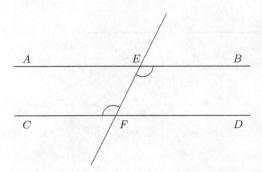

Figure A.6. Line EF intersecting lines AB and CD.

Proof: Suppose $\overline{AB}$ and $\overline{CD}$ intersect at some point G to the right of $\overline{EF}$ (Figure A.6). Then $\angle FEG$ is equal to $\angle CFE$, but $\angle FEG$ is an interior angle in $\triangle FEG$ and $\angle CFE$ is an exterior angle to $\triangle FEG$, which is a contradiction. So $\overline{AB}$ and $\overline{CD}$ do not intersect. $\qquad\square$

Next, we prove the converse.

Theorem A.7. *If two lines AB and CD are parallel, and a line intersects them at points E and F, respectively, then $\angle DFE \cong \angle AEF$. That is, if two lines are parallel, then their alternate interior angles are congruent.*

Proof: Suppose that $\angle DFE$ and $\angle AEF$ are not congruent. Let L be the line that intersects AB at the point E, and suppose that Q is another point on L such that $\angle DFE = \angle QEF$, as in Figure A.7.

Because there is a unique line through the point E parallel to the line CD, L is not equal to AB. But, by Theorem A.6, since the alternate

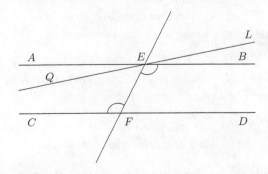

Figure A.7. Proof of Theorem A.7.

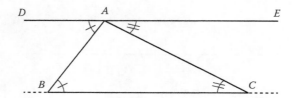

Figure A.8. Triangle ABC with line DE parallel to BC

interior angles are equal, L is parallel to CD, so L must equal AB. This is a contradiction. So, the alternate interior angles of lines AB and CD are congruent.   □

It follows from Theorem A.7 that if two lines are parallel, and a third line intersects them, then the sum of the interior angles on the same side of the third line is 180°.

Now we can prove Theorem A.8.

Theorem A.8. *The sum of the angles in a triangle is* 180°.

Proof: Through point A in $\triangle ABC$, draw a line $\overline{DE}$ parallel to $\overline{BC}$ (Figure A.8). Then, $\angle CBA \cong \angle DAB$ and $\angle BCA \cong \angle EAC$, so the sum of the angles in the triangle is a straight line, or 180°.   □

We now state some results about similar triangles.

Definition. Two triangles are *similar* if their corresponding angles are congruent and corresponding sides are proportional.

For triangles, it is true that if one of these conditions holds, then so does the other. That is, if corresponding sides are proportional, then corresponding angles are congruent, and conversely. There are two basic similarity theorems.

Theorem A.9 (AAA). *If two triangles have corresponding angles congruent, then the triangles are similar.*

Corollary A.10 (AA). *If two triangles have two pairs of corresponding angles that are congruent, then the triangles are similar.*

There are obvious theorems for similarity that are analogous to SAS, SSS, and ASA (Theorems A.1, A.3, and A.4) for congruent triangles.

Part 2

This part of the Appendix presents a set of figures (Figures A.9 and A.10) that illustrate some basic concepts of plane geometry. We start with Figure A.9, which is a quick reference to basic rules of geometry that also shows their relationships.

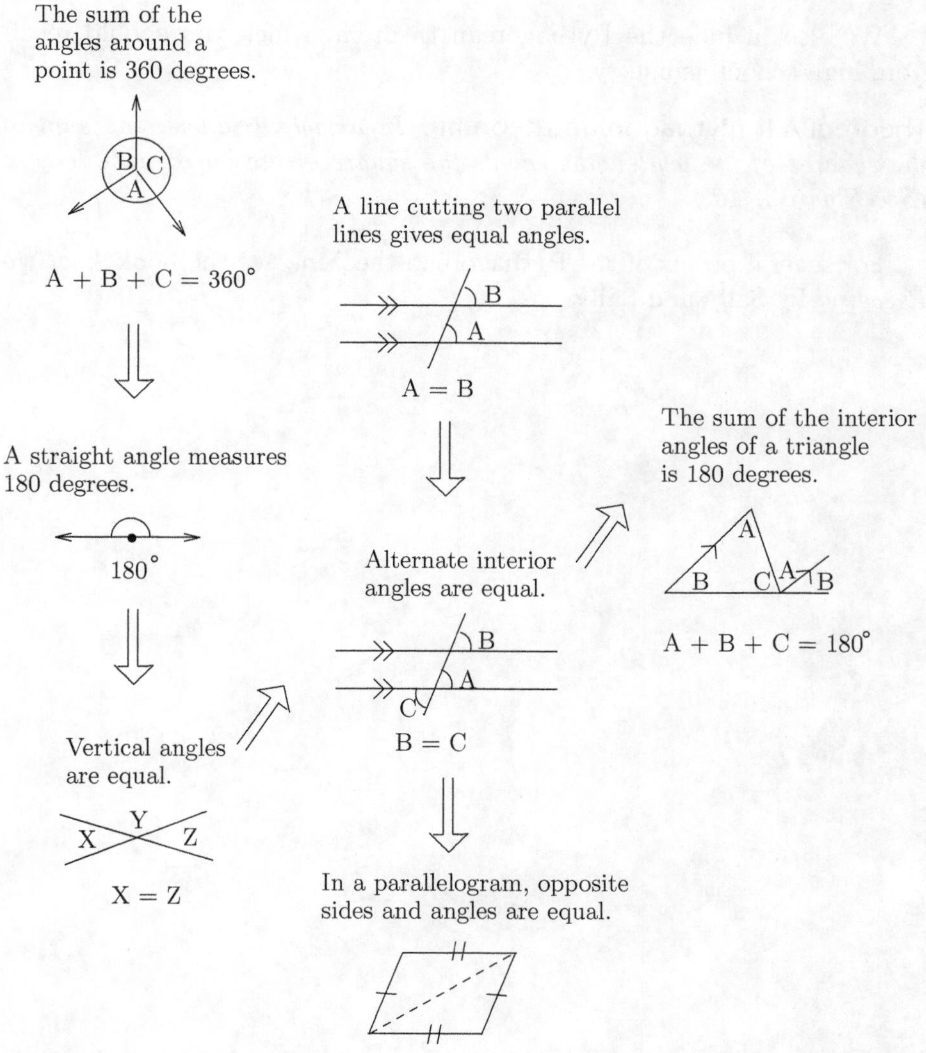
The sum of the angles around a point is 360 degrees.
B C A
A + B + C = 360°
A line cutting two parallel lines gives equal angles.
B A
A = B
A straight angle measures 180 degrees.
180°
The sum of the interior angles of a triangle is 180 degrees.
A B C A B
A + B + C = 180°
Alternate interior angles are equal.
B A C
B = C
Vertical angles are equal.
Y X Z
X = Z
In a parallelogram, opposite sides and angles are equal.

Figure A.9. Back-of-the-envelope geometry.

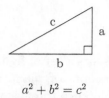

$$a^2 + b^2 = c^2$$

Figure A.10. The Pythagorean theorem.

We also include the Pythagorean theorem, which you should recall from high school geometry.

Theorem A.11 (Pythagorean theorem). *In a right triangle, the sum of the squares of the leg lengths equals the square of the hypotenuse length. (See Figure A.10.)*

For several proofs of the Pythagorean theorem, see the book *Roots to Research* by Sally and Sally.

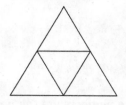

Glossary

abelian group *See* commutative group.

abundant An integer n is *abundant* if $\sigma(n) > 2n$.

adjacent In a graph, two vertices of an edge are *adjacent*. Also, two edges sharing a common vertex are *adjacent*.

Associative Property of Addition $(a + b) + c = a + (b + c)$.

binary An operation is *binary* if it combines two elements of a set to get an element of the universal set X.

Cancelation Law for $\mathbb{Z}$ If a, b, and c are in $\mathbb{Z}$, $a \neq 0$, and $ab = ac$, then $b = c$.

cardinal number *See* one-to-one correspondence.

Cartesian product The *Cartesian product* of sets A and B, written $A \times B$, is defined as the set $A \times B = \{(a, b) | a \in A \text{ and } b \in B\}$.

circuit A trail whose initial and final vertices are the same is a *circuit*.

closed If we start with a binary operation on a set, the set is *closed* under this binary operation if whenever we combine two elements from that set, the result is an element of the same set.

common divisor Two integers a and b have a *common divisor* d if $d \mid a$ and $d \mid b$.

commutative group If the group operation is commutative, the group is a *commutative group*. Such a group is also an *abelian* group, named after the Norwegian mathematician Niels Henrik Abel.

commutative ring A *commutative ring* is a set R closed under two binary operations, $+$ and $*$, that satisfy Axioms A1–A4, M1–M2, and D of the Rules of Arithmetic.

Commutative Property of Addition $a + b = b + a$.

complement of a graph The *complement* of graph G, written $\overline{G}$, is a graph that has the same vertices as G and two vertices of $\overline{G}$ are adjacent if and only if they are not adjacent in G; two vertices in G are adjacent if and only if they are not adjacent in $\overline{G}$.

complete graph A graph G is *complete* if every vertex of G is adjacent to every other vertex of G.

composite number Composite numbers are the positive integers greater than 1 that are not primes.

composition Suppose A, B, and C are sets and f and g are functions, where $f : A \rightarrow B$ and $g : B \rightarrow C$. The *composition* of f and g is a function $g \circ f : A \rightarrow C$ defined by $(g \circ f)(a) = g(f(a))$.

connected graph A graph G is *connected* if every pair of vertices is joined by a trail.

constructible number A positive number α is *constructible* if we can construct a line segment of length α in a finite number of steps, starting with a given segment of unit length and using only a straightedge and compass.

continued fraction An equation of the form

$$a_1 + \cfrac{1}{a_2 + \cfrac{1}{a_3 + \cfrac{1}{a_4 + \ldots + \frac{1}{a_n}}}}$$

is known as a *continued fraction*.

contradiction In a proof of contradiction, we assume that the conclusion of the theorem is false, and then we argue using clear, logical steps, and hope that we encounter a contradiction, or a statement that doesn't make sense.

convex polygon A *convex* polygon has the property that the line segment connecting any two vertices of the polygon will lie entirely within the polygon. Any polygon that does not have this property is called *nonconvex*.

convex polyhedron *See* polyhedron.

countably infinite The size of the set of natural numbers $\mathbb{N}$, labeled with the symbol aleph naught, or $\aleph_0$. This is also referred to as *countable*.

counting numbers *See* natural numbers.

cycle of length k A *cycle of length k*, where $a_1, a_2, a_3, \ldots, a_k$ are numbers from the set $\{1, 2, \ldots, n\}$, is an expression of the form $(a_1, a_2, a_3, \ldots, a_k)$. It indicates the permutation where a_1 goes to a_2, a_2 goes to a_3, and so on, until a_k goes to a_1.

cyclic group generated by a Suppose the binary operation on a group G is addition. If G has an element a so that every element of G can be written as a sum of a's and $-a$'s then, G is a *cyclic group generated by a*. If the group operation is multiplication instead of addition, then a generator of G will be an element a so that we can write every element of G as a product of a's and a^{-1}'s.

deficient An integer n is *deficient* if $\sigma(n) < 2n$.

degree The *degree* of a vertex v in G is the number of edges adjacent to v. The degree of v is written $\deg v$. A vertex is *odd* or *even* depending on whether its degree is odd or even, respectively.

diagonal of a polyhedron *See* polyhedron.

dihedral group D_6 The *dihedral group D_6* is the group made up of the six symmetric motions of the equilateral triangle. This group is a noncommutative group.

disjoint cycles Disjoint cycles have no elements in common.

Distributive Property $a \cdot (b + c) = (a \cdot b) + (a \cdot c)$.

divides For two integers a and b, a *divides* b if there is an integer k so that $ak = b$. We write this as $a \mid b$.

division If a and b are any two elements of a commutative ring, and if b has a multiplicative inverse, *division* of a by b is $\frac{a}{b} = a(b^{-1})$.

division algorithm For any two positive integers a and b, there are unique integers q and r satisfying $a = bq + r$, where $0 \le r < b$.

domain Suppose f is a function from a set A to a set B. The set A is the *domain* of f.

dual The *dual* of a convex polyhedron is the polyhedron obtained by joining the centers of adjacent faces of the original polyhedron.

edges of a graph *See* graph.

edges (or sides) of a polygon Edges, or sides, are the line segments of a polygon.

edges of a polyhedron *See* polyhedron.

empty set The *empty set* is the one special set that has no members at all, denoted by the symbol $\emptyset$.

endpoint A vertex with degree 1 is an *endpoint*.

Euclidean Algorithm Take positive integers a and b with $a > b$. Use the division algorithm to write

$$a = bq_1 + r_1, \text{ where } 0 \le r_1 < b.$$

Apply the division algorithm to b and r_1 and get

$$b = r_1 q_2 + r_2, \text{ where } 0 \le r_2 < r_1.$$

Continue this process to get

$$r_j = r_{j+1} q_{j+2} + r_{j+2}, \text{ where } 0 \le r_{j+2} < r_{j+1}$$

for $j = 0, 1, 2, \cdots, n-1$. If in the equation

$$r_{n-1} = r_n q_{n+1} + r_{n+1}$$

we have $r_{n+1} = 0$, then $(a, b) = r_n$. That is, (a, b) is the last nonzero remainder in this process.

Euler-ϕ function The *Euler-ϕ function* is a special function that counts the number of integers that are less than a positive integer m and relatively prime to m. It is written $\phi(m)$.

Euler's formula The *Euler formula* says that in convex polyhedra the number of vertices plus the number of faces minus the number of edges always equals 2, written symbolically as $V - E + F = 2$.

Eulerian trail, Eulerian circuit A trail that contains all the edges of a graph is called an *Eulerian trail*. A circuit that contains all the edges of a graph is an *Eulerian circuit*.

even degree *See* degree.

exponentiation The *exponentiation* of two nonzero integers is defined by $a \circ b = a^b$.

exterior angle An exterior angle is the angle formed by extending one edge of a convex polygon from a vertex. The exterior angle is supplementary to the interior angle at that vertex.

faces of a polyhedron *See* polyhedron.

Fermat number A *Fermat number* is any number of the form $2^{2^n} + 1$, where n is a nonnegative integer.

Fermat prime If a Fermat number is prime, it is a *Fermat prime*.

Fibonacci sequence The *Fibonacci sequence* $f_1, f_2, f_3, \cdots, f_n, \cdots$ is defined recursively by starting with $f_1 = 1, f_2 = 1$, and $f_n = f_{n-1} + f_{n-2}$ for $n \geq 3$.

final vertex *See* trail.

field A commutative ring that has a multiplicative identity is a *field* if every nonzero element is a unit.

finite set with n elements A *finite set with n elements* is one that can be put into one-to-one correspondence with the set of natural numbers $\{1, 2, 3, \ldots, n\}$.

First Principle of Mathematical Induction The *First Principle of Mathematical Induction* states: Suppose S is a subset of $\mathbb{N}$ with the following properties:

1. The number 1 is in S.
2. If n is in S, then $n + 1$ is in S.

function A *function* from A to B, where A and B are nonempty sets, is a subset of $A \times B$ such that each element of A occurs exactly once as a first coordinate.

Fundamental Theorem of Arithmetic Every positive integer greater than 1 can be written uniquely as a product of primes, with the prime factors in the product written in order of nondecreasing size.

Gaussian integers The set of *Gaussian integers*, written as $\mathbb{Z}[i]$, is the set $\mathbb{Z} \times \mathbb{Z}$ with the following two binary operations:

$$(a, b) \oplus (c, d) = (a + c, b + d),$$
$$(a, b) \odot (c, d) = (ac - bd, ad + bc).$$

glide reflection A *glide reflection* is the product of a translation and a horizontal reflection. This describes the way the pattern glides one space and then is reflected over a horizontal line.

Goldbach conjecture The *Goldbach conjecture* is a conjecture made by Christian Goldbach that every even integer larger than 2 can be written as the sum of two primes.

golden ratio If the ratio of the segment's entire length $(a + b)$ to the longer length (a) is the same as the ratio of the longer length (a) to the shorter length (b), then the ratio is *golden*.

graph A *graph* G is a finite nonempty set V together with a collection E of two-element subsets of V. The elements of V are called *vertices* and the elements of E are called *edges*.

greatest common divisor (GCD) The largest of all the common divisors of a and b is called the *greatest common divisor* of a and b, written as $(a, b) = d$ if d is the greatest common divisor of a and b.

group A *group* is a set of elements G closed under a binary operation $*$ that satisfies the following three axioms:

Axiom I. The operation $*$ is associative.

Axiom II. There is an element e in G that acts as the identity for the group. That is, if g is any element of the set G, then $e * g = g$ and $g * e = g$.

Axiom III. For every element g in the set G, there is an associated element g^{-1} that acts as the inverse for the element g. That is, $g^{-1} * g = e$ and $g * g^{-1} = e$.

Hamiltonian trail, Hamiltonian circuit In a graph G, a trail that includes each vertex exactly once is a *Hamiltonian trail*. If such a trail begins and ends at the same vertex, it is a *Hamiltonian circuit*.

Hilbert numbers The *Hilbert numbers* H are the set of positive integers that can be written in the form $4k + 1$, where $k \geq 0$ is an integer. In other words, $H = \{1, 5, 9, 13, 17, \ldots\}$.

Hilbert prime A number h in the set H is called a *Hilbert prime* if the only way it can be written as the product of two integers in H is $h \cdot 1$ or $1 \cdot h$.

identity An *identity* for an operation is an element that, when combined with any element, gives that second element back again.

image For a function f from a set A to a set B, the set of elements $f(A) = \{f(a) | a \in A\}$ is the *image* of f on B.

incident In a graph, vertices and edges that are joined are called *incident*.

induction *See* First Principle of Mathematical Induction; Second Principle of Mathematical Induction.

infinite order *See* order of an element.

infinite set A set is *infinite* if it can be put into one-to-one correspondence with a proper subset of itself.

initial vertex *See* trail.

integers The *integers* $\mathbb{Z}$ include 0 and the negatives of all the natural numbers.

integral domain An *integral domain* is a number system with two binary operations, $+$ and $\cdot$, in which Axioms A1–A4, M1–M3, D, and cancelation hold.

intersection of sets The *intersection of sets* A and B, written $A \cap B$, is the set of all elements that are elements of both A and B. That is, $A \cap B = \{x | x \in A \text{ and } x \in B\}$.

inverse For an element a, the *inverse* of a is the element that, combined with a by some operation, yields the identity for that operation.

isolated A vertex with degree 0 is called *isolated*.

isomorphic graphs Two graphs $G = (V, E)$ and $G' = (V, E')$ are called *isomorphic* if there exists a one-to-one correspondence between the vertex sets V and V' that preserves adjacency.

k-cycle *See* cycle of length k.

K_n If G has n vertices, the complete graph is called K_n.

least common multiple (LCM) The *least common multiple* of two positive integers a and b, written $[a, b]$ is the smallest positive integer that is divisible by both a and b.

lower bound A number is a *lower bound* for a set of solutions if all solutions are greater than or equal to that number.

matrix A 2×2 matrix (over $\mathbb{Q}$) is an array of the form $\left(\begin{smallmatrix} a & b \\ c & d \end{smallmatrix} \right)$, where a, b, c, and d are in $\mathbb{Q}$. We denote the collection of all such matrices as $M_2(\mathbb{Q})$.

magic square A *magic square* is an $n \times n$ grid in which each of the numbers $\{1, 2, 3, ..., n^2\}$ is used once, and the sum of each row, column, and diagonal is the same.

Mersenne number A *Mersenne number* is any number of the form

$$2^m - 1,$$

where m is a natural number.

Mersenne prime A *Mersenne prime* is a Mersenne number that is prime.

minimum A minimum, denoted $\min\{x, y\}$, is the smaller of the two numbers x and y.

modulo m For a positive integer $m \geq 2$ and two integers a and b, a is congruent to b modulo m, or $a \equiv b \pmod{m}$, if m divides $a - b$.

mutually relatively prime Numbers a, b, and c are *mutually relatively prime* if $((a, b), c) = 1$.

n-gon A polygon with n sides is an *n-gon*.

n-omino An *n-omino* is the shape of n squares joined together, edge-to-edge and vertex-to-vertex.

natural numbers The *natural numbers* $\mathbb{N}$ are the counting numbers: $1, 2, 3, 4, 5, ...$

nonconvex *See* convex polygon.

odd degree *See* degree.

onto For a function f from a set A to a set B, if $f(A) = B$, then f is *onto* B.

one-to-one For a function f from a set A to a set B, if each element in the image of f corresponds to exactly one element in the domain, f is *one-to-one*. This means that f is one-to-one if, whenever $f(a) = f(a')$, then $a = a'$.

one-to-one correspondence Two sets A and B are said to be in *one-to-one correspondence* if there is a function $f : A \to B$ that is one-to-one and onto. Then A and B have the same *cardinal number*.

order of a group The *order of a group* G, written $|G|$, is the number of elements in G. If G has n elements for some natural number n, G has finite order equal to n. If G has an infinite number of elements, then G is an *infinite group*.

order of an element If x is an element of a group G, the *order of element x* is the smallest positive integer n such that $x^n = e$, where $x^n = x \cdot x \cdot \cdots \cdot x$ (n times). In this case, x is said to be of order n. If G is written additively, the order of x is the smallest positive integer such that $nx = e$, where $nx = x + x + \cdots + x$ (n times). If such an n does not exist, x has *infinite order*.

parity check digit A *parity check digit* is the residue of the sum of the digits in a message modulo 2, inserted as an additional digit to the message.

path *See* trail.

perfect numbers The *perfect numbers* are those integers that are equal to half the sum of their positive divisors. For example, the positive divisors of 6 are 1, 2, 3 and 6, which add up to 12, and $6 = \frac{1}{2}(12)$.

period The *period* of a repeating decimal is the number of digits in the repeating part.

permutation A *permutation* of $S = \{1, 2, \ldots, n\}$ is a one-to-one function from S onto itself.

planar A graph G is *planar* if it can be diagrammed on a plane so that distinct edges do not cross.

Platonic solids The *Platonic solids*, the tetrahedron, cube, octahedron, dodecahedron, and icosahedron, have been known since the time of the Greeks in the sixth century B.C. (Plato was among those who first gave instructions on how to make models of them.)

polygon A *polygon* is a finite collection of line segments in a plane, each meeting exactly two others, one at each endpoint, such that no proper subset of the line segments has this same property.

polyhedron A *polyhedron* is the union of a finite set of polygons such that

1. Any pair of polygons meet at only their edges or vertices.
2. Each edge of each polygon meets exactly one other polygon along an edge.
3. It is possible to travel from the interior of any polygon to the interior of any other along the surface of the polyhedra

Furthermore,

▷ The polygons are called *faces* of the polyhedron.

▷ The intersection of two faces forms an *edge*.

▷ The intersection of two or more edges forms a *vertex*.

▷ A *diagonal* is a line joining any two vertices of a polyhedron that are not in the same face.

▷ If a line segment joining any two interior points of the polyhedron lies entirely inside the polyhedron, the polyhedron is said to be *convex*.

power set The *power set* of S, denoted $\mathcal{P}(S)$, consists of all subsets of the set S.

prime number A *prime number* p is a positive integer greater than 1 whose only positive divisors are 1 and p.

product of fractions The product of the two fractions, or rational numbers $\frac{a}{b}$ and $\frac{c}{d}$, is defined by

$$\frac{a}{b} \cdot \frac{c}{d} = \frac{ac}{bd}.$$

product of residue classes modulo m For a positive integer $m \geq 2$, and for residue classes A and B, modulo m, where a is an element of A and b is an element of B, the *product of residue classes module* m is AB, the residue class containing the element ab.

proper subset Set A is a *proper subset* of S if $A \subset S$ and $A \neq S$.

Ramsey number The *Ramsey number*, $r(m, n)$, is the smallest integer so that every graph G with $r(m, n)$ vertices contains a copy of K_m or $\overline{K}_n$.

Ramsey's theorem If G is a graph with six vertices, then either G or $\overline{G}$ must have a triangle.

range If f is a function from a set A to a set B, the set B is called the *range* of f.

rational numbers The *rational numbers* $\mathbb{Q}$ are those that can be expressed as the quotient of two integers a and b, written $\frac{a}{b}$, where $b \neq 0$ and the fraction is reduced to lowest terms.

real numbers The *real numbers* $\mathbb{R}$ include the rational numbers and irrational numbers.

reflection over a line l A *reflection over a line* l moves any point P to the point P' obtained by drawing a perpendicular from P to l and extending it the same distance to the other side of l. If P is on the line l, then $P = P'$. Every point on the line of reflection is a fixed point.

reflectional symmetry A figure has *reflectional symmetry* if it looks the same after you reflect it over a line.

regular polygon A *regular polygon* is one in which all the sides are of equal length and all the interior angles are equal.

regular polyhedron A *regular polyhedron* is a convex polyhedron whose faces are congruent regular polygons with the same number of faces meeting at each vertex.

regular prism A *regular prism* is a polyhedron made of two congruent regular n-gons, $n \geq 3$, in parallel planes, with squares for the remaining faces. A regular prism has $n + 2$ faces.

relatively prime If the greatest common divisor of two integers a and b is 1, then a and b are *relatively prime*.

residue class modulo m The set of numbers that correspond to the same remainder r when divided by m forms a *residue class modulo m*.

residue system modulo m The set of remainders $\{0, 1, 2, 3, \ldots, m-1\}$ is a *residue system modulo m*.

rotation A *rotation* about a center point O moves any point P to the point P' the same distance from O as P by moving P along the circle centered at O with radius OP counterclockwise to P'.

rotational symmetry A figure has *rotational symmetry* if it looks the same after you rotate it a certain number of degrees around a point.

Second Principle of Mathematical Induction The *Second Principle of Mathematical Induction* states: Suppose S is a subset of $\mathbb{N}$ with the following properties:

1. The number 1 is in S.
2. If n is in S and if every natural number k, where $k \leq n$, is in S, then $n + 1$ is in S.

semiregular polyhedron A *semiregular polyhedron* is a convex polyhedron with faces that comprise more than one kind of regular polygon. Every vertex is surrounded by the same kinds of polygons in the same cyclic order.

$\sigma(n)$ The symbol $\sigma(n)$ denotes the sum of the positive divisors of n.

subgraph A *subgraph H* of a graph G is a graph that has all its vertices and edges in G. If v is a vertex of the graph G, the graph $H = G - v$ is the subgraph of G consisting of all the vertices of G except for v, and all the edges of G not incident to v.

subgroup If G is a group with binary operation $*$, a nonempty subset H of G is a *subgroup* of G if H is also a group with the same binary operation $*$. Thus,

1. if a and b are in H, then $a * b$ is in H;
2. if e is the identity on G, then e is in H; and
3. if a is in H, then a^{-1} is in H.

subset Set A is a *subset* of S, written as $A \subset S$, if every element of A is also an element of S (that is, if $x \in A$, then also $x \in S$)

subtraction If a and b are any two elements of a commutative ring, $a - b = a + (-b)$.

sum of graphs If G_1 and G_2 are graphs with disjoint vertex sets V_1 and V_2 and edge sets E_1 and E_2, respectively, the *sum of graphs* $G_1 + G_2$ is the graph with vertices $V = V_1 \cup V_2$ and edges that join every vertex in V_1 with every vertex in V_2.

sum of residue classes modulo m For a positive integer $m \geq 2$, and for residue classes A and B, modulo m, where a is an element of A and b is an element of B, the *sum of residue classes modulo m* is $A + B$, the residue class containing the integer $a + b$.

symmetric group For any positive integer n, the *symmetric group* on n elements, called S_n, is the group of all the permutations of the elements designated by the set $\{1, 2, 3, \ldots, n\}$. There are $n! = n \cdot (n-1) \cdot (n-2) \cdots 3 \cdot 2 \cdot 1$ elements in the group.

$\tau(n)$ The symbol $\tau(n)$ denotes the number of positive divisors of n.

tessellation by one shape A *tessellation by one shape* is a covering of the plane by congruent copies of a particular shape with no overlaps and leaving no gaps.

trail In graph G, a *trail* (sometimes also called a *path*) is an alternating sequence of vertices and edges of the form

$$v_1, v_1v_2, v_2, v_2v_3, v_3, \ldots, v_{n-1}, v_{n-1}v_n, v_n,$$

where all the edges are distinct. Each edge is preceded and succeeded by the vertices adjacent to it. The vertex v_1 is the *initial vertex* and the vertex v_n is the *final vertex*.

translation A *translation* in a plane is a motion in the plane that moves every point in the plane a specified distance along a straight line.

transposition A *transposition*, or 2-cycle, is a permutation that interchanges two numbers.

tree A connected graph G is a *tree* if G has no circuits.

trigonal bipyramid A *trigonal bipyramid* is the polyhedron formed by two tetrahedrons glued along a face.

union of sets The *union of sets* A and B, written $A \cup B$, is the set of all elements of A and all elements of B. That is, $A \cup B = \{x | x \in A \text{ or } x \in B\}$.

union of graphs If G_1 and G_2 are graphs with disjoint vertex sets V_1 and V_2 and edge sets E_1 and E_2, respectively, the *union* $G_1 \cup G_2$ is the graph with vertices $V = V_1 \cup V_2$ and edges $E_1 \cup E_2$.

unit If a is a nonzero element in $\mathbb{Z}_m$ and a has a multiplicative inverse, a is a *unit* in $\mathbb{Z}_m$. The set of nonzero elements in $\mathbb{Z}_m$ that have multiplicative inverses are called the units of $\mathbb{Z}_m$, denoted by $\mathbf{U}(m)$.

upper bound A number is an *upper bound* for a set of solutions if all solutions are less than or equal to that number.

vertex of a polyhedron *See* polyhedron.

vertices of a graph *See* graph.

vertices of a polygon The *vertices of a polygon* are the endpoints or corners of the polygon.

weighted graph A *weighted graph* is a graph whose edges are assigned a "weight," such as the mileage between cities on the graphs representing a salesman's route.

well-ordering axiom for $\mathbb{Z}$ Any nonempty set of positive integers has a smallest element.

zero divisor A nonzero element a in a commutative ring R is called a *zero divisor* if there is a nonzero element b in R such that $ab = 0$.

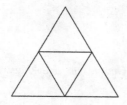

Bibliography

sarah-marie belcastro and Carolyn Yackel (editors). *Making Mathematics with Needlework: Ten Papers and Ten Projects.* Wellesley, MA: A K Peters, Ltd., 2007.

sarah-marie belcastro and Carolyn Yackel (editors). *Crafting by Concepts: Fiber Arts and Mathematics.* Natick, MA: A K Peters, Ltd., 2011.

H. S. M. Coxeter (editor). *M. C. Escher: Art and Science.* Amsterdam: Elsevier Science Ltd., 1987.

H. E. Dudeney. *The Canterbury Puzzles, and Other Curious Problems.* New York: E. P. Dutton and Company, 1908.

Joseph A. Gallian. *Contemporary Abstract Algebra,* Second Edition. Lexington, MA: DC Heath, 1990.

Martin Gardner. "On Tessellating the Plane with Convex Polygon Tiles." *Scientific American* (July 1975), 112–117.

Solomon W. Golomb. *Polyominoes,* Second Edition. Princeton, NJ: Princeton University Press, 1994.

George Markowsky. "Misconceptions about the Golden Ratio." *College Mathematics Journal* 23:1 (January 1992), 2–19.

Edwin Moise and Floyd L. Downs, Jr. *Geometry.* Menlo Park, CA: Addison-Wesley, 1991.

Kenneth H. Rosen. *Elementary Number Theory and Its Applications,* Fourth Edition. Menlo Park, CA: Addison Wesley, 2000.

Judith D. Sally and Paul J. Sally, Jr. *Roots to Research.* Providence, RI: American Mathematical Society, 2007.

Doris Schattschneider. "The Plane Symmetry Groups: Their Recognition and Notation." *American Mathematical Monthly* 85 (1978), 439–450.

Marjorie Senechal and George Fleck (editors). *Shaping Space: A Polyhedral Approach.* Boston: Birkhauser, 1988.

Michael Serra. *Discovering Geometry.* Berkeley, CA: Key Curriculum Press, 1997.

Joseph H. Silverman. *A Friendly Introduction to Number Theory*, Fourth Edition. Boston: Addison Wesley, 2012.

David Eugene Smith. *Essentials of Solid Geometry.* Boston: Ginn and Company, 1924.

Dorothy K. Washburn and Donald W. Crowe. *Symmetries of Culture: Theory and Practice of Plane Pattern Analysis.* Seattle: University of Washington Press, 1988.

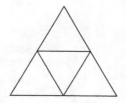

Index

In this index, bold-faced page numbers represent definitions.